Parallel Supercomputing in MIMD Architectures

R. Michael Hord

Advanced Technology Laboratories
General Electric Company
Moorestown, New Jersey

CRC Press
Taylor & Francis Group
Boca Raton London New York

CRC Press is an imprint of the
Taylor & Francis Group, an **informa** business

First published 1993 by CRC Press
Taylor & Francis Group
6000 Broken Sound Parkway NW, Suite 300
Boca Raton, FL 33487-2742

Reissued 2018 by CRC Press

© 1993 by CRC Press, Inc.
CRC Press is an imprint of Taylor & Francis Group, an Informa business

No claim to original U.S. Government works

Library of Congress Cataloging-in-Publication Data

Hord, R. Michael, 1940–
    Parallel supercomputing in MIMD architectures / author, R. Michael Hord.
        p.   cm.
    Includes bibliographical references and index.
    ISBN 0-8493-4417-4
    1. Parallel processing (Electronic computers).   2. Supercomputers.   3. Computer architecture.   I. Title.
QA76.58.H67 1993
004'.35—dc20   92-45596

A Library of Congress record exists under LC control number: 92045596

Publisher's Note
The publisher has gone to great lengths to ensure the quality of this reprint but points out that some imperfections in the original copies may be apparent.

Disclaimer
The publisher has made every effort to trace copyright holders and welcomes correspondence from those they have been unable to contact.

ISBN 13: 978-1-315-89623-6 (hbk)
ISBN 13: 978-1-351-07533-6 (ebk)

Visit the Taylor & Francis Web site at http://www.taylorandfrancis.com and the
CRC Press Web site at http://www.crcpress.com

For Susan

TABLE OF CONTENTS

FOREWORD

We live in a golden age of computer architecture innovation. The advent of parallel processing has produced a flood of architecture paradigms, some more successful than others.

The most well known taxonomy for parallel computers was proposed by M. J. Flynn in 1966. It is based on the multiplicity of the instruction and data streams, which identifies four classes of computers: (1) single instruction stream single data stream (SISD) computers correspond to regular von Neumann nonparallel computers that can execute one instruction at a time on one data item at a time; (2) single instruction multiple data (SIMD) computers are based on a central program controller that drives the program flow and a set of processing elements that all execute the instructions from the central controller on their individual data items; (3) multiple instruction single data (MISD) computers are based on a pipeline principle. A given data item passes from processor to processor and is acted on differently at each stage of this assembly line; (4) multiple instruction multiple data (MIMD) computers consist of a number of SISD computers configured to communicate among themselves in the course of a program.

Examples of successful SIMD computers are too numerous to list. Instances that were discussed in the previously published companion to this book included the Illiac IV (I4), the Massively Parallel Processor (MPP), the GAPP, the Distributed Array Processor (DAP), and the Connection Machine (CM); other examples currently on the market include the MasPar, Wavetracer, and the Princeton Engine.

In this book we explore the MIMD category, but of course, this category is not monolithic; it exhibits great diversity, and we can only examine the high-performance subset of this rich and thriving group of architectures. It is only a snapshot since new computing paradigms are conceived and produced with maddening frequency, a phenomenon that can only be explained by the handsome economic rewards that accrue to anyone who can devise a way to achieve more instructions per dollar.

Using these parallel machines effectively can be a considerable challenge. Just as instructing a single bricklayer exercises less managerial skill than being the foreman of a crew constructing a high rise building, programming an MIMD supercomputer requires a great deal more planning, coordination, and talent than developing an application on a sequential machine. In time, practitioners acquire the ability to "think parallel", but to remain competitive, programmers need this skill more and more because the movement toward parallel processing is inexorable.

Much of the material included here is based on or adapted from work previously published in government reports, journal articles, books, and vendor literature. A listing of the major sources is provided following the text.

PREFACE

Parallel Supercomputing in MIMD Architectures is a survey book providing a thorough review of multiple instruction multiple data machines, a type of parallel processing computer that has grown to importance in recent years. It was written to describe a technology in depth including the architectural concepts, a variety of hardware implementations, both commercial and research, major programming concepts, algorithmic methods, representative applications, and an assessment of benefits and drawbacks.

The book is intended for a wide range of readers. Computer professionals will find sufficient detail to incorporate much of this material into their own endeavors. Program managers and applications system designers may find the solution to their requirements for high computational performance at an affordable cost. Scientists and engineers will find sufficient processing speed to make interactive simulation a practical adjunct to theory and experiment. Students will find a case study of an emerging and maturing technology. The general reader is afforded an opportunity to appreciate the power of advanced computing and some of the ramifications of this growing capability.

Although there are numerous books on MIMD parallel processing, this is the first volume devoted to supercomputing on a wide variety of parallel machines of the MIMD class. The reader already familiar with lock-step parallel processing, as addressed in the prior book on SIMD architectures, will discover an alternative philosophy — the widespread program parallel paradigm instead of the data parallel scheme.

The contents of the book are organized into the Background section and three major parts, rich with illustrations and tables. Part 1 covers MIMD computers, both commercial and research. The commercial machines reviewed include Connection Machine CM-5, NCUBE, Butterfly, Meiko, Intel iPSC, iPSC/2, Paragon, and iWarp, DSP-3, Multimax, Sequent, and Teradata; the research machines are the J-Machine, PAX, Concert and ASP. Part 2 deals with MIMD software, including operating systems, languages, translating sequential programs to parallel, and semiautomatic parallelizing. Part 3 addresses MIMD issues, such as scalability, partitioning, processor utilization, and heterogeneous networks.

LIST OF FIGURES

Partitioning

Heterogeneous Networks

THE AUTHOR

R. Michael Hord is presently Director of High Performance Computing at the General Electric Advanced Technology Laboratories, Moorestown, New Jersey. In this capacity he directs research and development activities employing the Connection Machine, the Butterfly, and the Warp advanced architecture computers. The current emphasis is on acoustic signal analysis, military data/information systems, and future architectures.

Until the end of 1989, Mr. Hord was Head of the Advanced Development Center at MRJ, Inc., Oakton, Virginia, where he directed diverse computer applications using parallel architectures including two Connection Machines. Application areas included image processing, signal processing, electromagnetic scattering, operations research and artificial intelligence. Mr. Hord joined MRJ in 1984 where he also directed the corporate research and development program.

For 5 years (1980 to 1984) Mr. Hord was the Director of Space Systems for General Research Corporation. Under contract to NASA and the Air Force, he and his staff assessed technology readiness for future space systems and performed applications analysis for innovative on-board processor architectures.

SIMD parallel processing was the focus of his efforts as the Manager of Applications Development for the Institute for Advanced Computation (IAC). IAC was the joint DARPA/NASA sponsored organization responsible for the development of the Illiac IV parallel supercomputer at Ames Research Center. Projects included computational fluid dynamics, seismic simulation, digital cartography, linear programming, climate modeling and diverse image and signal processing applications.

Prior positions at Earth Satellite Corporation, Itek Corporation and Technology Incorporated were devoted to the development of computationally intensive applications such as optical system design and natural resource management.

Mr. Hord's six prior books and scores of papers address advanced parallel computing, digital image processing and space technology. He has long been active in the applied imagery pattern recognition community, has been an IEEE Distinguished Visitor, and is a frequent guest lecturer at several universities. His B.S. in physics was granted by Notre Dame University in 1962 and in 1966 he earned an M.S. in physics from the University of Maryland.

INTRODUCTION

As demands emerge for ever greater computer processing speeds and capacities, traditional serial processors have begun to encounter physical laws that inhibit further speed increases. One impediment is the speed of light. Signals cannot propagate faster than about a foot in a nanosecond. Hence, computer components commanded by another component ten feet away cannot respond in less than 10 ns. To defeat this limit, designers try to make computers smaller. This effort encounters limits on the allowed smallness of chip feature sizes and the need to dissipate heat.

The most promising strategy to date for overcoming these limits is the abandonment of serial processing in favor of parallel processing. Parallel processing is the use of multiple processors simultaneously working on one problem. The hope is that if a single processor can generate X floating point operations per second (FLOPS), then ten of these may be able to produce 10X FLOPS, and, in the case of massively parallel processing, a thousand processors may produce 1000X FLOPS.

Problems of obvious interest for parallel processing because of their computational intensity include

- Matrix inversion
- Artificial vision
- Data base searches
- Finite element analysis
- Computational fluid dynamics
- Simulation
- Optical ray trace
- Signal processing
- Optimization

However, the range of applications for parallel processing has proven to be much broader than expected. This volume will examine one type of parallel processing termed MIMD (multiple instruction multiple data) and describe by example the wide variety of application areas that have shown themselves to be well suited to parallel processing.

Parallel processing, or concurrent computing as it is sometimes termed, is not conceptually new. For as long as there have been jobs that can be broken into multiple tasks which in turn can be handed out to individual workers for simultaneous performance, team projects have been an effective way to achieve schedule speedup. In the realm of computation, one recalls the WPA projects of the 1930s to generate trigonometric and logarithmic tables that employed hundred of mathematicians, each calculating a small portion of the total work. Lenses were designed the same way, with each optical engineer tracing one ray through a candidate design.

The recent excitement for parallel computer architectures results from the rising demand for supercomputer performance and the simultaneous maturing of constituent computer technologies that make parallel processing supercomputers a viable possibility.

The term supercomputer enjoys an evolving definition. It has been facetiously defined as those computers that exhibit throughput rates 50% greater than the highest rate currently available. The advent of the term occurred in the 1975 time frame, when it was variously applied to the CDC-7600, the Illiac IV, and other high-performance machines of the day. Upon the arrival of the Cray-1, the usage became firmly established. Today, with the need for high performance computing greater than ever, the supercomputer identifier is commonplace. For the purpose of this book the term supercomputer means that class of computers that share the features of high speed and large capacity compared with the average of what is available at any given time. Both elements are important; high speed on small problems is insufficient. With this definition supercomputers over time cease being supercomputers and retire to the category "former supercomputers".

Parallel processing supercomputers haven't always been technically feasible. They require interprocessor communication to perform sufficiently well that multiple processors can execute an application more quickly than a single processor acting alone can execute that application. Even today we see cases where 32 processors are slower than 16 processors working the same problem, not because the problem is insufficiently parallel, but because the interprocessor communications overhead is too high. These cases are becoming less common as the various constituent technologies mature.

Another technology maturity issue making parallel processing supercomputing feasible today is that the cost of implementation is more and more affordable. VLSI chip technology is revolutionizing the cost-performance characteristics of recent systems.

Parallel computers have been evolving for 25 years. Today they have become an essential and undeniable force in large-scale computing. This book explores their design, their programming methods, and a selection of their applications in some depth.

Background

SUPERCOMPUTING

Supercomputers address the big problems of their time. Because they are expensive, they need a computationally intensive application to warrant their use, but for such problems they are economically justified because they are less costly than their conventional competitors on that class of problems; and, of course, they solve problems that are so big that conventional computers are hopelessly inadequate to address, the so-called grand challenge problems.

Today there are hundreds of supercomputers in the world because there are many applications that offer a very high payoff despite the large costs of supercomputing methods. For example, consider the case described by W. Ballhaus of the NASA Ames Research Center, Mountainview, CA: Supercomputers were used to design the A-310 Airbus Airliner. It is estimated that the use of the supercomputer resulted in a fuel efficiency improvement of 20%. For jet fuel price at \$1.30/gal, over a 15-year life expectancy with 400 flights per year per plane, the fleet life fuel savings due to the improved efficiency is \$10 billion.

Supercomputers achieve their status by virtue of their speed and capacity. In part, their speed is derived from faster circuits; but as Dr. E. W. Martin, Vice President of Boeing Aerospace, said, as quoted in *Photonics Spectra*, January 1991: "In 40 years, electronic logic-element switching speeds have increased only three orders of magnitude; however, computer speeds have increased by nine orders of magnitude. The key to this gain is architectures that provide increased parallelism."

The *New York Times*, November 20, 1990, reported: "The long running debate about how best to make computers thousand of times more powerful than they are today appears to be ending in a stunning consensus. The nation's leading computer designers have agreed that a technology considered the underdog as recently as two years ago has the most potential. The winning technology is known as massively parallel processing. For example, Cray and Convex announced that their companies are embarking on designs for massively parallel computers."

A report prepared for the Department of Energy and the President's office of Science and Technology Policy estimates that sales of massively parallel computers will exceed vector-based supercomputer sales as early as 1996.

MIMD VS. SIMD

Parallel processors fall mainly into two general classes as described in Table 1. The fundamental distinction between the two classes is that one class is MIMD (multiple instruction multiple data), while the other is SIMD (single instruction multiple data). This distinction can be summarized as follows:

Multiple Instruction Multiple Data
Each processor runs its own instruction sequence
Each processor works on a different part of the problem
Each processor communicates data to other parts
Processors may have to wait for other processors or for access to data

Single Instruction Multiple Data
All processors are given the same instruction
Each processor operates on different data
Processors may "sit out" a sequence of instructions

In MIMD architectures several processors operate in parallel in an asynchronous manner and generally share access to a common memory. Two features are of interest to differentiate among designs: the coupling of processor units and memories, and the homogeneity of the processing units.

In tightly coupled MIMD multiprocessors, the number of processing units is fixed, and they operate under the supervision of a strict control scheme. Generally, the controller is a separate hardware unit. Most of the hardware controlled tightly coupled multiprocessors are heterogeneous in the sense that they consist of specialized functional units (e.g., adders, multipliers) supervised by an instruction unit which decodes instructions, fetches operands, and dispatches orders to the functional units.

In SIMD architectures a single control unit (CU) fetches and decodes instructions. Then the instruction is executed either in the CU itself (e.g., a jump instruction) or it is broadcast to a collection of processing elements (PEs). These PEs operate synchronously, but their local memories have different contents. Depending on the complexity of the CU, the processing power and the address-

TABLE 1
Two General Classes of Parallel Processing

MIMD	SIMD
Relatively few powerful processors	Many simple processors
Control level parallelism which assigns a processor to a unit of code	Data level parallelism which assigns a processor to a unit of data
Typically shared memory so can have memory contention	Typically distributed memory so can have data communication problem
Needs good task scheduling for efficiency	Needs good processor utilization for efficiency

ing method of the PEs, and the interconnection facilities between the PEs, we can differentiate among various subclasses of SIMD machines, e.g., array processors, associative processors, and processing ensembles.

FINE GRAIN VS. COARSE GRAIN

There is some confusion about the terms fine grain and coarse grain as applied to parallel computers. Some use the terms to characterize the power of the individual processors. In this sense a fine-grain processor is rather elemental, perhaps operating on a one-bit word and having very few registers. This is contrasted with powerful processing elements in coarse-grain computers, each a full computer in its own right with a 32- or 64-bit word size.

Others have used the terms to differentiate between computers with a small number of processors and those with a large number. More recently those with a small number have been termed simply "parallel", while those with a large number are termed "massively parallel". The dividing line between these classes has been set at 1000 so that computers with more than 1000 processors are termed massively parallel.

SHARED VS. DISTRIBUTED MEMORY

Another distinction between MIMD and SIMD architecture classes is whether the memory is shared or distributed. In shared memory designs all processors have direct access to all of the memory; in distributed memory computers each processor has direct access only to its own local memory. Typically MIMD computers use shared memory, while SIMD computers use distributed memory.

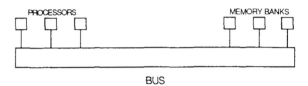

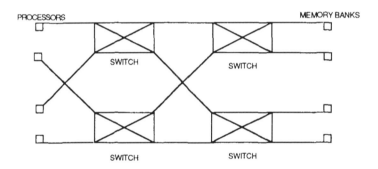

FIGURE 1. Shared memory architecture.

INTERCONNECT TOPOLOGY

The final major distinguishing characteristic among current parallel computers is the topology of the interconnections, i.e., which processors have direct interconnection with which other processors. Diagrams of various examples are shown in Figures 1 and 2. Figure 1 shows variations in connectivity for the shared memory class of architecture. Figure 2 illustrates five kinds of connectivity for the distributed memory architecture class.

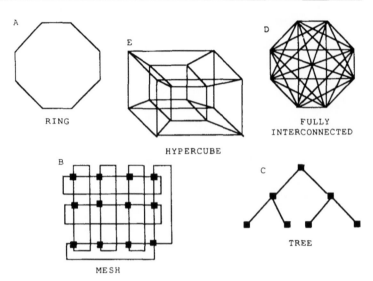

FIGURE 2. Distributed memory architecture. (A) Ring; (B) mesh;
(C) tree; (D) fully interconnected; (E) hypercube.

MIMD CONCEPTS

It is possible to create a vast variety of MIMD parallel programs with just five library functions: one for creating processes; another for destroying processes; a third for sharing memory; and two more for interprocess synchronization. The complexity of parallel programming in a real application is due to the interactions among these primitives.

The five functions are: (1) forks, to create a parallel process; (2) joins, to destroy a parallel process; (3) memory shares; (4) memory locks; and (5) barriers, which are used to synchronize parallel programs.

One way to spread a single calculation among several different processors is to make each part of the calculation into an independent program, or appear to be an independent program to the operating system. In reality, of course, they are not independent because they share intermediate results and they may share data. In MIMD parallel programming, these "independent" programs are termed processes, a generalization of the program concept and the logical foundation of shared memory multiprocessing.

In a time-shared uniprocessor, two users can be running programs at the same time. For parallel processing, this scenario is extended in three ways: the two users are one person, the programs are processes that can communicate, and the two programs do not time share a single processor, but instead each has their own dedicated processor. Naturally this extension from time-shared uniprocessing applies to three, four, and more users. A parallel program may have hundreds or thousands of processes.

SHARED MEMORY

When a calculation has been subdivided among two or more processes, each must be able to carry out its portion of that calculation. For this to be the case, each process must have some part of the total memory that is private to itself to be certain that the value it previously stored will still be there when it goes back to get it. However, there must also be part of memory which all processes can access. These memory locations that can be accessed by all of the processes are termed shared.

7

FORKS

Processes are created by the operating system. The user can command the operating system to create a new process by calling the library function process_fork. Some libraries call this function by a different name; if the function is not in a library at all, the function can be constructed based, for example, on UNIX®* system calls.

The function process_fork is used as follows:

```
integer id, nproc, process_fork
id = process_fork (nproc)
```

The function process_fork (nproc), when called from an executing process, creates nproc-1 additional processes. Each additional process is an exact copy of the caller. The original process, which is called process_fork, is still running, so there are now nproc processes in all. Each of the nproc processes has a private copy of each of the variables of the parent, unless the variable is explicitly shared. The process that made the call is termed the parent, and the newly created processes are termed the children. If nproc = 1 in the original call, no new processes are created, i.e., there are no children.

While the process_fork function is invoked by only one process, it returns nproc times to nproc identical processes of which nproc-1 are new. After the return, each process continues executing at the next executable instruction following the process_fork function call.

The function process_fork returns an integer, called the process-id, which is unique to each process. It returns 0 to the parent and the integers 1,2,3...nproc-1 to the children, each child getting a different number. That each child gets a different number allows the different processes to identify themselves and do a different calculation.

JOINS

The join is the opposite of a fork. It destroys all the children, leaving only the parent running. The join is executed by the statement

```
call process_join()
```

The subroutine process_join has no arguments, so the parentheses are optional. The user's library may have a program with the functionality of process_join but, with a different name.

If a child process makes the process_join call, it is destroyed. If the calling process is the parent (id = 0), it waits until all the children are destroyed, then it

* UNIX is a registered trademark of AT&T.

Portions of the text taken from *Introduction to Parallel Programming*, S. Brawer, Ed., Academic Press, San Diego, CA (Copyright © 1989 by Academic Press, Inc.)

continues with the statement following the call to process_join. When a process is destroyed, the memory it occupied is released for other uses. The operating system no longer has a record of the destroyed process. There is no way to access the data of the destroyed process.

RANDOMLY SCHEDULED PROCESSES

In analyzing shared memory parallel programs it is essential to assume that all processes are randomly scheduled. Any process can be idled and restarted at any time. It can be idled for any length of time. It is incorrect to assume that two processes will proceed at the same rate or that one process will go faster because it has less work to do than another process. The operating system can intervene at any time to idle a process.

This random scheduling of processes can cause the order of the output of different processes to be different in each run. It is also possible for processes to incorrectly alter something in shared memory so that the final value of a shared variable will be different depending on the order in which the variable is altered.

SELF-SCHEDULING AND SPIN-LOCKS

One common way to share a calculation among nproc processors is loop splitting. The general form of loop splitting is

```
    do 1 i=1+id, n, nproc
              (work)
  1 continue
```

On a sequential uniprocessor the loop would start with i = 1 and end n passes later with i = n. On a multiprocessor with nproc processors, process 0 would perform the work of loops i = 1, i = 1 + nproc, i = 1 + 2 * nproc, ..., while process 1 would perform the work of loops i = 2, i = 2 + nproc, i = 2 + 2 * nproc, ..., etc.

The work section of this loop can be long for some values of i, but short for other values of i. Consider an example where several processes are exploring a maze. Some processes may reach a dead end in a short time; others may take a long time. Some processes may explore short paths, while others may explore long paths.

The alternative is self-scheduling; each process chooses its next index value only when it is ready to begin executing with it. Self-scheduling allows some processes to execute only a few iterations, while others execute many. This permits the ensemble of processes to finish more closely at the same time, keeping all processors in use. Self-scheduling is invoked by

```
    call shared (next_index, ...)
```

In the call to shared, next_index has the next available value of the index i that will be used by the process. After the process assigns the value of the next_index to i, it decrements next_index so that the next process can assign the next lower value of the index. The intention of the program is to allocate the values of i to processes on a first-come, first-served basis.

Difficulty arises in the event of contention, i.e., if two or more processes try to alter the value of next_index at the same time. If process 0 assigns i = next_index and process 1 assigns i = next_index before process 0 can decrement next_index, a program can perform incorrectly. To prevent this, the assignment and the decrement operations must be treated as a unit so that there is no interruption and, once started, both operations complete. This section of operations is therefore "protected" in that while a process is executing protected code, all other processes are locked out.

In order to enforce protection, it is necessary that one process communicate with all other processes that it is in a protected portion of the program and all other processes must not interfere. Such communication is an example of synchronization.

In the case of self-scheduling, the required software structure for implementing this synchronization is the spin-lock. The spin-lock is normally supplied by the operating system.

The use of the spin-lock requires three function calls and a shared variable. The statement:

```
spin_lock_init (lock_word, lock_condition)
```

initializes the spin-lock. A spin-lock only has two conditions, locked and unlocked. The integer lock_word is shared by all the processes and describes the state of the spin-lock. For example, when the state of the spin-lock is unlocked, then lock_word is zero, which means that there is no process executing instructions in a protected region of the code. When lock_word is one, some process is in a protected region and the spin-lock is locked. The lock_condition value in the call to initialize indicates whether the initial condition is for the spin-lock to be locked or unlocked.

Besides the initialization, there are two functions that actually perform the work of protecting a group of instructions:

```
call spin_lock (lock_word)
call spin_unlock (lock_word)
```

A process calls spin-lock when it is about to enter a protected region of its code. When called, the function first checks to see whether the lock is unlocked. If it is, the lock is locked and the process proceeds into the protected region. However, if the function finds the lock is already locked, then the calling process must wait until the lock becomes unlocked. Thus, the process that seeks to enter a protected region cannot proceed until the other process that caused the lock to go to the locked condition completes its protected code and unlocks the lock. It is this behavior that gives rise to the name spin-lock. The call to spin-lock puts the process into a loop

```
100 continue
        if (lock_word.eq.1) go to 100
```

that spins around and around until lock_word has the value 0, indicating that the lock has been unlocked. When this happens, only one of the waiting processes can enter the protected region; the remaining processes keep spinning while they wait their turn. The order in which waiting processes enter the protected region is undefined. When properly employed, a spin-lock eliminates contention, i.e., only one process is able to update a shared variable at a time.

The spin-lock is the most basic synchronization mechanism. Practical parallel programming requires a generous use of locks. However, an inefficient implementation of locks can make a parallel program run slower than the sequential version.

BARRIERS

A barrier causes processes to wait and allows them to proceed only after a predetermined number of processes are waiting at the barrier. It is used to ensure that one stage of a calculation has been completed before the processes proceed to a next stage, which requires the results of the previous stage. Barriers are used to eliminate race conditions, in which the result of a calculation depends on the relative speed at which the processes execute.

Part 1
MIMD Computers

In this part we examine a representative sample of MIMD computers, both those that are commercially available and the research machines not yet available to the general user community. Each is addressed in its own chapter, some with application examples.

In the rapidly evolving world of MIMD computers, particularly at the high-performance end of the range, the more recent machines are more advanced than earlier implementations. This is reflected in the selection of computers covered here; for example, the BBN Butterfly and Encore Multimax both played important roles in the development of the technology, but are no longer offered in the marketplace.

The evolutionary progress of MIMD computers is not as advanced as SIMD computers. The 1992 Gorden Bell Prize for parallel processing achievement went, in the MIMD arena, to the first application to exceed one GigaFLOPS. The SIMD part of the price was awarded to a 14-GigaFLOPS application. However, the MIMD paradigm is believed to be richer in application opportunities.

Commercial Machines

1 THINKING MACHINES CORPORATION CM-5

OVERVIEW

In October 1991, Thinking Machines Corporation (TMC) introduced the Connection Machine model CM-5. The design of the CM-5 surprised the parallel high-performance computing community because many of the features are distinct departures from the pattern that characterized prior TMC products. Prior Connection Machine models were SIMD, but the CM-5 has MIMD capabilities. They were hypercube connected, but the CM-5 is not. They had custom chips, but the CM-5 uses a standard off-the-shelf RISC microprocessor. The CM-5 has vector pipes, four per node, while prior machines did not. It has a Unix-like operating system, new to TMC.

TMC emphasizes the scalability of the design. Since one can always increase performance by adding more processors, the question becomes does the I/O, communications and reliability grow in proportion with the size of the system. Danny Hillis, TMC's chief scientist, believes the CM-5 architecture answers that question and can grow to teraFLOPS throughput rates.

PROCESSORS

A CM-5 (Figure 1) system may contain hundreds or thousands of parallel processing nodes. Each node has its own memory. Nodes can fetch from the same address in their respective memories to execute the same (SIMD-style) instruction, or from individually chosen addresses to execute independent (MIMD-style) instructions.

The processing nodes are supervised by a control processor, which runs an enhanced version of the UNIX operating system. Program loading begins on the control processor; it broadcasts blocks of instructions to the parallel processing nodes and then initiates execution. When all nodes are operating on a single control thread, the processing nodes are kept closely synchronized and blocks are broadcast as needed. (There is no need to store an entire copy of the program at each node.) When the nodes take different branches, they fetch instructions

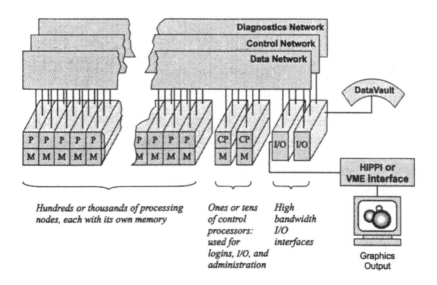

FIGURE 1. Organization of the CM system.*

independently and synchronize only as required by the algorithm under program control.

To maximize system usefulness, a system administrator may divide the parallel processing nodes into groups, known as *partitions*. There is a separate control processor, known as a *partition manager*, for each partition. Each user process executes on a single partition, but may exchange data with processes on other partitions. Since all partitions utilize UNIX time-sharing and security features, each allows multiple users to access the partition while ensuring that no user's program interferes with another's.

Other control processors in the CM-5 system manage the system's I/O devices and interfaces. This organization allows a process on any partition to access any I/O device and ensures that access to one device does not impede access to other devices. Figure 2 shows how this distributed control works with the CM-5 interprocessor communication networks to enhance system efficiency.

Functionally, the CM-5 is divided into three major areas. The first contains some number of partitions, which manage and execute user applications; the second contains some number of I/O devices and interfaces; and the third contains the two interprocessor communication networks that connect all parts of the first two areas. (A fourth functional area, covering system management and diagnostics, is handled by a third interprocessor network and is not shown in Figure 2.)

Because all areas of the system are connected by the Data Network and the Control Network, all can exchange information efficiently. The two networks provide high bandwidth transfer of messages of all sorts: downloading code from

---

* Figures and portions of the text were taken from The Connection Machine CM-5 Technical Summary, Thinking Machines Corp., Cambridge, MA, Oct. 1991.

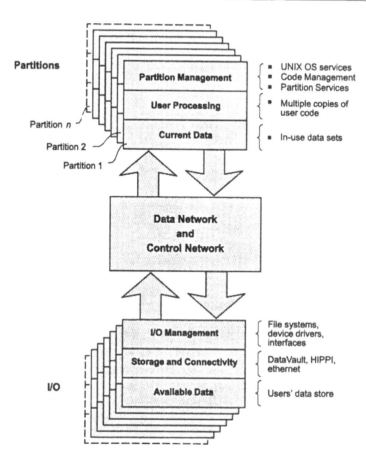

FIGURE 2. Distributed control on the CM-5.

a control processor to its nodes, passing I/O requests and acknowledgments between control processors, and transferring data, either among nodes (whether in a single partition or in different partitions) or between nodes and I/O devices.

NETWORKS

Every control processor and parallel processing node in the CM-5 is connected to two scalable interprocessor communication networks, designed to give low latency combined with high bandwidth in any possible configuration a user may wish to apply to a problem. Any node may present information, tagged with its logical destination, for delivery via an optimal route. The network design provides low latency for transmissions to near neighboring addresses, while preserving a high, predictable bandwidth for more distant communications.

The two interprocessor communications networks are the Data Network and the Control Network. In general, the Control Network is used for operations that involve all the nodes at once, such as synchronization operations and broadcasting; the Data Network is used for bulk data transfers where each item has a single source and destination.

A third network, the Diagnostics Network, is visible only to the system administrator; it keeps tabs on the physical well-being of the system.

External networks, such as Ethernet and FDDI, may also be connected to a CM-5 system via the control processors.

I/O

The CM-5 runs a UNIX-based operating system; it provides its own high-speed parallel file system and also allows full access to ordinary NFS file systems. It supports both high-performance parallel interface (HIPPI) and VME interfaces, thus allowing connections to a wide range of computers and I/O devices, while using standard UNIX commands and programming techniques throughout. A CMIO interface supports mass storage devices such as the DataVault and enables sharing of data with CM-2 systems.

I/O capacity may be scaled independently of the number of computational processors. A CM-5 system of any size can have the I/O capacity it needs, whether that be measured in local storage, in bandwidth, or in access to a variety of remote data sources. Communications capacity scales both with processors and with I/O.

Just as every partition is managed by a control processor, every I/O device is managed by an input/output control processor (IOCP), which provides the software that supports the file system, device driver, and communications protocols. Like partitions, I/O devices and interfaces use the Data Network and the Control Network to communicate with processes running in other parts of the machine. If greater bandwidth is desired, files can be spread across multiple I/O devices: a striped set of eight DataVaults, for example, can provide eight times the I/O bandwidth of a single DataVault.

The same hardware and software mechanisms that transfer data between a partition and an I/O device can also transfer data from one partition to another (through a named UNIX pipe) or from one I/O device to another.

A UNIVERSAL ARCHITECTURE

The architecture of the CM-5 is optimized for data parallel processing of large, complex problems. The Data Network and Control Network support fully general patterns of point-to-point and multiway communication, yet reward patterns that exhibit good locality (such as nearest-neighbor communications) with reduced latency and increased throughput. Specific hardware and software support improve the speed of many common special cases.

Two more key facts should be noted about the CM-5 architecture. First, it depends on no specific types of processors. As new technological advances

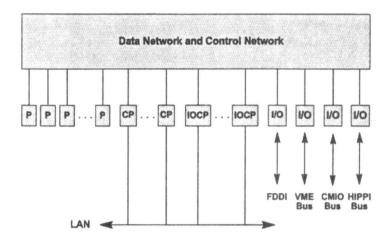

FIGURE 3. System components.

arrive, they can be moved with ease into the architecture. Second, it builds a seamlessly integrated system from a small number of basic types of modules. This creates a system that is thoroughly scalable and allows for great flexibility in configuration.

ARCHITECTURE DETAILS

A Connection Machine Model CM-5 system contains thousands of computational processing nodes, one or more control processors, and I/O units that support mass storage, graphic display devices, and VME and HIPPI peripherals. These are connected by the Control Network and the Data Network. (For a high-level sketch of these components, see Figure 3.)

A CM-5 system contains tens, hundreds, or thousands of processing nodes, each with up to 128 Mflops of 64-bit floating-point performance. It also contains a number of I/O devices and external connections. The number of I/O devices and external connections is independent of the number of processing nodes. Both processing and I/O resources are managed by a relatively small set of control processors. All these components are uniformly integrated into the system by two internal communications networks, the Control Network and the Data Network.

PROCESSORS

Every processing node is a general-purpose computer that can fetch and interpret its own instruction stream, execute arithmetic and logical instructions, calculate memory addresses, and perform interprocessor communication. The processing nodes in a CM-5 system can perform independent tasks or collaborate

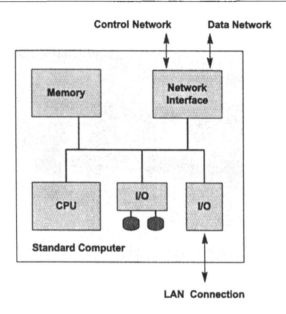

FIGURE 4. Control processor.

on a single problem. Each processing node has 8, 16, or 32 Mbytes of memory; with the high-performance arithmetic accelerator, it has the full 32 Mbytes of memory and delivers up to 128 Mips or 128 Mflops.

The control processors are responsible for administrative actions such as scheduling user tasks, allocating resources, servicing I/O requests, accounting, enforcing security, and diagnosing component failures. In addition, they may also execute some of the code for a user program. Control processors have the same general capabilities as processing nodes, but are specialized for performing managerial functions rather than computational functions. For example, control processors have additional I/O connections and lack the high-performance arithmetic accelerator. (See Figure 4.)

The basic CM-5 control processor consists of a RISC microprocessor, memory subsystem, I/O (including local disks and Ethernet connections), and a CM-5 Network Interface (NI), all connected to a standard 64-bit bus. Except for the NI, this is a standard off-the-shelf workstation-class computer system. The NI connects the control processor to the rest of the system through the Control Network and Data Network. Each control processor runs CMost, a UNIX-based operating system with extensions for managing the parallel-processing resources of the CM-5. Some control processors are used to manage computational resources and some are used to manage I/O resources.

In a small system, one control processor may play a number of roles. In larger systems, individual control processors are often dedicated to particular tasks and referred to by names that reflect those tasks. Thus, a control processor that manages a partition and initiates execution of applications on that partition is referred to as a partition manager (PM), while a processor that controls an I/O device is called an I/O control processor (IOCP).

NETWORKS AND I/O

The Control Network provides tightly coupled communications services. It is optimized for fast response (low latency). Its functions include synchronizing the processing nodes, broadcasting a single value to every node, combining a value from every node to produce a single result, and computing certain parallel prefix operations.

The Data Network provides loosely coupled communications services. It is optimized for high bandwidth. Its basic function is to provide point-to-point data delivery for tens of thousands of items simultaneously. Special cases of this functionality include nearest-neighbor communication and FFT butterflies. Communications requests and data delivery need not be synchronized. Once the Data Network has accepted a message, it takes on all responsibility for its eventual delivery; the sending processor can then perform other computations while the message is in transit. Recipients may poll for messages or be notified by interrupt on arrival. The Data Network also transmits data between the processing nodes and I/O units.

A standard NI connects each node or control processor to the Control Network and Data Network. This is a memory-mapped control unit; reading or writing particular memory addresses will access network control registers or trigger communication operations.

The I/O units are connected to the Control Network and Data Network in exactly the same way the processors, using the same NI. Many I/O devices require more data bandwidth than a single NI can provide; in such cases multiple NI units are ganged. For example, a HIPPI channel interface contains six NI units, which provide access to six Data Network ports. (At 20 Mbytes/s apiece, six NI units provide enough bandwidth for a 100-Mbyte/s HIPPI interface with some to spare.)

Individual I/O devices are controlled by dedicated IOCPs. Some I/O devices are interfaces to external buses or networks; these include interfaces to VME buses and HIPPI channels. Noteworthy features of the I/O architecture are that I/O and computation can proceed independently and in parallel, that data may be transferred between I/O devices without involving the processing nodes, and that the number of I/O devices may be increased completely independently of the number of processing nodes.

In the background is a third network, the Diagnostic Network. It can be used to isolate any hardware component and to test both the component itself and all connections to other components. The Diagnostic Network pervades the hardware system but is completely invisible to the user; indeed, it is invisible to most of the control processors. A small number of the control processors include command interfaces for the Diagnostic Network; at any given time, one of these control processors provides the System Console function.

THE USER-LEVEL VIRTUAL MACHINE

The virtual machine provided by the hardware and operating system to a single-user task consists of a control processor acting as a PM, a set of processing nodes, and facilities for interprocessor communication. Each node is an ordinary, general-purpose microprocessor capable of executing code written in C, Fortran, or assembly language. The processing nodes may also have optional vector units for high arithmetic performance.

The operating system is CMost, a version of SunOS enhanced to manage CM-5 processor, I/O, and network resources. The PM provides full UNIX services through standard UNIX system calls. Each processing node provides a limited set of UNIX services.

A user task consists of a standard UNIX process running on the PM and a process running on each of the processing nodes. Under time-sharing, all processors are scheduled *en masse*, so that all are processing the same user task at the same time. Each process of the user task, whether on the PM or on a processing node, may execute completely independently of the rest during their common time slice.

The Control Network and Data Network allow the various processes to synchronize and transfer data among themselves. The unprivileged control registers of the network interface hardware are mapped into the memory space of each user process, so that user programs on the various processors may communicate without incurring any operating system overhead.

COMMUNICATIONS FACILITIES

Each process of a user task can read and write messages directly to the Control Network and the Data Network. The network used depends on the task to be performed.

The Control Network is responsible for communications patterns in which many processors may be involved in the processing of each datum. One example is broadcasting, where one processor provides a value and all other processors receive a copy. Another is reduction, where every processor provides a value and all values are combined to produce a single result. Values may be combined by summing them, finding the maximum input value, or taking the logical OR or exclusive OR of all input values; the combined result may be delivered to a single processor or to all processors. (Software provides minimum-value and logical AND operations by inverting the inputs, applying the hardware maximum-value or logical OR operation, then inverting the result.) Note that the control processor does not play a privileged role in these operations; a value may be broadcast from, or received by, the control processor or any processing node with equal facility.

The Control Network contains integer and logical arithmetic hardware for carrying out reduction operations. This hardware is distinct from the arithmetic hardware of processing nodes; Control Network operations may be overlapped with arithmetic processing by the processors themselves. The arithmetic hardware of the Control Network can also compute various forms of parallel prefix operations, where every processor provides a value and receives a result; the nth result is produced by combining the first n input values. Segmented parallel prefix operations are also supported in hardware.

The Control Network provides a form of two-phase barrier synchronization (also known as "fuzzy" or "soft" barriers). A processor can indicate to the Control Network that it is ready to enter the barrier. When all processors have checked in, the Control Network relays this fact to all processors. A processor can thus overlap unrelated processing with the possible waiting period between the time it has checked in and the time it has been determined that all processors have checked in. This allows thousands of processors to guarantee the ordering of certain of their operations without ever requiring that they all be exactly synchronized at one given instant.

The Data Network is responsible for reliable, deadlock-free point-to-point transmission of tens of thousands of messages at once. Neither the senders nor the receivers of a message need be globally synchronized. At any time, any processor may send a message to any processor in the user task. This is done by writing first the destination processor number, and then the data to be sent, to control registers in the NI. Once the Data Network has accepted the message, it assumes all responsibility for eventual delivery of the message to its destination. In order for a message to be delivered, the processor to which it was sent must accept the message from the Data Network. However, processor resources are not required for forwarding messages. The operation of the Data Network is independent of the processing nodes, which may carry out unrelated computations while messages are in transit.

There is no separate interface for special patterns of point-to-point communication, such as nearest neighbors within a grid. The Data Network presents a uniform interface to the software. The hardware implementation, however, has been tuned to exploit the locality found in commonly used communication patterns.

Data Network performance follows a simple model. The Data Network provides enough bandwidth for every NI to sustain data transfers at 20 Mbytes/s to any other NI within its group of 4; at 10 Mbytes/s to any other NI within its group of 16; or at 5 Mbytes/s to any other NI in the system. (Two NIs are in the same group of $2k$ if their network addresses differ on the k lowest-order bits.) These figures are for maximum sustained network hardware performance, which is sufficient to handle the transfer rates sustainable by node software. Note that worst-case performance is only a factor of four worse than best-case performance. Other network designs have much larger worst/best ratios.

To see the consequences of this performance model, consider communication within a two-dimensional grid. If, say, the processors are organized so that each group of 4 represents a 2×2 patch of the grid, and each group of 16 processors represents a 4×4 patch of the grid, then nearest-neighbor communication can be sustained at the maximum rate of 20 Mbytes/s per processor. For within each group of four, two of the processors have neighbors in a given direction (north, east, west, south) that lie within the same group and therefore can transmit at the maximum rate. The other two processors have neighbors outside the group of four. However, the Data Network provides bandwidth of 40 Mbytes/s out of that group, enough for each of the 4 processors to achieve 10 Mbytes/s within a group of 16. The same argument applies to the 4 processors in a group of 16 that have neighbors outside the group: not all processors have neighbors outside the group, so their outside-the-group bandwidth can be borrowed to provide maximum bandwidth to processors that do have neighbors outside the group.

There are two mechanisms for notifying a receiver that a message is available. The arrival of a message sets a status flag in a NI control register; a user program can poll this flag to determine whether an incoming message is available. The arrival of a message can also optionally signal an interrupt. Interrupt handling is a privileged operation, but the operating system converts an arrived-message interrupt into a signal to the user process. Every message bears a 4-bit tag; under operating system control, some tags cause message-arrival interrupts and others do not. (The operating system reserves certain of the tag numbers for its own use; the hardware signals an invalid-operation interrupt to the operating system if a user program attempts to use a reserved message tag.)

The Control Network and Data Network provide flow control autonomously. In addition, two mechanisms exist for notifying a sender that the network is temporarily clogged. Failure of the network to accept a message sets a status flag in a NI control register; a user program can poll this flag to determine whether a retry is required. Failure to accept a message can also optionally signal an interrupt.

Data can also be transferred from one user task to another, or to and from I/O devices. Both kinds of transfer are managed by the operating system using a common mechanism. An intertask data transfer is simply an I/O transfer through a named UNIX pipe.

DATA PARALLEL COMPUTATIONS

While the user may code arbitrary programs for the various processors and put the general capabilities of the network interface to any desired use, the CM-5 architecture is designed to support especially well the data parallel model of programming. Parallel programs are often structured as alternating phases of local computation and global communication. Local computation consists of operations by each processor on the data in its own memory. Global communications includes any transfer of data between or among processors, possibly with arithmetic or logical computation on the data as it is transferred. By managing data transfers globally and coherently rather than piecemeal, the data parallel model often realizes economies of scale, reducing the overhead of synchronization for interprocessor communication. Frequently used patterns of communication are captured in carefully tuned compiler code generators and run-time library routines; they are presented as primitive operators or intrinsic functions in high-level languages so that the programmer need not constantly reinvent them.

The following sections discuss various aspects of the data parallel programming model and sketch the ways in which each is supported by the CM-5 architecture and communications structure.

ELEMENTAL AND CONDITIONAL COMPUTATIONS

Elemental computations, which involve operating on corresponding elements of arrays, are purely local computations if the arrays are divided in the same way among the processors. If two such matrices are to be added together, for example, every pair of numbers to be added resides together in the memory of a single processing node, and that node takes responsibility for performing the addition.

Because each processing node executes its own instruction stream as well as processing its own local data, conditional operations are easily accommodated. For example, one processing node might contain an element on the boundary of an array, while another might contain an interior element; certain filtering operations, while allowing all elements to be processed at once, require differing computations for boundary elements and interior elements. In the CM-5 data parallel architecture, some processors can take one branch of a conditional and others can take a different branch simultaneously with no loss of efficiency.

REPLICATION

Replication consists of making copies of data. The most important special case is broadcasting, in which copies of a single item are sent to all processors. This is supported directly in hardware by the Control Network.

Another common case is spreading, in which copies of elements of a lower dimensional array are used to fill out the additional dimensions of a higher dimensional array. For example, a column vector might be spread into a matrix, so that each element of the vector is copied to every element of the corresponding row of the matrix. This case is handled by a combination of hardware mechanisms.

If the processors are partitioned into clusters of differing size, such that the network addresses within each cluster are contiguous, then one or two parallel prefix operations by the Control Network can copy a value from one processor in each cluster to all others in that cluster with particular speed.

REDUCTION

Reduction consists of combining many data elements to produce a smaller number of results. The most important special case is global reduction, in which every processor contributes a value and a single result is produced. The operations of integer summation, finding the integer maximum, logical OR, and logical exclusive OR are supported directly in hardware by the Control Network. Floating-point reduction operations are carried out by the nodes with the help of the Control Network and Data Network.

A common operation sequence is a global reduction immediately followed by a broadcast of the resulting value. The Control Network supports this combination as a single step, carrying it out in no more time than a simple reduction.

The cases of reduction along the axes of a multidimensional array correspond to the cases of spreading into a multidimensional array and have similar solutions. The rows of a matrix might be summed, for example, to form a column matrix. This case is handled by a combination of hardware mechanisms.

If the processors are partitioned into clusters of differing size, such that the network addresses within each cluster are contiguous, then one or two parallel prefix operations by the Control Network can reduce values from all processors within each cluster and optionally redistribute the result for that cluster to all processors in that cluster.

PERMUTATION

The Data Network is specifically designed to handle all cases of permutation, where each input value contributes to one result and each result is simply a copy of one input value. The Data Network has a single, uniform hardware interface and a structure designed to provide especially good performance when the pattern of exchange exhibits reasonable locality. Both nearest-neighbor and nearest-but-one-neighbor communication within a grid are examples of patterns with good locality. These particular patterns also exhibit regularity, but regularity is not a requirement for good Data Network performance. The irregular polygonal tessellations of a surface or a volume that are typical of finite-element methods lead to communications patterns that are irregular but local. The Data Network performs as well for such patterns as for regular grids.

PARALLEL PREFIX

Parallel prefix operations embody a very specific, complex yet regular, combination of replication and reduction operations. A parallel prefix operation produces as many results as there are inputs, but each input contributes to many results and each result is produced by combining multiple inputs. Specifically, the inputs and results are linearly ordered; suppose there are n of them. Then result j is the reduction of the first j inputs; it follows that input j contributes to the last $n - j + 1$ results. (For a reverse parallel prefix operation—also called a parallel suffix operation — these are reversed: result j is the reduction of the last $n - j + 1$ inputs, and input j contributes to the first j results.)

The Control Network handles parallel prefix (and parallel suffix) operations directly, in the same manner and at the same speed as reduction operations, for integer and logical combining operations. The input values and the results are linearly ordered by network address.

The Control Network also directly supports segmented parallel prefix operations. If the processors are partitioned into clusters of differing size, such that the network addresses within each cluster are contiguous, then a single Control Network operation can compute a separate parallel prefix or suffix within each cluster.

More complex cases of parallel prefix operations, such as on the rows or columns of a matrix or on linked lists, are variously handled through the Control Network or Data Network in cooperation with the nodes.

VIRTUAL PROCESSORS

Data parallel programming provides the high-level programmer with the illusion of as many processors as necessary; one programs as if there were a processor for every data element to be processed. These are often described as *virtual processors*, by analogy with conventional virtual memory, which provides the illusion of having more main memory than is physically present.

The CM-5 architecture, rather than implementing virtual processors entirely in firmware, relies primarily on software technology to support virtual processors.

CM-5 compilers for high-level data parallel languages generate control-loop code and run-time library calls to be executed by the processing nodes. This provides the same virtual processor functionality made available by the Paris instruction set on the Connection Machine Model CM-2, but adds further opportunities for compile-time optimization.

LOW-LEVEL USER PROGRAMMING

Low-level programs may be written for the CM-5 in C or Fortran 77. Assembly language is also available, though C should be adequate for most low-level purposes; all hardware facilities are directly accessible to the C programmer. A special assembler allows hand coding of individual vector instructions for the processing nodes.

One writes low-level programs as two pieces of code: one piece is executed in the control processor, and the other is replicated at program start-up and executed by each processing node. One speaks of writing a program in "C & C" (a C program for the control processor and a C program for the nodes); one may also write in "Fortran & Fortran" ("F & F") or in "C & assembler", etc.

A package of macros and run-time functions supports common communications operations within a message-passing framework. Such low-level communications access allows the user to experiment with MIMD program organizations other than data parallel, to port programs easily from other MIMD architectures, and to implement new primitives for use in high-level programs.

CONTROL PROCESSOR ARCHITECTURE

A control processor (CP) is essentially like a standard high-performance workstation computer. It consists of a standard RISC microprocessor, associated memory and memory interface, and perhaps I/O devices such as local disks and Ethernet connections. It also includes a CM-5 NI, providing access to the Control Network and Data Network.

A CP acting as a PM controls each partition and communicates with the rest of the CM-5 system through the Control Network and Data Network. For example, a PM initiates I/O by sending a request through the Data Network to a second CP, an I/O CP. A PM initiates task-switching by using the Control Network to send a broadcast interrupt to all processing nodes; privileged operating system support code in each node then carries out the bulk of the work. To access the Control Network and Data Network, each CP uses its NI, a memory-mapped device in the memory address space of its microprocessor.

The microprocessor supports the customary distinction between user and supervisor code. User code can run in the CP at the same time that user code for the same job is running in the processing nodes. Protection of the supervisor, and of one user from another, is supported by the same mechanisms used in workstations and single-processor time-shared computers, namely memory address mapping and protection and the suppression of privileged operations in user mode. In particular, the operating system prevents a user process from performing privileged NI operations; the privileged control registers simply are not mapped into the user address space.

The initial implementation of the CM-5 CP uses a SPARC microprocessor. However, it is expected that, over time, the implementation of the CP will track the RISC microprocessor technology curve to provide the best possible functionality and performance at any given point in time; therefore it is recommended that low-level programming be carried out in C as much as possible, rather than in assembly language.

PROCESSING NODE ARCHITECTURE

The CM-5 processing node is designed to deliver very good cost performance when used in large numbers for data parallel applications. Like the CP, the node makes use of industry-standard RISC microprocessor technology. This microprocessor may optionally be augmented with a special high-performance hardware arithmetic accelerator that uses wide datapaths, deep pipelines, and large register files to improve peak computational performance.

The node design is centered around a standard 64-bit bus. To this node bus are attached a RISC microprocessor, a CM-5 NI, and memory. Note that all logical connections to the rest of the system pass through the NI.

The node memory consists of standard DRAM chips and a 2-Kbyte boot ROM; the microprocessor also has a 64-Kbyte cache that holds both instructions

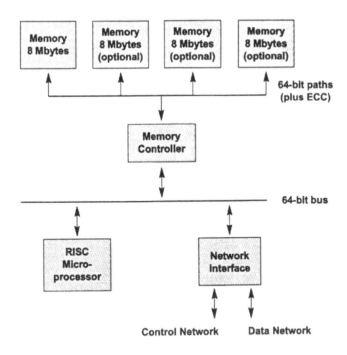

FIGURE 5. Processing node.

and data. All DRAM memory is protected by ECC checking, which corrects single-bit failure and detects 2-bit errors and DRAM chip failures. The boot ROM contains code to be executed following a system reset, including local processor and memory verification and the communications code needed to download further diagnostics or operation system code.

The memory configuration depends on whether the optional high-performance arithmetic hardware is included. Without the arithmetic hardware, the memory is connected by a 72-bit path (64 data bits plus 8 ECC bits) to a memory controller that in turn is attached to the node bus. (See Figure 5.) In this configuration the memory size can be 8, 16, or 32 Mbytes. (This assumes 4-Mbit DRAM technology. Future improvements in DRAM technology will permit increases in memory size. The CM-5 architecture and chip implementations anticipate these future improvements.)

The basic CM-5 processing node consists of a RISC microprocessor, memory subsystem, and a CM-5 Network Interface all connected to a standard 64-bit bus. The RISC microprocessor is responsible for instruction fetch, instruction execution, processing data, and controlling the NI. The memory subsystem consists of a memory controller and either 8, 16, or 32 Mbytes of DRAM memory. The path from each memory back to the memory controller is 72 bits wide, consisting of 64 data bits and 8 bits of ECC code. The ECC circuits in the memory controller can correct single-bit errors and detect double-bit errors as

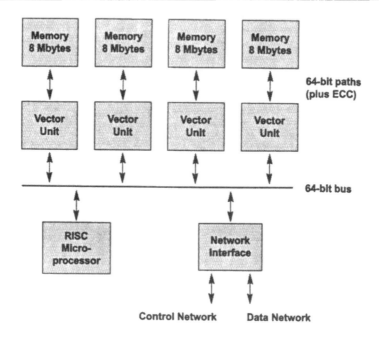

FIGURE 6 . Processing node with VUs.

well as failure of any single DRAM chip. The NI connects the node to the rest of the system through the Control Network and Data Network.

If the high-performance arithmetic hardware is included, then the node memory is divided into four independent banks, each with a 72-bit (64 data bits plus 8 ECC bits) access path. (See Figure 6.)

A CM-5 processing node may optionally contain an arithmetic accelerator. In this configuration the node has a full 32 Mbytes of memory, four banks of 8 Mbytes each. The memory controller is replaced by four vector units (VUs). Each VU has a dedicated 72-bit path to its associated memory bank, providing peak memory bandwidth of 128 Mbytes/s per VU, and performs all the functions of a memory controller, including generation and checking of ECC bits. Each VU has 43 MFLOPS peak 64-bit floating-point performance and 32 FLOPS peak 64-bit integer performance. The VUs execute vector instructions issued to them by the RISC microprocessor. Each vector instruction may be issued to a specific VU (or pair of units), or broadcast to all four VUs at once. The microprocessor takes care of such "housekeeping" computations as address calculation and loop control, overlapping them with vector instruction execution. Together, the VUs provide 512 Mbytes/s memory bandwidth and 128 MFLOPS peak 64-bit floating-point performance. A single CM-5 node with vector units is a supercomputer in itself.

The special arithmetic hardware consists of four VUs, one for each memory bank, connected separately to the node bus. (See Figure 7.) In this configuration

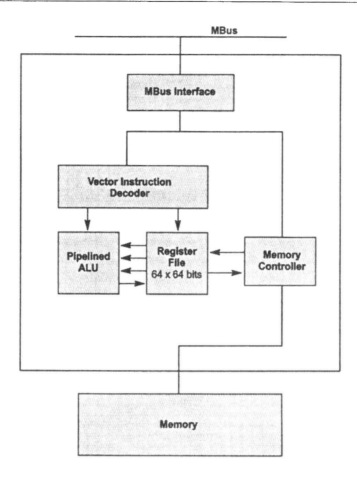

FIGURE 7. VU functional architecture.

the memory size is 8 Mbytes per VU for a total of 32 Mbytes per node. (Again, this figure assumes 4-Mbit DRAM technology and will increase as industry-standard memories are improved.) Each VU also implements all memory controller functions, including ECC checking so that the entire memory appears to be in the address space of the microprocessor exactly as if the arithmetic hardware were not present.

The memory controller or vector unit also provides a word-based interface to the system Diagnostics Network. This provides an extra communications path to the node; it is designed to be slow but reliable and is used primarily for hardware fault diagnosis.

As with the CPs, the implementation of the CM-5 processing node is expected to track the RISC microprocessor technology curve to provide the best possible functionality and performance at any given point in time; therefore, it is recommended that low-level programming be carried out in C as much as possible, rather than in assembly language. The initial implementation of the CM-5 node uses a SPARC microprocessor.

VU ARCHITECTURE

Each VU is a memory controller and computational engine controlled by a memory-mapped control-register interface. When a read or write operation on the node bus addresses a VU, the memory address is further decoded. High-order bits indicate the operation type:

- For an ordinary *memory transaction*, the low-order address bits indicate a location in the memory bank associated with the VU, which acts as a memory controller and performs the requested memory read or write operation.
- For a *control register access*, the low-order address bits indicate a control register to be read or written.
- For a *data register access*, the low-order address bits indicate a data register to be read or written.
- For a *VU instruction*, the node memory bus operation must be *write* (an attempt to *read* from this part of the address space results in a bus error). The data on the memory bus is not written to memory, but is interpreted as an instruction to be executed by the vector execution portion of the VU. The low-order address bits indicate a location in the memory bank associated with the VU; the instruction will use this address if it includes operations on memory. A VU instruction may be addressed to any single VU (in which case the other three VUs will ignore it), to a pair of VUs, or to all four VUs simultaneously.

The first types of operation are identical to those performed by the memory controller when VUs are absent. The third type permits the microprocessor to read or write the register file of any VU. The fourth type of operation initiates high-performance arithmetic computation. This computation has both vector and parallel characteristics: each VU can perform vector operations, and a single instruction may be issued simultaneously to all four. If the vector length is 16, then issuing a single instruction can result in as many as 64 individual arithmetic operations (or 128 if the instruction specifies a compound operation such as *multiply-add*).

VUs cannot fetch their own instructions; they merely react to instructions issued to them by the microprocessor. The instruction format, instruction set, and maximum vector length (VL) have been chosen so that the microprocessor can keep the VUs busy while having time of its own to fetch instructions (both its own and those for the VUs), calculate addresses, execute loop and branch instructions, and carry out other algorithmic bookkeeping.

Each VU has 64 64-bit registers, which can also be addressed as 128 32-bit registers. There are some other control registers as well, most notably the 16-bit vector mask (VM) and the 4-bit VL registers. The VM register controls certain conditional operations and optionally receives single-bit status results for each vector element processed. The VL register specifies the number of elements to be processed by each vector instruction.

The VU actually processes both vector and scalar instructions; a scalar-mode instruction is handled as if it were a vector-mode instruction of length 1. Thus, scalar-mode instructions always operate on single registers; vector-mode instructions operate on sequences of registers. Each register operand is specified by a 7-bit starting register number and a 7-bit stride. The first element for that vector operand is taken from the starting register; thereafter the register number is incremented by the stride to produce a new register number, indicating the next element to be processed. Using a large stride has the same effect as using a negative stride, so it is possible to process a vector in reverse order. Most instruction formats use a default stride of 1 for 32-bit operands or 2 for 64-bit operands, so as to process successive registers, but one instruction format allows arbitrary strides to be specified for all operands, and another allows one vector operand to take its elements from an arbitrary pattern of registers by means of a mechanism for indirect addressing of the register bit.

Each VU includes an adder, a multiplier, memory load/store, indirect register addressing, indirect memory addressing, and population count. Every VU instruction can specify at least one arithmetic operation and in independent memory operation. Every instruction also has four register-address fields: three for the arithmetic operation and one for the memory operation. All binary arithmetic operations are fully three-address; an addition, for example, can read two source registers and write into a third destination register. The memory operation can address a completely independent register. If, however, a load operation addresses a register that is also a source for the arithmetic operation, then load-chaining occurs, so that the loaded memory data is used as an arithmetic operand in the same instruction. Indirect memory addressing supports scatter/gather operations and vectorized pointer indirection.

Two mechanisms provide for conditional processing of vector elements within each processing node. Each VU contains a VM register; vector elements are not processed in positions where the corresponding VM bit is zero. Alternatively, a VM enumeration mechanism may be used in conjunction with the scatter/gather facility to pack vector elements that require similar processing; after bulk application of unconditional vector operations, the results are then unpacked and scattered to their originally intended destinations.

VU instructions come in five formats. (See Figure 8.) The 32-bit short format allows many common scalar and vector operations to be expressed succinctly. The four 64-bit long formats extend the basic 32-bit format to allow additional information to be specified: a 32-bit immediate operand, a signed memory stride, a set of register strides, or additional control fields (some of which can update certain control register with no additional overhead).

Each instruction issued by the RISC microprocessor to the VUs is 32 or 64 bits wide. The 32-bit format is designed to cover the operations and register access

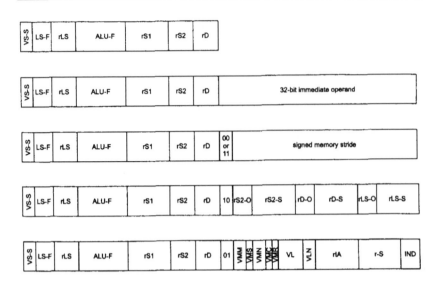

FIGURE 8. VU instruction formats.

patterns most likely to arise in high-performance compiled code. The 32 high-order bits of the 64-bit format are identical to the 32-bit format. The 32 low-order bits provide an immediate operand, a signed memory stride, or specifications for more complex or less frequent operations.

The short format includes an arithmetic opcode (8 bits), a load/store opcode (3 bits), a vector/scalar mode specifier (2 bits), and four register fields called rLS, rD, rS1, and rS2 that designate the starting registers for the load/store operation and for the arithmetic destination, first source, and second source, respectively. The vector/scalar specifier indicates whether the instruction is to be executed once (scalar mode) or many times (vector mode). It also dictates the expansion of the 4-bit register specifiers into full 7-bit register addresses. The short format is designed to support a conventional division of the uniform register file into vector registers of length 16, 8, or (for 64-bit operands only) 4, with scalar quantities kept in the first 16 registers. For a scalar-mode instruction, the 4-bit register field provides the low-order bits of the register number (which is then multiplied by 2 for 64-bit operands); for a vector-mode instruction, it provides the high-order bits of the register number. The rS1 field is 7 bits wide; in some cases these specify a full 7-bit register number for the arithmetic source 1, and in other cases 4 bits specify a vector register and the other 3 bits convey stride information.

A short scalar-mode instruction can therefore access the first 16 32- or 64-bit elements of the register file, simultaneously performing an arithmetic operation and loading or storing a register. (The memory address that accompanies the issued instruction indicates the memory location to be accessed.) One of the arithmetic operands (S1) may be in any of the 128 registers in the register file.

A short vector-mode instruction can conveniently treat the register file as a set of vector registers:

```
16   4 × 64-bit vector registers
 8   8 × 64-bit vector registers
 4  16 × 64-bit vector registers
16   8 × 32-bit vector registers
 8  16 × 32-bit vector registers
```

Many options are available for vector-mode instructions. These include a choice between a default memory stride and the last explicitly specified memory stride, as well as a choice of register stride for the S1 operand (last specified, 1, or 0 — stride 0 treats the S1 operand as a scalar to be combined with every element of a vector).

The long instruction formats are all compatible extensions of the short format: the most significant 32 bits of a 64-bit instruction are decoded as a 32-bit instruction, and the least significant 32 bits specify additional operations or operands. If the rS2 field of a long instruction is zero, then the low-order 32 bits of the instruction constitute an immediate scalar value to be used as the S2 operand. If the arithmetic operation requires a 64-bit operand, then the immediate value is zero-extended left if an unsigned integer is required, sign-extended left for a signed integer, or zero-extended right for a floating-point number.

If the rS2 field of a long instruction is not zero, then the two high-order bits of the low 32 are decoded. If the two bits match, then the low-order 32 bits are an explicit signed memory stride. (Note that it is possible to specify such a stride even in a scalar-mode long instruction in order to latch the stride in preparation for a following vector-mode instruction that might need to use another of the long formats.) Code 01 indicates additional register number and register stride information, allowing specification of complete 7-bit register numbers and register strides for the rLS, rD, and rS2 operands. This enables complex regular patterns of register access. Code 10 indicates a variety of control fields for such mechanisms as changing the VL, controlling use of the VM, indirect addressing, S1 operand register striding, and population count.

The arithmetic operations that can be specified by the ALU-F instruction field are summarized in Table 1. Note the large set of three-operand multiply-add instructions. These come in three different addressing patterns: accumulative, which adds a product into a destination register (useful for dot products); inverted, which multiplies the destination by one source and then adds in the other (useful for polynomial evaluation and integer subscription computations); and full triadic, which takes one operand from the load/store register so that the destination register may be distinct from all three sources. The triadic multiply-add operations are provided for signed and unsigned integers as well as for floating-point operands, in both 32- and 64-bit sizes. Unsigned 64-bit multiply-boolean operations are also provided. (Note that multiplying by a power of 2 has the effect of a shift.)

TABLE 1
Summary of VU Arithmetic Instructions

imove	dimove	umove	dumove	fmove	dfmove	Move: D = S1+0
itest	ditest	utest	dutest	ftest	dftest	Move and generate status
icmp	dicmp	ucmp	ducmp	fcmp	dfcmp	Compare
iadd	diadd	uadd	duadd	fadd	dfadd	Add
isub	disub	usub	dusub	fsub	dfsub	Subtract
isubr	disubr	usubr	dusubr	fsubr	dfsubr	Subtract reversed
imul	dimul	umul	dumul	fmul	dfmul	Multiply (low 64 bits for integers)
	dimulh		dumulh			Integer multiply (high 64 bits)
				fdiv	dfdiv	Divide
				finv	dfinv	Invert: D = 1.0/S1
				fsqrt	dfsqrt	Square root
				fisqt	dfisqt	Inverse square root: D = 1.0/SQRT(S2)
ineg	dineg			fneg	dfneg	Negate
iabs	diabs			fabs	dfabs	Absolute value
iaddc	diaddc	uaddc	duaddc			Integer add with carry
isubc	disubc	usubc	dusubc			Integer subtract with borrow
isbrc	disbrc	usbrc	dusbrc			Integer subtract reversed with borrow
		ushl	dushl			Integer shift left
		ushlr	dushlr			Integer shift left reversed
		ushr	dushr			Integer shift right logical
		ushrr	dushrr			Integer shift right logical reversed
ishr	dishr					Integer shift right arithmetic
ishrr	dishrr					Integer shift right arithmetic reversed
		uand	duand			Bitwise logical AND
		uandc	duandc			Bitwise logical AND with Complement
		unand	dunand			Bitwise logical NAND
		uor	duor			Bitwise logical OR
		unor	dunor			Bitwise logical NOR
		uxor	duxor			Bitwise logical XOR
		unot	dunot			Bitwise logical NOT
		umrg	dumrg			Merge: D = (if mask then S2 else S1)
		uffb	duffb			Find first 1-bit
imada	dimada	umada	dumada	fmada	dfmada	rD = (rS1*rS2)+rD
imsba	dimsba	umsba	dumsba	fmsba	dfmsba	rD = (rS1*rS2)-rD
imsra	dimsra	umsra	dumsra	fmsra	dfmsra	rD = −(rS1*rS2)+rD
inmaa	dinmaa	unmaa	dunmaa	fnmaa	dfnmaa	rD = −(rS1*rS2)-rD
imadi	dimadi	umadi	dumadi	fmadi	dfmadi	rD = (rS2*rD)+rS1
imsbi	dimsbi	umsbi	dumsbi	fmsbi	dfmsbi	rD = (rS2*rD)−rS1

TABLE 1 (continued)

imsri	dimsri	umsri	dumsri	fmsri	dfmsri	rD = –(rS2*rD)+rS1
inmai	dinmai	unmai	dunmai	fnmai	dfnmai	rD = –(rS2*rD)–rS1
imadt	dimadt	umadt	dumadt	fmadt	dfmadt	rD = (rS1*rLS)+rS2
imsbt	dimsbt	umsbt	dumsbt	fmsbt	dfmsbt	rD = (rS1*rLS)–rS2
imsrt	dimsrt	umsrt	dumsrt	fmsrt	dfmsrt	rD = –(rS1*rLS)+rS2
inmat	dinmat	unmat	dunmat	fnmat	dfnmat	rD = –(rS1*rLS)–rS2
			dumsa			rD = lower(rS1*rS2) AND rD
			dumhsa			rD = upper(rS1*rS2) AND rD
			dumma			rD = lower(rS1*rS2) AND NOT rD
			dumhma			rD = upper(rS1*rS2) AND NOT rD
			dumoa			rD = lower(rS1*rS2) OR rD
			dumhoa			rD = upper(rS1*rS2) OR rD
			dumxa			rD = lower(rS1*rS2) XOR rD
			dumhxa			rD = upper(rS1*rS2) XOR rD
			dumsi			rD = lower(rS2*rD) AND rS1
			dumhsi			rD = upper(rS2*rD) AND rS1
			dummi			rD = lower(rS2*rD) AND NOT rS1
			dumhmi			rD = upper(rS2*rD) AND NOT rS1
			dumoi			rD = lower(rS2*rD) OR rS1
			dumhoi			rD = upper(rS2*rD) OR rS1
			dumxi			rD = lower(rS2*rD) XOR rS1
			dumhxi			rD = upper(rS2*rD) XOR rS1
			dumst			rD = upper(rS1*rLS) AND rS2
			dumhst			rD = upper(rS1*rLS) AND rS2
			dummt			rD = lower(rS1*rLS) AND NOT rS2
			dumhmt			rD = upper(rS1*rLS) AND NOT rS2
			dumot			rD = lower(rS1*rLS) OR rS2
			dumhot			rD = upper(rS1*rLS) OR rS2
			dumxt			rD = lower(rS1*rLS) XOR rS2
			dumhxt			rD = upper(rS1*rLS) XOR rS2
				fclas	dfclas	Classify operand
				fexp	dfexp	Extract exponent
				fmant	dfmant	Extract mantissa with hidden bit
		uend	duenc			Make float from exponent (s1) and mantissa (s2)
				fnop		No arithmetic operation
cvtfi						Convert integer to float*

TABLE 1 (continued)

cvtf	Convert float to float*
cvtir	Convert float to integer (round)*
cvti	Convert float to integer (truncate)*
trap	Generate debug trap
etrap	Generate trap on enabled exception
ldvm	Load vector mask
stvm	Store vector mask

(*) The rS2 field encodes the source and result sizes and formats for these instructions.

The LS-F instruction field specifies one of five load/store operations:

- No operation
- 32-bit load
- 64-bit load
- 32-bit store
- 64-bit store

The load/store size (32 or 64 bits) need not be the same as the arithmetic operand size. They should be the same, however, if load chaining is used. There is no distinction between integer and floating-point loads and stores. A 64-bit load or store may be used to load or store an even-odd 32-bit register pair.

EXECUTING VECTOR CODE

All instruction fetching and control decisions for the VUs are made by the node microprocessor. When VUs are present, all instructions and data reside in the memory banks associated with the VUs. A portion of each memory bank is conventionally reserved for instruction and data areas for the microprocessor. The memory management hardware of the microprocessor is used to map pages from the four memory banks so as to make them appear contiguous to the microprocessor.

While the microprocessor does not have its own memory, it does have a local cache that is used for both instruction and data references. Thus, the microprocessor and VUs can execute concurrently so long as no cache misses occur.

When a cache block must be fetched from memory, the associated VU may be in one of three states. If it is not performing any local operations, then the cache block is fetched immediately. If it is performing a local load or store operation, then the block fetch is delayed until the operation completes. If the VU is doing an operation that does not require the memory bus, then the block fetch proceeds immediately, concurrently with the executing vector operation.

The microprocessor issues VU instructions by storing to a specially constructed address: the microprocessor fetches the instruction itself from its data memory, calculates the special VU destination address for issuing the instruction, and executes the store. The time it takes the microprocessor to do this is generally less than the time it takes a VU to execute an instruction with a VL of 4. Moreover, the tail end of one vector instruction may be overlapped in time with the beginning of the next, thus eliminating memory latency and vector start-up overhead. With careful programming, therefore, the microprocessor can sustain delivery of vector instructions so as to keep the VUs continuously busy.

The VU is optimally suited for a VL of 8; with vectors this long, the timing requirements are not so critical, and the microprocessor has time to spare for bookkeeping operations. The short-VU instruction format supports addressing of length-8 register blocks for either 32- or 64-bit operands. This provides 8 vector registers for 64-bit elements or 16 vector registers for 32-bit elements, with the first 2 such register blocks also addressable as 16 scalar registers. This is only a conventional arrangement, however; long-format instructions can address the registers in arbitrary patterns.

Flow control of instructions to the VUs is managed using the hardware protocol of the node bus. When a vector instruction is issued by the microprocessor, an addressed VU may stall the bus if it is busy. A small write buffer and independent bus controller within the microprocessor allows it to continue local execution of its own instructions while the bus is stalled by a VU. If the microprocessor gets far enough ahead, the small write buffer becomes full, causing the microprocessor to stall until the VU(s) catch up.

Each vector instruction either completes successfully or terminates in a hard error condition. Exceptions and other nonfatal conditions are signaled in sticky status registers that may be either polled or enabled to signal interrupts. Hard errors and enabled exception conditions are signaled to the microprocessor as interrupts via the NI.

The memory addresses on the node bus are physical addresses resulting from memory map translation in the microprocessor. The memory map provides the necessary protection to ensure that the addressed location itself is in fact within a user's permitted address space, but cannot prevent accesses to other locations by execution of vector instructions that use indirect addressing or memory strides. Additional protection is provided in each VU by bounds-checking hardware that signals an interrupt if specified physical address bounds are exceeded.

Certain privileged VU operations are reserved for supervisor use. These include the interrupt management and memory management features. The supervisor can interrupt a user task at any time for task-switching purposes and can save the state of each VU for transparent restoration at a later time.

GLOBAL ARCHITECTURE

A single-user process "views" the CM-5 system as a set of processing nodes plus a PM, with I/O and other extrapartitional activities being provided by the operating system.

To support such processes, however, requires that the underlying system software make appropriate use of the global architecture provided by the communications networks of the CM-5.

All the computational and I/O components of a CM-5 system interact through two networks, the Control Network and the Data Network. Every such component is connected through a standard CM-5 NI. The NI presents a simple, synchronous 64-bit bus interface to a node or I/O processor, decoupling it both logically and electrically from the details of the network implementation.

The Control Network supports communication patterns that may involve all the processors in a single operation; these include broadcasting, reduction, parallel prefix, synchronization, and error signaling. The Data Network supports point-to-point communications among the processors, with many independent messages in transit at once.

THE NETWORK INTERFACE (NI)

The CM-5 NI provides a memory-mapped control-register interface to a 64-bit processor memory bus. All network operations are initiated by writing data to specific addresses in the bus address space.

Many of the control registers appear to be at more than one location in the physical address space. When a control register is accessed, additional information is conveyed by the choice of which of its physical addresses was used for the access; in other words, information is encoded in the address bits. For example, when the Control Network is to be used for a combining operation, the first — and perhaps only — bus transaction writes the data to be combined, and the choice of address indicates which combining operation is to be used. One of the address bits indicates whether the access has supervisor privileges; an error is signaled on an attempt to perform a privileged access using an unprivileged address. (Normally the operating system maps the unprivileged addresses into the address space of the user process, thereby giving the user program zero-overhead access to the network hardware while prohibiting user access to privileged features.)

The logical interface is divided into a number of functions units. Each functional unit presents two first-in-first-out (FIFO) interfaces, one for outgoing data and one for incoming data. A processor writes messages to the outgoing FIFO and pulls messages from the incoming FIFO, using the same basic protocol for each functional unit. Different functional units, however, respond in different ways to these messages. For example, a Data Network unit treats the first 32 bits of a message as a destination address to which to send the remainder of the message; a Control Network combining unit forwards the message to be summed (or otherwise combined) with similar messages from all the other processors.

Data are kept in each FIFO in 32-bit chunks. The memory-bus interface accepts both 32- and 64-bit bus transactions. Writing 64 bits thus pushes two 32-bit chunks onto an output FIFO; reading 64 bits pulls two chunks from an output FIFO.

For outgoing data, there are two control registers called *send* and *send first.* Writing data to the *send first* register initiates an outgoing message; address bits encode the intended total length of the message (measured in 32-bit chunks). Any additional data for that message are then written to the *send* register. After all the data for that message have been written, the program can test the *send ok* bit in the third control register. If the bit is 1, then the network has accepted the message and bears all further responsibility for handling it. If the bit is 0, then the data were not accepted (because the FIFO overflowed) and the entire message must be repushed into the FIFO at a later time. The *send space* control register may be checked before starting a message to see whether there is enough space in the FIFO to hold the entire message; this should be treated only as a hint, however, because the supervisor operations (such as task switching) might invalidate it. In many situations throughput is improved by pushing without checking first, in the expectation that the FIFO will empty out as fast as new data are being pushed. It is also permissible to check the *send ok* bit before all the data words for the message have been pushed; if it is 0, the message may be retried immediately.

For incoming data, a processor can poll the *receive ok* bit until it becomes 1, indicating that a message has arrived; alternatively, it can request that certain types of messages trigger an interrupt on arrival. In either case, the program can then check the *receive-length left* field to find out how long the message is and then read the appropriate number of data words from the *receive* control register.

The supervisor can always interrupt a user program and send its own message; this is done by deleting any partial user message, sending the supervisor message, and then forcing the *send ok* bit for that unit to 0 before resuming the user program. To the user program it merely appears that the FIFO was temporarily full; the user program should then retry sending the message. The supervisor can also lock a send-FIFO, in which case it appears always to be full, or disable it, in which case user access will cause an interrupt. The supervisor can save and transparently restore the contents of any receive-FIFO.

Each NI records interrupt signals and error conditions generated within its associated processor; exchanges error and interrupt information with the Control Network; and forwards interrupt and reset signals to its associated processor.

THE CONTROL NETWORK

Each NI contains an assortment of functional units associated with the Control Network. All have the same dual-FIFO organization but differ in detailed function.

Every Control Network operation potentially involves every processor. A processor may push a message into one of its functional units at any time; shortly after all processors have pushed messages, the result becomes available to all processors. Messages of each type may be pipelined; a number of messages may be sent before any results are received and removed. (The exact depth of the pipeline varies from one functional unit to another.) The general idea is that every processor should send the same kinds of messages in the same order. The Control

Network, however, makes no restrictions about when each processor sends or receives messages. In other words, processors need not be exactly synchronized to the Control Network; rather, the Control Network is the very means by which processors conduct synchronized communication *en masse*.

There are exceptions to the rule that every processor must participate. The functional units contain mode bits for *abstaining*. A processor may set the appropriate mode bit in its NI in order to abstain from a particular type of operation; each operation of that type will then proceed without input from that processor or without delivering a result to that processor. A *participating* processor is one that is not abstaining from a particular kind of Control Network operation.

BROADCASTING

The *broadcast* unit handles broadcasting operations. There are actually three distinct broadcasting units: one for user broadcast, one for supervisor broadcast, and one for interrupt broadcast. Access to the supervisor broadcast unit or interrupt broadcast unit is a privileged operation.

Only one processor may broadcast at a time. If another processor attempts to send a broadcast message before completion of a previous broadcast operation, the Control Network signals an error.

A broadcast message is 1 to 15 32-bit words long. Shortly after a message is pushed into the broadcast send-FIFO, copies of the message are delivered to all participating processors. The user broadcast and supervisor broadcast units are identical in function, except that the latter is reserved for supervisor use.

An interrupt broadcast message causes every processor to receive an interrupt or reset signal. A processor can abstain from receiving interrupts, in which case it ignores interrupt messages when it receives them; but a processor cannot abstain from a reset signal (which causes the receiving NI and its associated processor to be reset).

As an example of the use of broadcast interrupts, consider a PM coordinating the task-switching of user processes. When it is time to switch tasks, the PM uses the Control Network to send a broadcast interrupt to all nodes in the partition. This transfers control in each node to supervisor code, which can then read additional supervisor broadcast information about the task-switch operation (such as which task is up next).

COMBINING

The *combine* unit handles reduction and parallel prefix operations. A combine message is 32 to 128 bits long and is treated as a single integer value. There are four possible message types: reduction, parallel prefix, parallel suffix, and router-done. The router-done operation is simply a specialized logical OR reduction that assists the processors in a protocol to determine whether Data Network communications are complete. Reduction, parallel prefix, and parallel suffix may combine the messages using any one of five operators: bitwise logical OR, bitwise logical XOR, signed maximum, signed integer addition, and unsigned integer addition. (The only difference between signed and unsigned addition is in the reporting of overflow.) The message type and desired combining operation are encoded by address bits when writing the destination address to the *send first* register.

As an example, every processor might write a 64-bit integer to the combine interface, specifying signed integer addition reduction. Shortly after the last participating processors write their input values, the signed sum is delivered to every participating processor, along with an indication of whether overflow occurred at any intermediate step.

As another example, every processor might write a 32-bit integer to the combine interface, specifying signed maximum parallel prefix. Shortly after the last participating processors write their input values, every participating processor receives the largest among the values provided by itself and all lower numbered processors.

The combine interface also supports segmented parallel prefix (and parallel suffix) operations. Each combine unit contains a *scan start* flag; when this flag is 1, that NI is considered to begin a new segment for purposes of parallel prefix operations. Such an NI will always receive the very value that was pushed.

Every participating processor must specify the same message type and combining operation. If, in the course of processing combine requests in order, the Control Network encounters different combine requests at the same time, it signals an error.

GLOBAL OPERATIONS

Global bit operations produce the logical OR of 1 bit from every participating processor. There are three independent global operation units, one synchronous and two asynchronous, which may be used completely independently of each other and of other Control Network functions. This makes them useful for signaling conditions and exceptions.

The synchronous global unit is similar to the combine unit except that the operation is always logical OR reduction and each message consists of a single bit. Processors may provide their values at any time; shortly after the last participating processors have written their input bits, the logical OR is delivered as a single-bit message to every participating processor.

Each synchronous global unit produces a new value any time the value of any input is changed. Input values are continually transported, combined, and delivered throughout the Control Network without waiting for all processors to participate. Processors may alter their input bits at any time. These units are best used to detect the transition from 0 to 1 in any processor or to detect the transition from 1 to 0 in all processors. (The NI will signal an interrupt, if enabled, whenever a transition from 0 to 1 is observed.)

There are two asynchronous global units, one for the user and one for the supervisor. Access to the supervisor asynchronous global unit is a privileged operation.

SYNCHRONIZATION

Both the synchronous global unit and the combine unit may be used to implement barrier synchronization: if every processor writes a message and then waits for the result, no processor will pass the barrier until every processor has reached the barrier. The hardware implementation of this function provides extremely rapid synchronization of thousands of processors at once. Note that the router-done combine operation is designed specifically to support barrier synchronization during a Data Network operation, so that no processor abandons its effort to receive messages until all processors have indicated that they are no longer sending messages.

FLUSHING THE CONTROL NETWORK

There is a special functional unit for clearing the intermediate state of combine messages, which may be required if an error or task switch occurs in the middle of a combine operation. A flush message behaves very much like a broadcast message: shortly after one processor has sent such a message, all processors are notified that the flush operation has completed. Access to the flush functional unit is a privileged operation.

ERROR HANDLING

The Control Network is responsible for detecting certain kinds of communications errors, such as an attempt to specify different combining operations at the same time. More important, it is responsible for distributing error signals throughout the system. Hard error signals are collected from the Data Network and all NIs: these error signals are combined by logical OR operations and the result is redistributed to every NI.

THE DATA NETWORK

Each NI contains one Data Network function unit. The first 32-bit chunk of a message is treated as a destination address; it must be followed by one to five additional 32-bit chunks of data. These data are sent through the Data Network and delivered to the receive-FIFO of the NI at the specified destination address. Each message also bears a 4-bit tag, which is encoded by address bits when writing the destination address to the *send first* register. The tag provides a cheap way to differentiate among a small number of message types. The supervisor can reserve certain tags for its own use; any attempt by the user to send a message with a reserved tag signals an error. The supervisor also controls a 16-bit interrupt mask register; when a message arrives, an interrupt is signaled to the destination processor if the mask bit corresponding to the tag value of the message is 1.

A destination address may be physical or relative. A physical address specifies a particular NI that may be anywhere in the system and is not checked for validity. Using a physical address is a privileged operation. A relative address is bounds checked, passed through a translation table, and added to a base register. A relative destination address is thus very much like a virtual memory address: it provides to a user program the illusion of a contiguous *address space* for the nodes running from 0 to one less than the number of processing elements. Access to the bounds register, translation table, or base register is a privileged operation; thus the supervisor can confine user messages within a partition.

While programs may use an interrupt protocol to process received messages, data parallel programs usually use a receiver-polls protocol in which all processors participate. In the general case, each processor will have some number of messages to send (possibly none). Each processor alternates between pushing outgoing messages onto its Data Network send-FIFO and checking its Data Network receive-FIFO. If any attempt to send a message fails, that processor should then check the receive-FIFO for incoming messages. Once a processor has sent all its outgoing messages, it uses the Control Network combine unit to assert this fact; it then alternates between receiving incoming messages and checking the Control Network. When all processors have asserted that they are done sending messages and all outstanding messages have been received, the Control Network asserts the *router-done* signal to indicate to all the processors that the communications step is complete and they may proceed.

For task-switching purposes, the supervisor can put the Data Network into All Fall Down (AFD) mode. Instead of trying to route messages to their intended destinations, the Data Network drops each one into the nearest node. The advantage of this strategy is that no node will receive more than a few hundred bytes of AFD messages, even if they were all originally intended for a single destination. The supervisor can then read them from the Data Network receive-FIFO and save them in memory as part of the user task state, resending them when that user task is resumed.

2 NCUBE

OVERVIEW

The NCUBE hardware architecture (see Figure 1) is based on a scalable network of between 32 and 8192 general-purpose processing nodes and one or more UNIX-based hosts. The hosts are connected to the NCUBE network by I/O nodes, which are themselves built around general-purpose processors. Both distributed and shared I/O are supported. I/O services can be provided directly by the I/O nodes through the host.

The NCUBE system uses a MIMD architecture. Each processing node has up to 64 Mbytes of local DRAM memory. For the 8K processor configuration, this corresponds to 512 Gbytes aggregate memory. Data and control messages are passed on high-speed DMA communication channels supported by hardware routing. This routing method is faster than store-and-forward methods. Any node can communicate via DMA with any other node on the network, including I/O nodes.

The host workstations serve as operator interfaces to the NCUBE network. For large problems, a single host can use all the processors and memory in the NCUBE system. Alternatively, the NCUBE can be divided into subsets of processing nodes that are dedicated to individual hosts in a multihost environment; in such configurations, computations and message passing by one user are protected from other users on the network.

An NCUBE system, with its maximum configuration of 8192 processors, can perform 60 billion instructions per second and 27 GFLOPS (billion floating-point operations per second). Because NCUBE systems use many processors, each with its own distributed memory, a failure at one processor has only local (not global) effects. Each processor is a custom-designed 64-bit VLSI chip that integrates many parts found in traditional computers, including a 64-bit IEEE floating point unit, a memory management unit, a message-routing unit, and 14 DMA communication channels.

The topology of the NCUBE network is a hypercube — an *n-dimensional* cube. The processors in the hypercube lie at the nodes (vertices) of the cube, and neighboring nodes are linked along the edges of the cube by the message-passing

Portions reprinted with permission from *SIAM J. Sci. Stat. Comput.* 9(4), 609-638, © 1988 by the Society for Industrial and Applied Mathematics, Philadelphia. All rights reserved

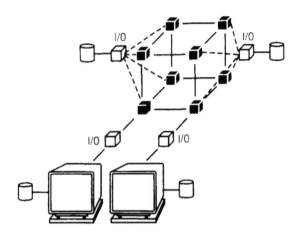

FIGURE 1. NCUBE interconnections.

communication channels. Each node is therefore linked to its n-neighboring processors, and the entire hypercube contains 2^n processors.

NCUBE systems are expandable. Two NCUBE networks of the same dimension can be merged to form an NCUBE network of the next higher dimension. A single system clock drives all the processors.

NCUBE software includes software development tools, including C and Fortran compilers, parallel debuggers, performance monitors, and subroutine libraries adapted to the NCUBE environment.

Sandia National Laboratories achieved record-breaking performance using an NCUBE first-generation supercomputer to process U.S. Department of Defense and Department of Energy radar simulation application in 1989.

The radar simulation application permits users to preview the appearance of objects detected with radar. When run on the NCUBE/ten, a single view that took CRAY X-MP 15 min to compute was created and the image was drawn in 2 min, performing 6.8 times faster than a CRAY Y-MP and 7.9 times faster than a CRAY X-MP.

The NCUBE/ten is well suited for parallel supercomputing. Since each processor has 512 Kbytes of memory and a complete environment, it is possible to run a fixed-sized problem of practical size on a single processor. Each NCUBE node has fast enough 32- and 64-bit floating-point arithmetic for the 1024-node ensemble to be competitive with conventional supercomputers on an absolute performance basis.

All memory is distributed in the hypercube architecture. Information is shared between processors by explicit communications across I/O channels (as opposed to the shared-memory approach of storing data in a common memory). Therefore, the best parallel applications are those that seldom require communications which must be routed through nodes. The applications presented here use these point-to-point paths exclusively. The NCUBE provides adequate bandwidth for moving data to and from I/O devices such as host, disk, and graphics display. The operating system can allocate *subcubes* to multiple users with very little interference between subcubes. In fact, much benchmarking is performed while sharing the cube with various-sized jobs and various applications.

The NCUBE processor node is a proprietary, custom VLSI design. It contains a complete processor (similar in architecture to a VAX-11/780 with floating-point accelerator), bidirectional direct memory access (DMA) communications channels, and an error-correcting memory interface, all on a single chip. Both 32- and 64-bit IEEE floating-point arithmetics are integral to the chip and to the instruction set. Each node consists of the processor chip and six 1-Mbit memory chips (512 Kbytes plus error-correction code).

Of relevance to this discussion is the ratio of computation time to communication time in such a processor node, as actually measured. Currently, a floating-point operation takes between 7 and 15 µs to execute on one node, using the Fortran compiler and indexed memory-to-memory operations (peak observed floating point performance is 0.17 MFLOPS for assembly code kernels with double-precision arithmetic). Computationally intensive, single-node Fortran programs fall within this range (0.07 to 0.13 MFLOPS). This performance is expected to improve as the compiler matures. Integer operations are much faster, averaging a little over 1 µs when memory references are included.

The time to move data across a communications channel can sometimes be overlapped, either with computations or with other communications. However, using subroutine calls from Fortran shows that a message requires about 0.35 ms to start and then continues at an effective rate of 2 µs/byte. It is then possible to estimate just how severe a constraint on speed-up one faces when working a fixed-sized problem using 1024 processors: suppose that an application requires 400 Kbytes for variables on one node (50K 64-bit words). If distributed over 1024 processors, each node will only have 50 variables in its domain. For a typical time-stepping problem, each variable might involve ten floating-point operations (120 µs) per time step, for a total of 6 milliseconds before data must be exchanged with neighbors. This computational granularity excludes the effective overlap of communication with computation that is achieved for larger problems. Data exchange might involve four reads and four writes of 80 bytes each, for a worst-case time of $(4+4) \times (350 + 80 \times 2)$ µs, or about 4 ms. Therefore, when a single-node problem is distributed on the entire 1024-processor ensemble, the parallel overhead on the NCUBE will be about 40%. This estimate is validated by the experimental results. The cost of synchronization and load imbalance appears secondary to that of message transfers (for interprocessor communications and I/O).

Even when the computation appears perfectly load balanced on the ensemble, there can be load imbalance caused by data-dependent differences in arithmetic times on each node. For example, the NCUBE processor does not take a fixed amount of time for a floating-point addition. The operands are shifted to line up their binary points at a maximum speed of 2 bits per clock prior to the actual addition, or normalized at a similar rate if the sum yields a number smaller than the operands. The sum of 3.14 and 3.15 executes at maximum speed, but he sum of 3.14 and 0.0003, or 3.14 and −3.15, takes additional cycles. The microprocessor also does a check, in parallel, of whether either operand is zero and shortcuts

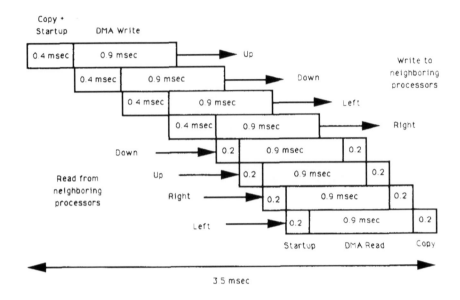

FIGURE 2. Overlapped communications.

the calculation if true. This timing variability is typical of the VAX-type architectures, but is very unlike that of machines like the CRAY that have pipelined arithmetic. Statistically, the NCUBE nodes have nearly the same amount of work to do; however, variation caused by data introduces a slight load imbalance in the large jobs.

Communication Overlap

The communications channels for each node are, in principle, capable of operating concurrently with the processor itself and with each other up to the point where memory bus bandwidth (7.5 Mbytes/s) is saturated. However, the DMA channels are managed by software running on the processor. The software creates overhead that limits the extent to which communications can be overlapped.

Careful ordering of reads and writes can yield considerable overlap and economy, halving the time spent on interprocessor communications. As an example, Figure 2 shows the pattern used for a two-dimensional wave mechanics problem.

The messages sent and received in Figure 2 are 768 bytes long. The actual DMA transfers require 1.20 μs per byte, or 0.9 milliseconds for each message. Before a message can be written, it is first copied to a location in system buffer memory where messages are stored in a linked-list format. For a 768-byte message, the copy and start-up time for writing a message is about 0.4 ms. It is best to arrange the writes as in Figure 2 rather than alternate writes and reads. This arrangement reduces the number of synchronizations from four to one; it also ensures, as much as possible, that messages have arrived by the time the corresponding reads are executed.

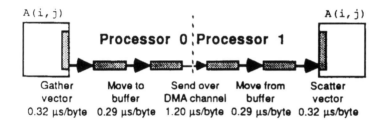

FIGURE 3. Quadruple buffering.

When subroutine calls to read messages are issued in the same order that corresponding messages are written, the probability of an idle state (waiting for a message) is reduced. Therefore, if the first write is in the "up" direction, the first read should be from the "down" direction. About 0.2 ms of processor time is needed to prepare to receive the message. If there is no waiting (i.e., all connected processors are ready to write), and there is little contention for the memory bus, then the read operations proceed with overlapped DMA transfers. As shown in Figure 2, four channels can operate simultaneously for modest message lengths, even when startup overhead is taken into account. In Figure 2, an average of 2.5 DMA channels are operating simultaneously.

After each message is read into a system buffer, it is copied back to the Fortran array so it can again be used in the program. For the 768-byte message, this requires about 0.2 ms. The total time for a complete set of four writes and four reads is less than 4 ms for this example. This time compares with a computation time of about 3 s for the interior points. Thus, with careful management, computation time can be made almost three orders of magnitude greater than parallel communication overhead.

Note also that the total transfer in Figure 2 is 6144 bytes in 3.5 ms, or 0.6 μs/byte; because of overlap, this is less than the theoretical time required to send 6144 bytes over one channel: 7.3 ms. Hence, one cannot use simple parametric models of communication speed and computation speed to accurately predict ensemble performance.

Message Gather-Scatter

In sending edge data to nearest neighbors in a two-dimensional problem, two of the four edges in the array have a nonunit stride associated with their storage. Since the communications routines require contiguous messages, it is first necessary to gather the data into a message with a Fortran loop for those two edges. The result is *quadruple buffering* of messages, as shown in Figure 3.

The NCUBE Fortran compiler produces considerable loop overhead for the gather and scatter tasks, and studies show that a hand-coded assembly routine for these tasks is actually 11 times faster than Fortran. On the parallel version of *WAVE* (see below), for example, the assembly code routine is used to gather the edge data and to scatter the message back to the edge. (Figure 4 shows times using assembly code.) This change improved parallel speed-up by as much as 20%.

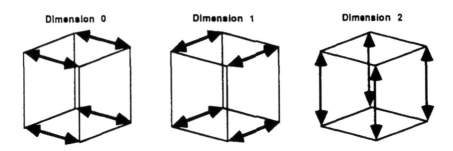

FIGURE 4. Global exchange for inner products.

Message Organization

Considerable communication time can be saved by judicious reorganization of data and computation within the application. In particular, it is very important to avoid *message start-up time* by coalescing individual messages wherever possible. The total overhead for every message is about 0.35 ms, which limits the fine-grained parallel capability of the hypercube. In spreading problems thinly across processors for the purpose of measuring fixed-sized problem speed-up, message start-up time dominates the parallel overhead. To mitigate this effect, we have structured the algorithms so that communications are grouped rather than alternated with computation. Data structures were organized so that successive communications can be changed into a single communication of concatenated data.

As an example, the first attempt at a hypercube version of a program descended from a vector uniprocessor version required over 400 nearest-neighbor read-write message pairs per time step. Reorganization of data structures and computation reduced the nearest-neighbor communication cost to 48 message pairs per time step. This reduction primarily involved the reorganization of dependent variable arrays into one large structure with one more dimension. The restructuring placed nearest-neighbor data into a contiguous array for two of the communicated edges, and a constant stride array on the other two edges. The constant stride arrays are gathered into a contiguous array by an optimized routine at a cost of about 0.3 µs/byte.

Global Exchange

Many of the kernels generally thought of as "serial" (order P time complexity for P processors) can actually be performed in $\log_2 P$ using a series of exchanges across the dimensions of the cube. For example, the accumulation of *inner products* is performed efficiently by means of bidirectional exchanges of values along successive dimensions of the hypercube, interspersed with summation of the newly acquired values (Figure 4). This algorithm requires the optimal number of communication steps, $\log_2 P$. Note that we do *not* perform a "global collapse" that condenses the desired scalar to one processor which must then be broadcast to all nodes. The exchange does more computations and messages than a collapse, but requires half the passes to produce the desired sum on each

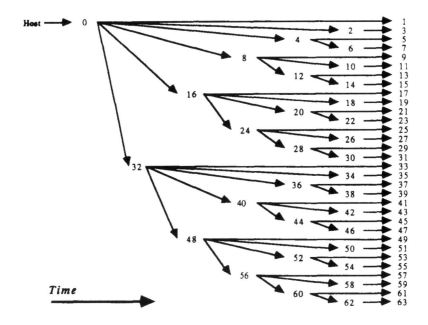

FIGURE 5. Logarithmic fanout.

processor. A similar pattern can be used to perform such "serial" operations as finding global maxima, global minima, and global sums, in time proportional to $\log_2 P$ rather than P.

Logarithmic-Cost Fanout

Figure 5 shows the *fanout* algorithm used to load 64 processors in an order 6 hypercube. By using a tree to propagate redundant information, the time for operations such as loading the applications program or the node operating system is greatly reduced. This pattern is used in reverse to collect output. In contrast to the global exchange technique, the tree method is most useful when information must be sent to or received from the host.

As an example of the savings provided by this technique, a 12,500-byte executable program is loaded onto a 512-node hypercube in 1.33 s using a pattern similar to the one shown in Figure 5, compared with 61.18 s using a serial loading technique. The performance ratio is 46 to 1. For a 1024-node hypercube, the logarithmic technique is 83 times faster. The logarithmic technique can disseminate a block of 480 Kbytes (the maximum user space in each node) in less than 1 min, whereas the serial technique requires 1 h, 20 min.

APPLICATION: WAVE MECHANICS

Application Description

The WAVE program calculates the progress of a two-dimensional surface (acoustic) wave through a set of deflectors and provides a graphic display of the resulting heavily diffracted wavefront. The program is capable of handling reflectors of any shape (within the resolution of the discretized domain).

Mathematical Formulation

The wave equation is

$$c^2 \nabla^2 \phi = \phi_{tt} \tag{1}$$

where ϕ and c are functions of the spatial variables. In general, ϕ represents the deviation of the medium from some equilibrium (pressure, height, etc.), and c is the speed of wave propagation in the medium (assumed to be isotropic). For nonlinear waves, c is also a function of the wave state.

A discrete form of Equation 1 for a two-dimensional problem on $[0, 1] \times [0, 1]$ is

$$c^2 \left[F(i, j+1) + F(i, j-1) + F(i+1, j) + F(i-1, j) - 4F(i, j) \right] / h^2$$
$$= \left[F_{\text{new}}(i, j) - 2F(i, j) + F_{\text{old}}(i, j) \right] / (\Delta t)^2 \tag{2}$$

where $F(i,j) = \phi(ih, jh)$, $h = 1/N$. Equation (2) can be rearranged into a scheme where F_{new} is computed explicitly from F and F_{old}. Hence, only two time steps need to be maintained in memory simultaneously ("leapfrog" method).

There is ample literature regarding the convergence of this method as a function of c^2, h, and Δt. For example, it is necessary (but not sufficient) that $(\Delta t)^2 \leq (h/c)^2/2$ (CFL condition). We use constant c and $(\Delta t)^2 = (h/c)^2/2$ in our benchmark.

Test Problem

To demonstrate the capability of the algorithm, a test problem is used that includes internal reflecting boundaries (Figure 6). The reflecting barrier is a rectangle that is one sixth by one third the domain size. A diagonal wave of width one sixth the domain size impinges on the barrier, creating reflections and diffractions difficult to compute by analytic means. Feature proportions of one sixth and one third allow discretizations as coarse as 6 by 6 for the domain to correspond in a simple way to discretizations of size 12 by 12, 24 by 24, and so forth.

Communication Cost

The equations describing communication overhead are

$$C_p(N) \quad = 32N / P \tag{3a}$$

$$M_p(N) \quad = 8 \tag{3b}$$

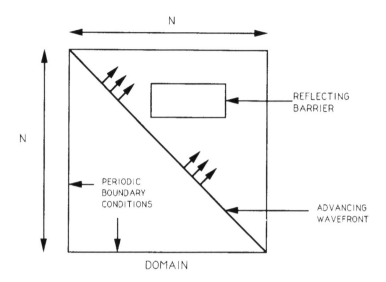

FIGURE 6. Wave mechanics problem.

where $C_p(N)$ is the number of bytes sent and received per time step per processor, $M_p(N)$ is the number of messages sent and received per time step per processor, N is the number of global grid points in the x and y directions, and P is the number of processors ($P > 1$). For this application, the expressions for communication cost are simple because all messages are to nearest-neighbor processors in the two-dimensional topology. For the smallest problems studied (6 by 6 grid points per processor), the communication time per timestep is dominated by the 350-μs-per-message start-up time.

Computation Cost

The wave mechanics problem does relatively few arithmetic calculations per time step. One consequence of this is that Fortran loop overhead dominates when the subdomains are very small. For example, the 6-by-6 subdomain initially ran at about 40 kFLOPS on a single processor, whereas the 192-by-192 subdomain ran at about 80 kFLOPS on a single processor. Besides degrading absolute performance, this overhead introduced another efficiency loss for the fixed-size case, since a 50% efficiency loss resulting from spatial decomposition had nothing to do with interprocessor communication.

To mitigate this efficiency loss, the kernel of the WAVE time step update was coded in assembly language. This refinement raised the absolute performance while also flattening performance across the whole range of subdomain sizes. With an assembly version, the 6-by-6 problem achieved 83 kFLOPS, and larger problems quickly approach a plateau of 111 kFLOPS. Thus, a 25% loss of efficiency for the fixed-sized case is the result of loop start-up time within a single processor, and absolute performance is improved in all cases.

TABLE 1
MFLOPS for the Wave Mechanics Problem (32-bit Arithmetic)

Problem size per node	Hypercube dimension					
	0	2	4	6	8	10
192 by 192	0.111	0.442	1.77	7.07	28.3	113.0
96 by 96	0.111	0.442	1.77	7.06	28.2	113.0
48 by 48	0.111	0.439	1.76	7.02	28.1	112.0
24 by 24	0.111	0.431	1.72	6.87	27.4	109.0
12 by 12	0.106	0.400	1.59	6.32	25.0	98.1
6 by 6	0.083	0.314	1.23	4.82	18.8	70.6

TABLE 2
Time in Seconds for the Wave Mechanics Problem

Problem size per node	Hypercube dimension					
	0	2	4	6	8	10
192 by 192	12,780.0	12,806.0	12,809.0	12,817.0	12,822.0	12,824.0
96 by 96	3,194.0	3,206.0	3,206.0	3,208.0	3,211.0	3,212.0
48 by 48	796.5	805.7	805.9	806.5	807.3	808.7
24 by 24	199.3	205.5	205.8	206.1	206.6	207.8
12 by 12	52.1	55.3	55.6	56.0	56.5	57.7
6 by 6	16.7	17.6	18.0	18.4	19.1	20.0

We have used 32-bit precision for this application. The numerical method error is of order $(h + \Delta t)$, which dominates any errors introduced by the finite precision of the arithmetic. The parallel speed-up benefits from the larger subdomains permitted by reducing memory requirements from 64-bit words to 32-bit words. Absolute performance is also increased by about 50%, with virtually no additional truncation error.

Measured Performance

By keeping the number of time steps constant, the resulting performance charts would ideally show constant MFLOPS as a function of problem size, and constant time in seconds as a function of the number of processors. It is interesting to compare Tables 1 and 2 against this ideal. Note in particular that time in seconds varies little with hypercube dimension, as one would hope, except for the loss in going from the serial version to the parallel version.

The physical problem being solved is identical only along diagonal lines in the above charts, from top left to bottom right. For example, the 12-by-12-per-node problem on a serial processor is the same as a 6-by-6 problem on each of

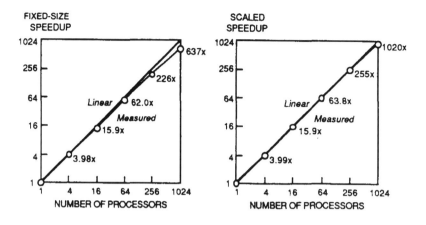

FIGURE 7. Wave mechanics problem speed-ups.

4 processors. In contrast, the 192-by-192-per-node problem on an 8-dimensional hypercube is the same as the 96-by-96-per-node problem on a 10-dimensional hypercube; both perform a very high resolution (3072 by 3072 global grid points) simulation, but the algorithmic time steps now represent a smaller interval of physical time for the high-resolution simulation than for the low-resolution simulation, since Δt is proportional to $1/N$. By fixing the number of algorithmic time steps, we are able to study the effects of parallel overhead across rows of the table, and the effect of loop overhead along columns.

The job with 192 by 192 global grid points, the largest problem that spans the entire range of hypercube sizes, yields a fixed-sized speed-up of 637. If we scale the problem to fit the processor, always assigning a 192-by-192 subdomain to each node, the overall efficiency never drops below 99.6%. In particular, the 1024-node job executes 1020 times as fast as it would on one identical processor with similar access times to 0.5 Gbytes of memory. Alternatively, the 1024-node job executes at an MFLOPS rate of 1020 times that of a single-node job (see Table 1). Both speed up curves are shown in Figure 7.

This application revealed a series of subtle hardware problems with specific nodes that initially caused a *spurious load imbalance* of up to 10% of the 256- and 1024-node jobs. By partitioning the cube and comparing times on identical subcubes, we identified "slow nodes" that were performing memory error correction, communication error recovery, or overly frequent memory refresh, all of which diminished performance without causing incorrect results.

The fixed-sized problem speed up of 637 is 62.2% of the linear speed-up of 1024. Ideally, the fixed-size 12,780-s serial case would have been reduced to 12.5 s; instead, the measured time is 20.0 s. Of the 7.5-s difference, 4.2 s is due to the reduced MFLOPS rate caused by the shortening of Fortran loop counts on each node (from 192 to 6). This MFLOPS reduction can be observed directly in the leftmost column of Table 1. Another 0.7 s is lost in program loading and non-overlapped I/O. The remaining time, 2.6 s, is lost in interprocessor communication; the sum of the latter two effects is visible in the bottom row of Table 2.

The scaled problem speed-up of 1020 is 99.66% of the ideal. Of the 0.34% loss, 0% is caused by loss of operation efficiency, 0% is lost in loop overhead (since further improvements in serial MFLOPS were negligible beyond the 48-by-48 subdomain size; see Table 1, leftmost column), 0.01% is lost in program loading, 0.17% is incurred in going from the serial version to the parallel version of the program, and the remaining 0.16% is from load imbalance induced by data-dependent MFLOPS rates. Based on the top row of Table 2, the extrapolated uniprocessor execution time for this problem (6144 by 6144 grid points) is approximately 13.1 million s (5 months).

3 iWARP

iWarp is a product of a joint effort between Carnegie Mellon University and Intel Corporation. The goal of the effort is to develop a powerful building block for various distributed memory parallel computing systems and to demonstrate its effectiveness by building actual systems. The building block is a custom VLSI single-chip processor, called iWarp, which consists of approximately 600,000 transistors.

The iWarp component contains both a powerful computation processor (20 MFLOPS) and a high-throughput (320 Mbytes/s), low-latency (100 to 150 ns) communication engine. Using nonpipelined floating-point units, the computation processor will sustain high computation speed for vectorizable as well as nonvectorizable codes.

An iWarp component connected to a local memory forms an iWarp cell; up to 64 Mbytes of memory are directly addressable. A large array of iWarp cells will deliver an enormous computing bandwidth. Because of the strong computation and communication capabilities and because of its commercial availability, iWarp is expected to be an important building block for a diverse set of high-performance parallel systems.

The iWarp architecture evolved from the Warp machine,[1] a programmable systolic array developed at Carnegie Mellon and produced by General Electric. All applications of Warp, including low-level vision, signal processing, and neural network simulation,[2,18] can run efficiently on iWarp. However, systems made of the iWarp building block can achieve at least one order of magnitude improvement over Warp in cost, reliability, power consumption, and physical size. Much larger arrays can be easily built. The clock speed of iWarp is twice as high as Warp; the increase in computation throughput is matched by a similar increase in I/O bandwidth. Therefore, we expect iWarp to achieve the same high efficiency as Warp. For example, the NETtalk neural network benchmark[20] runs at 16.5 million connections per second and 70 MFLOPS on a 10-cell Warp array; the same benchmark runs at 36 million connections per second and 153 MFLOPS on an iWarp array of the same number of cells.

Although the design of the iWarp architecture profited greatly from programming and applications experiences gained from many Warp machines in the field, iWarp is not just a straightforward VLSI implementation of Warp. iWarp is intended to have a much more expanded domain of applications than Warp. The following summarizes the goals of iWarp as a system building block.

- iWarp is useful for the implementation of both special-purpose arrays, which require high computation and I/O bandwidth, and general-purpose arrays where programmability and programming support are essential.
- iWarp is useful for both high-performance processors attached to general-purpose hosts and autonomous processor arrays capable of performing all the computation and I/O by themselves. That is, iWarp can be used for both "host-centric" and "array-centric" processing.
- iWarp supports both tightly and loosely coupled parallel processing, and both systolic[12] and message-passing models of communication.
- iWarp can implement a variety of processor interconnection topologies including one-dimensional (1D) arrays, rings, two-dimensional (2D) arrays, and tori.
- iWarp is intended for systems of various sizes ranging from several processors to thousands of processors.

Besides conventional high-level languages such as C and Fortran, the programming of iWarp arrays will be supported by programming tools such as parallel program generators. Previous experience and research on Warp indicates that parallel program generators are one of the most promising approaches to programming distributed memory parallel computers. In this approach, a specialized, machine-independent language is created which embodies a particular parallel computation model (for example, input partitioning, domain partitioning, or task queuing).[13] The compiler for that language then maps the program onto a target parallel architecture. This approach can allow efficient parallel programs to be generated automatically for large processor arrays.

Automatic parallel program generators have been developed for iWarp in two applications areas: scientific computing and image processing. The scientific computing language, called array language (AL),[21] incorporates the domain partitioning model and allows programmers to transfer data between a common space and a partitioned space, perform computation in parallel in the partitioned space, and then transfer data back. For scientific routines such as those found in LINPACK,[5] the AL compiler generates efficient code for iWarp and Warp, as well as for uniprocessors. The image-processing language, called Apply,[9] incorporates the input partitioning model: the input images are partitioned among the processors, each of which generates part of the corresponding output image. Apply compilers exist for iWarp, Warp uniprocessors, and the Meiko Computing Surface, as well as several other computer architectures. Benchmark comparisons on Apply programs have validated the above claims.[22]

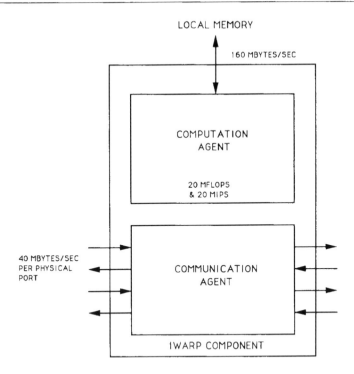

FIGURE 1. iWarp component overview.

The architecture and logic designs for iWarp have been completed. In the software area, an optimizing compiler developed for Warp,[8,16] has been retargeted to generate code for iWarp.

iWARP OVERVIEW

An iWarp system is composed of a collection of iWarp cells, each of which consists of an iWarp component and its local memory.

iWarp Component

The iWarp component has a *communication agent* and a *computation agent*, as depicted in Figure 1. The computation agent can carry out computations independently from the operations being performed at the communication agent. Therefore, a cell may perform its computation while communication through the cell from and to other cells is taking place, and the cell program does not need to be involved with the communication. While separating the control of the two agents makes programming easy, having the two agents on the same chip allows them to cooperate in a tightly coupled manner. The tight coupling allows several communication models to be implemented efficiently.

In the following we summarize the major features in the two agents and their interface. The performance numbers are based on the clock speed of 20 MHz, i.e., a clock is 50 ns.

Communication Agent
Four input and four output ports

- 40 Mbytes/s data bandwidth per port.
- Word by word, hardware flow control at each port.
- An output port can be connected to an input port of another iWarp cell via a point-to-point *physical bus*.

Multiple logical busses multiplexed on each physical bus

- Maintaining up to 20 incoming pathways simultaneously in an iWarp component.
- Idle logical busses do not consume any bandwidth of the physical bus.

Pathway unit

- Routing for 1D and 2D configurations.
- Capable of implementing wormhole and street sign routing schemes.

Both message passing and systolic communication are supported for coarse-grain and five-grain parallel computation.

Computation Agent
Computational units

- Floating-point adder
  - 10 and 5 MFLOPS for 32- and 64-bit additions (IEEE 754 standard), respectively.
  - Nonpipelined
- Floating-point multiplier
  - 10 and 5 MFLOPS for 32- and 64-bit multiplications (IEEE 754 standard), respectively.
  - Nonpipelined.
  - Full divide, remainder, and square root support.
- Integer/logical unit
  - 20 MIPS peak performance on 8/16/32-bit integer/ordinal data.
  - Arithmetic, logical, and bit operations.

All the above three units may be scheduled to operate in parallel in one instruction, generating a peak computing rate of 20 MFLOPS plus 20 MIPS.

Internal data storage and interconnect

- A shared, multiported, 128 word register file
- Special register file locations for local memory and communication agent access

Memory units

- Off-chip local memory for data and instructions
  - Separate address and data busses (24-bit word address bus, 64-bit data bus).
  - 20 million memory accesses per second peak performance.
  - 160 Mbytes/s peak memory bandwidth.
  - Read, write, and read/modify/write support.
- On-chip program store
  - 256-word cache RAM.
  - 2K-word ROM (built-in functions).
  - 32- and 96-bit instructions.

Communication and computation interface

- Communication agent notifies computation agent on message arrival.
- Dynamic flow control: Computation agent spins when reading from an empty queue or writing to a full queue in communication agent.
- Hardware spools data between queues and local memory.

FORMING iWARP SYSTEMS

Various iWarp systems can be constructed with the iWarp cell. We describe how copies of the iWarp cell can be connected together, and how an iWarp cell can connect to peripherals to form these systems.

There are two ways that an iWarp cell, consisting of an iWarp component and its local memory, physically interfaces with the external world. Recall that the iWarp component has four input ports and four output ports. The first interface method is to use a physical *bus* to connect an output port of an iWarp to an input port of another. The former and latter port can write to and read from the bus, respectively. Thus, this is a unidirectional bus between the two components Usually another unidirectional bus in opposite direction is also provided, so that bidirectional data communication between the two components is possible.

The second interface method is via the local memory of the iWarp cell. Using this interface, the iWarp cell can reach peripherals such as standard busses, disks, graphics devices and sensors. Therefore the iWarp cell's connection with peripherals uses the local memory, while its intercell connection uses ports of iWarp. Since these two functions are different physical resources of the iWarp cell, they can be implemented independently from each other. This implies, for example, that peripherals can be attached to any set of iWarp cells in an array of iWarp cells, independently from the array interconnection topology. With these two interface methods many system configurations can be implemented, as will be shown.

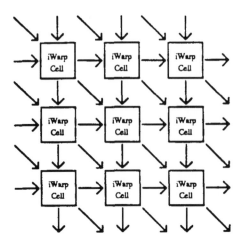

FIGURE 2. Hexagonal array with unidirectional physical busses between cells.

iWARP USAGES AND SYSTEM CONFIGURATIONS

The iWarp cell, consisting of the iWarp component and local memory, is a building block for a variety of system configurations. These systems can be used as general and special-purpose computing engines.

General Purpose Arrays

With its four pairs of input and output ports, the iWarp cell is a convenient building block for a 2D array or torus. Peripherals can be attached to any of the iWarp cells via its local memory. The initial demonstration iWarp system is an 8 × 8 torus, with a total of 32 Mbytes SRAM. It has a peak performance of 1280 MFLOPS. The memory of each cell can be expanded up to 1.5 Mbytes, and with different memory components, a memory space of up to 64 Mbytes per cell is possible. The same system design can be extended to a 32 × 32 torus, giving an aggregate peak performance of 20,480 MFLOPS.

The iWarp cell is a building block of 1D arrays or rings as well. A 1D array or ring of a moderate number of iWarp cells, delivering on the order of hundreds of MFLOPS, can be an effective attached processor to a workstation. This has been demonstrated by the ten-cell Warp array. Using the same approach with iWarp, there is an order of magnitude improvement in cost performance over Warp. To meet the requirement of lower cost (and lower performance) applications, one or a few iWarp cells can also form a single-board accelerator for low-end workstations or PCs.

Special-Purpose Arrays

Many systolic algorithms can make effective use of large arrays for applications such as signal processing and graphics.[10] With the iWarp cell, various special-purpose arrays that execute only a predetermined set of these algorithms can easily be built. For example, a hexagonal array (as depicted in Figure 2) with unidirectional physical busses between cells can be built to execute some classical systolic algorithms for matrix operations.[15] For such an array, sensors and array output ports may be connected to the local memories of a number of cells, so that I/O can be carried out in parallel. In areas such as high-speed signal processing, special-purpose arrays can effectively use hundreds or even thousands of iWarp cells. In some systolic algorithms, cells on the array boundary may execute a different function from cells inside the array.[7] In this case, individual iWarp cells can be programmed to perform different functions according to their locations in the array.

In general, the performance of special-purpose arrays made of iWarp cells will be comparable to that of those arrays made of custom hardware using similar VLSI technology. Although the iWarp array will probably have a larger physical size, it can be readily programmed to implement the target algorithms and will incur a much shorter development time.

COMMUNICATION MODELS

Interprocessor communication is an integral part of parallel computing on a distributed memory processor array. To balance the high numerical processing capability on the processor, iWarp must be equally efficient in communication. The development of efficient parallel software is simplified if the communication cost is low and can be estimated reliably.

We have identified two communication models used for distributed memory parallel systems: *message passing* and *systolic*. They differ primarily in the granularity of communication and computation. In message passing mode, as in computer networks, the unit of processing is a complete message. That is, a message is accumulated in the source cell memory, transmitted (as a unit) to the destination cell, and only when the full message is available in the local memory of the destination cell is it ready to be operated upon. Conversely, in systolic mode,[12] the unit of communication and processing can be as fine grained as a single word in a message.

Message Passing

Message passing is a commonly used model for coarse-grain parallel computation. Processes at each cell operate independently on the cell's local data and only occasionally communicate with other cells. The timing, order, and even the communication partner are often determined at run time. The dynamic nature makes certain communication overheads unavoidable, such as routing the message across the array and asynchronously invoking the answering party. To efficiently support message passing, we need the capabilities described below.

Hardware Support for 1D and 2D Configurations

1D arrays and rings are easier to build than 2D arrays and tori. However, computations on a 2D configuration can be more efficient than those on a 1D configuration for large systems with many cells. Suppose that there are n cells in the system. Using a 2D configuration, not only is the distance between cells reduced from $O(n)$ to $O(\sqrt{n})$ hops, but also the effective bandwidth of the communication network is increased since a transfer takes up fewer hops.

Spooling

Suppose a process wants to send a message to a destination cell. Since the communication network is shared, a process does not have guaranteed instantaneous access. Ideally, the sender process can simply specify the destination and message location, and continue with its processing regardless of the availability of the network, and a separate thread of control spools the data out of the memory. Similarly, another spooling process can take the data from the network and store them into the receiver's memory, with minimal interference with the computational process in progress.

Separate Communication Support Hardware

In an iWarp array, a message may be routed through intermediate cells before reaching the final destination. The routing of data through a cell is logically unrelated to the process local to the cell. It can be supported in dedicated hardware to handle the high bandwidth of the array.

Word-Level Synchronization

Although the granularity of communication is a message, this does not mean that the intermediate hops should forward the data at the same grain size. In wormhole routing,[3] the routing information in the header of the message can be used to set up the next leg of the communication path even before the rest of the data arrive. The contents of the message can be forwarded word by word, without having to be buffered in entirety on intermediate cells. Wormhole routing reduces the latency of communication and does not take up any of the memory bandwidth of the intermediate cells. Since the communication path is built link by link, a word-by-word handshake is necessary to throttle the data flow in case the next link is temporarily unavailable.

Multiplexing Messages on a Physical Bus

Since wormhole routing may use up multiple links at the same time, prevention of deadlock is necessary in ring or torus architectures. One scheme to prevent deadlock is the virtual channel method, which uses another set of links when routing beyond a certain cell.[4,19] If in each direction only one physical bus is available between two connecting cells, this deadlock prevention scheme requires that multiple communication paths be multiplexed on a physical bus. Multiplexing can also be used to keep a long message from monopolizing the physical bandwidth for an indefinitely long time.

Door-to-Door Message Passing

When a message arrives at the destination cell, it is generally first buffered in a system memory space and then copied into the user's memory space. This extra copy can be eliminated if the data are stored directly into the desired memory location. Using the data throttling mechanism above, the receiving process can first examine the message header to determine the memory address for the message. We call this scheme of shipping data directly from a sender's data structures to a receiver's data structures, without any system memory buffering, *door-to-door* message passing.

Systolic Communication

Systolic communication supports efficient, fine-grain parallelism. In this model, the source cell program sends data items to the destination cell as it generates them, and the destination cell program can start processing the data as soon as the first word of input has arrived. For example, the outputs of an adder in one cell can be used as operands to the adder of another, without going through the memories of either cells, in a matter of several clocks. This mode provides tight coupling and synchronization between cooperating processes.

Systolic algorithms rely on the ability to transfer long streams of intermediate data between processes at high throughput and with low latency. More importantly, the communication cost must be consistently small, because cost variations can greatly increase delays in the overall computation. This implies that dedicated communication paths are desirable, which may be neighboring or nonneighboring paths depending on communication topologies of the algorithm.

Raw data words are sent along a communication path, identified only by their ordering in the data stream. The sender appends data to the end of the data stream, and the receiver must access the words in the order they arrive. Experience with the Warp systolic array[1] shows that FIFO queuing along a communication path is useful in relaxing the coupling between the sender and the receiver. A sender does not need to wait for the receiver unless the queue is full; similarly, the receiver can process the queued data until they run out. Word-level synchronization is provided by stalling a process that tries to read from an empty queue or write to a full queue.

A special-purpose systolic array can be tailored to a specific algorithm by implementing the dedicated communication paths directly in hardware and providing long enough queues to ensure a steady flow of data. As a programmable array, iWarp processors should implement the common systolic algorithms well, but can also degrade gracefully to cover other algorithms. The following are requirements for efficient support of systolic communication.

Hardware Support for 1D and 2D Configurations

Many systolic algorithms in signal and image processing and in scientific computing use 1D and 2D processor arrays.[10] iWarp can directly support such configurations in hardware.

Multiplexing Communication Paths on a Physical Bus

iWarp can also support other configurations, with degraded performance, if necessary. A systolic algorithm may call for more communication paths between a pair of connecting cells than those provided directly by hardware. It may require extra communication paths for configurations such as a hexagonal array, or to implement deadlock avoidance schemes.[14] Divide-and-conquer algorithms may require communication between powers of two distances away at different times of the algorithm. All these considerations motivate the need to multiplex multiple communication paths on a physical bus.

Coupling of Computation and Communication

The computation part of a cell needs to access the communication part directly without going through the cell's local memory. This extra source of data is a key to systolic algorithm efficiency. For example, fine-grained systolic algorithms for important matrix operations can consume and produce up to four data words per clock. Memory bandwidth cannot match this high communication bandwidth.

Spooling

Regardless of the size of the hardware queue available on each cell, there is always some systolic algorithm that requires deeper queues. For example, a systolic algorithm for convolving a kernel with a 2D image requires some cells to store the entire row of the image.[11] Therefore, it is desirable to provide an automatic facility to overflow the data to the cell's local memory if necessary.

Reserving a Communication Subnetwork

In both message passing and systolic communication models, there is a need to multiplex multiple communication paths onto a physical bus. Efficient support of communication paths requires dedicated hardware resources; thus, only a small number of paths can be provided. This resource limitation raises the issue of resource management.

The need for managing the communication resource is more pronounced in the systolic communication model. This is because the production and consumption rates of a data stream are tied directly to the computation rates of the cells. As the computation on a cell can stall and even deadlock while waiting for data, the lifetime of a communication path can be arbitrarily long. Although an idle communication path does not consume any communication bandwidth, if all the multiplexed paths on a physical bus are occupied, no other traffic can get through.

In message passing, an entire message is first prepared and buffered in the sender cell's local memory, and the message is stored into the destination cell's local memory directly. Once a communication path becomes available, the data can be spooled in and out of the memories. With a proper routing scheme to avoid deadlocks, a message can always get through, although it may have to wait for a while if the network is backed up with long messages.

Both models can benefit from a mechanism to reserve a set of communication paths for a class of messages. For instance, we can reserve a set of communication paths for system messages for purposes such as synchronization, program debugging, and code downloading. First of all, this guarantees that the system can reach all the cells even if the user uses up all other paths. Moreover, since the system has full control over all messages on the reserved network, the behavior of the network is more predictable, and attributes such as a guaranteed response time are possible.

Messages and Pathways

Message passing and systolic communication are two very different kinds of communication. The former supports coarse-grain parallelism where processes at different cells behave independently, and the latter supports fine-grain parallelism where processes at different cells cooperate synchronously. However, on examining the requirements to make message passing *efficient* and systolic communication *general*, they are not that dissimilar. For example, wormhole routing uses up multiple hardware links simultaneously, much like a communication path in systolic communication that connects two non-neighboring processes. On iWarp, both communication models can be unified and supported efficiently by the same programming abstractions of a *pathway* and a *message*, defined below.

A *pathway* is a direct connection from a cell (called the source cell) to another cell (called the destination cell). Each segment of the pathway that connects the communication agent of a cell to the computation agent of the same cell or to the communication agent of another cell is called a *pathway segment*.

A *message* consists of a header, a sequence of data words, and a marker denoting the end of the message. Messages can be sent from the source cell to the destination cell over a pathway. The pathway is initiated and terminated by the source cell. It assembles a header containing a destination address and additional routing information and hands it to the communication agent. The source cell closes the pathway by sending a special marker to signal the end.

Normally, one pathway is set up for each individual message. The source cell opens a pathway to the destination cell, sends its message, and then closes the pathway. In the message-passing model, the sending process dynamically creates a new pathway and message for each data transfer. Not all intermediate links of a pathway need to exist at the same time. In wormhole routing, the marker denoting the end of the pathway may have reached an intermediate cell even before the header and data reach the destination. In the systolic communication model, the cells typically set up required pathways for a longer duration. The sending program transmits individual data words along a pathway as they are generated without sending any additional headers or markers. On termination, the cell programs close the message and the pathway.

However, it is possible that a pathway is set up for multiple messages. That is, the sender cell does not take down the pathway immediately after the first message has passed through, so the sender can send further messages over the same pathway. The sender cell has reserved the pathway for its future use.

Reservation of multiple pathways is also possible on iWarp. Two pathways are said to be connected if the destination cell of one is the source of the other. The sender of the first pathway can send messages to the destination of the second using both pathways. In this way, a cell can send messages to multiple destinations through a set of reserved pathways.

iWARP COMMUNICATION

The functionality of the communication agent falls into four categories.

1. *Physical communication network.* Hardware support for 1D and 2D configurations.
2. *Logical communication network.* A mechanism to multiplex multiple pathway segments on a physical bus on a word level basis.
3. *Pathway unit.* A mechanism to establish pathways.
4. *Streaming and spooling unit.* Direct access of communication agent from the computation agent, and a spooling mechanism to transfer data from and to local memory.

Physical Communication Network

The communication network of iWarp is based on a set of high-bandwidth, point-to-point physical busses, linking the input and output ports of a pair of cells. Each cell has four input and four output ports, allowing cells to be connected in various topologies.

Each physical bus can transmit one 32-bit word data every 100 ns. The VLSI custom chip implementation made it possible to have fine-grain, word-level hand shaking without any synchronization delay. Thus, each physical bus has a data bandwidth of 40 Mbytes/s, giving an aggregate data transfer rate of 320 Mbytes/s.

Logical Communication Network

Besides the four input and four output *external* busses described above for connecting to other cells, a communication agent is also connected to the computation agent in its cell through two input and two output *internal* busses. Each of these busses can be multiplexed on a word-level basis to support a number of *logical* busses in the same direction, whereas each logical bus can implement one pathway segment of a pathway at a time.

To the communication agent on a cell, there are four kinds of logical busses:

1. Incoming busses from the communication agent of a neighboring cell.
2. Incoming busses from computation agent of the same cell.
3. Outgoing busses to communication agent of a neighboring cell.
4. Outgoing busses to computation agent of the same cell.

The mapping of the logical busses to physical busses is performed statically, under software control. The hardware allows the total number of incoming logical busses in the communication agent of each cell to be as large as 20. For example, in a 2D array, the logical busses can be evenly distributed among the four neighbors and the computation agent. In this case, the heart of the communication agent is a 20×20 crossbar that links incoming logical busses to outgoing logical busses. Logical busses are managed by the source, i.e., the

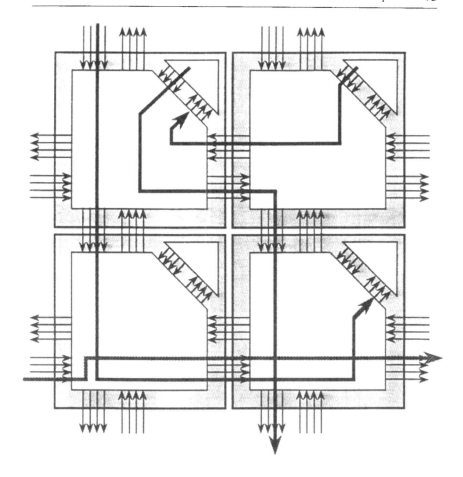

FIGURE 3. Pathways in a 2D array.

sending cell. The sender can initiate communication using any of its preallocated free logical busses without consulting the receiver. This design minimizes the time needed to set up a pathway between cells.

The Pathway Unit

A pathway is formed by connecting a sequence of pathway segments together. Figure 3 contains an example of three pathways through some cells in a 2D array. Pathway 1 connects the computation agent of B to that of A through a pathway segment between the communication agents of the two cells. Pathway 2 passes through cell A, turns a corner at cell C, and finally reaches the destination, D. Lastly, pathway 3 passes through both cells C and D. Two pathway segments are multiplexed on the physical bus from cell C to cell D.

A new pathway is established by the use of a special *open pathway marker*. As the communication agents pass the open pathway marker along from cell to cell, they allocate resources to form the pathway.

Open pathway markers carry *addresses* to tell how to route them from their source to their destination, using *street sign* routing (e.g., "go to Jones and stop"). There are two components of such a street sign: the *street name* (e.g., Jones) and the associated *action* (e.g., stop). The controller provides special hardware support for address recognition. For example, a pathway route might consist of "go to Jones, turn right, go to Smith, turn left, go to Johnson, and stop". Each pathway unit is responsible for recognizing addresses of open pathway markers requiring service or attention at that cell. Given the sequential nature of street sign routing interpretation, a given communication agent needs to deal with only the "next" street name on passing open pathway markers.

Upon the arrival of an open pathway marker, the pathway unit interprets the address to see if it is addressed to this cell, and, if so, posts an event to the computation agent to invoke the appropriate routine. Otherwise, it finds a free, outgoing logical bus along the route given by the header and connects it to the incoming logical bus. The pathway is dismantled, one link at a time, by the flow of a *close pathway marker* along the pathway, cell to cell, from the source to the destination.

The latency of communication through a cell is 100 ns normally, and 150 ns in the case of corner turning. The interpretation of addresses and the establishment of pathways are completely performed by hardware. Creating a new pathway segment does not incur any additional time delay.

The Streaming and Spooling Unit

The computation agent can get access to the communication data by (1) directly accessing the communication agent a word at a time, or (2) spooling the data in and out of local memory using special hardware support.

Programs can read data from a message or write data to a message via the side effects of special register references. These special registers are called *streaming gates*, because they provide a "gating" or "windowing" function allowing a stream of data to pass, word by word, between the communication agent and the computation agent. There are two input gates and two output gates. These gates can be bound to different logical busses dynamically. A read from the gate will consume the next word of the associated input message; correspondingly, a write to an output gate will generate the next word of the associated output message. Data word-at-a-time synchronization is expressed in algorithms by the side effects of gate register references (e.g., a read of an input gate at which no data are available causes the instruction to spin until the data arrive).

iWarp also provides a transparent, low-overhead mechanism for transferring data between the pathway unit and the local memory via *spooling gates*. Spooling has low overhead to avoid significant reduction of the efficiency of any ongoing or parallel computation. Spooling is transparent except for delays incurred due to either cycle stealing (i.e., for address computation) or local memory access interference from other memory references (i.e., due to concurrent cache or instruction activities).

iWARP COMPUTATION

The iWarp processor is designed to execute numerical computations with a high, sustained floating-point arithmetic rate. The iWarp cell has a high peak computation rate of 20 MFLOPS for single precision and 10 MFLOPS for double precision. More importantly, iWarp can attain a high computation rate consistently. This is because the multiple functional units in the computation agent are directly accessible through a long instruction word (LIW) instruction. By translating user's code directly into these long instructions using an optimizing compiler,[16] a high computation rate can be achieved for all programs, vectorizable or not.

THE COMPUTATION AGENT

The computation agent has been optimized for LIW controlled, parallel operation of multiple-functional units. Chief among these optimizations are

- Nonpipelined floating-point arithmetic units.
- Inter- and intra-unit, output-to-input, operand bypassing.
- Parallel, hardware-supported, zero-overhead looping.
- Large, shared, multiported register file.
- A high-bandwidth, low-latency (no striding penalty) memory.
- High-bandwidth, low-latency interface with the communication agent.

The LIW workhorse instruction of iWarp is called the *ComputeAndAccess* (C&A) instruction. As an example of the parallelism available, a loop with code

```
FOR     i := 0 TO n-1 DO BEGIN
        f:=(A[i]*B[3*i])+f;
END;
```

is compiled into a loop that initiates one iteration every cycle surrounded by a loop prologue and epilogue to get the iterations started. Similarly, a loop body that reads a value V1 from one message, V2 from another message, computes V1 * V2 + C[i] and sends the result as well as V2 on to the next processor is also translated into a single C&A instruction in the loop body. The single precision C&A instruction executes in two clocks, and the double-precision C&A instruction executes in four clocks, so both loops execute at the peak computation rate of the processor.

A C&A instruction requires up to eight operands and produces up to four results. Memory accesses may produce or consume up to two of those operands: either a read and a write or two reads. Each memory reference includes an address computation (e.g., an indexing operation with a nonunit stride). The C&A instruction employs a read-ahead/write-behind pipeline that makes memory read operands from one instruction available for use in the next. Conversely, computational results of one instruction are written to memory during the next.

Those operand references that are not satisfied by the memory read operation or read from a gate must be to the register file. These operands may themselves be the results of previous operations (e.g., intermediate results held in the register file). To avoid any interinstruction latencies, the results of the integer unit or a floating-point unit may be "bypassed" directly back to that unit as an input operand, without waiting for the destination register file location to be updated. Also, the results of either floating-point unit may be "bypassed" directly to the other floating-point unit (e.g., to support multiply-accumulate sequences).

Thus, the execution of a single C&A instruction can include up to one floating-point multiplication, one floating-point addition, two memory accesses (including two integer operations for addressing), four gate accesses, several more register accesses (enough to provide the rest of the required operands), and branching back to the beginning of the loop.

Incremental to the single, "long" C&A instruction, the iWarp computation agent provides a full complement of "short" instructions. They can be thought of as two- and three-address RISC-like instructions. These "short" instructions are provided to make iWarp a generally programmable processor. They usually control only a single functional unit.

LIVERMORE LOOPS PERFORMANCE

The Livermore Loops,[6] a set of computational kernels typically found in scientific computing, have been used since the 1960s as a benchmark for computer systems. The loops range from having no data dependence between iterations (easily vectorizable) to having only a single recurrence (strictly sequential). This combination of vector and scalar code provides a good measure on the performance of a machine across a spectrum of scientific computing requirements.

The Livermore Loops were manually translated from Fortran to W2. (W2 is a Pascal-like language developed for the Warp machine. The retargeted W2 compiler[8,16,17] has been used as a tool in developing and evaluating the iWarp architecture.) The translation into W2 was straightforward, preserving loop structures and changing only syntax, except for kernels 15 and 16, which were translated from Feo's restructured loops.[6]

The performance of Livermore Loops (double precision) on a single iWarp processor is presented in Table 1. The unweighted mean is 4.2 MFLOPS, the standard deviation is 2.6 MFLOPS, and the harmonic mean is 2.7 MFLOPS. Since the peak double-precision performance of the machine is 10 MFLOPS, these numbers demonstrate a highly effective use of the raw computation power of iWarp.

The iWarp cell is a scalar processor and does not require that loops be vectorizable for full utilization of its floating-point units. This is why it does not

TABLE 1
Double-Precision Performance of Livermore
Loops on a Single iWarp Cell

Kernel	MFLOPS	Kernel	MFLOPS
1	8.2	13	1.8
2	3.3	14	3.2
3	9.8	15	1.6
4	3.2	16	0.8
5	3.3	17	1.6
6	5.0	18	6.6
7	9.9	19	4.0
8	7.4	20	3.1
9	6.7	21	4.1
10	2.0	22	3.1
11	2.0	23	6.3
12	2.5	24	0.9

exhibit the same tremendous disparity in MFLOPS rates for the different loops as do vector machines. Nonetheless, the variation between the MFLOPS rates obtained is still significant. Near peak performance can be achieved (using a high-level language and an optimizing compiler), as demonstrated by kernels 3 and 7. On the other hand, performance of near 1 MFLOPS is also observed. The factors that limit iWarp performance are data dependency and the critical resource bottleneck.

Data Dependency
Consider kernel 5:

```
FOR    i := 0 TO n-1 DO BEGIN
       X[i] := Z[i] * (Y[i] - X[i-1]);
END;
```

The multiplications and additions are serialized because of the data dependencies. Just by this consideration alone, iWarp is limited to a peak performance of 5 MFLOPS on this loop. However, iWarp still executes data-dependent code better than vector machines. The floating-point units are not pipelined, and there is no penalty on nonunit stride memory accesses. More importantly, not all data dependencies force the code to be serialized. As long as the loop contains other independent floating-point operations, the floating-point units can still be utilized. This is a unique advantage an LIW architecture has over vector machines.

Critical Resource Bottleneck
The execution speed of a program is limited by the most heavily used resource. Unless *both* the floating-point multiplier and adder are the most critical resources, the peak MFLOPS rate cannot be achieved. Programs containing no

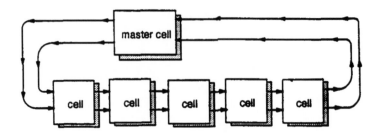

FIGURE 4. Master cell and cell communication configuration.

multiplications cannot run faster than 5 MFLOPS since the multiplier is idle all the time. For example, for kernel 13, since the integer unit is the most heavily used resource, the MFLOPS measure is naturally low.

PROGRAMMING AN iWARP SYSTEM

The following program gives an example of what the user can do by programming the array in C. This approach requires that the programmer be familiar with the functioning of the cell array, but it offers great programming flexibility. As an alternative, the program developer can use the Apply language, which is specialized for image processing applications. Such programs are fairly easy to write in Apply.

The following simple example shows segments of an iWarp C program that evaluates a polynomial. It demonstrates the solution of an *n-th* order polynomial at *m* data points using an $n+1$ cell ring array of iWarp cell processors. Each iWarp cell processor is connected to its right neighbor with two paths. Each of *n* cells computes one step of the Horner algorithm and passes the result to the right. The $n+1th$ cell serves as a master cell. The master cell provides data and receives, stores, and prints the results. In this example, we are computing the following polynomial using six cells:

$$P(z) = c_0 z^4 + c_1 z^3 + c_2 z^2 + c_3 z + c_4 \tag{1}$$

Figure 4 illustrates how data flows from the master cell around the array of cells.

There are two parts to the program: one that runs on the master cell and one part that runs one each cell. The #include statements and the master cell program segment that reads in the data from the host and writes back the results have been omitted for brevity.

```
static float c[5]; /*coefficients*/
static float z[5]; /*data points*/
static float p[5];/*results*/
static int nc; /*number of coefficients*/
static int nz; /*number of data points*/
main()
```

```
{
        register int i, tmp, error = 0;
        register float ftmp, fzero = 0.0;
        /*host input*/
        if( pathway_init() ) return(1);
        _sendi( GATE0, nc );
        tmp = _receivei( GATE0 );
        _sendi( GATE0, nz );
        tmp = _receivei( GATE0 );
        /* send coefficients */
        for ( i = 0; i  nc; i++)
                _sendf( GATE0, c[i] );
        /* send z and receive p */
        for ( i = 0; i < nz; i++ ) {
                _sendf( GATE0, z[i] );
                -sendf( GATE1, fzero );
                ftmp = _receivef( GATE0 );
                p[i] = _receivef( GATE1 );
        }
        if( pathway_close() ) return(1);
        /* write results back to host */
        return 0;
}
```

This portion of the program runs on each cell.

```
main()
{
register int i, nc, nz;
register float temp, coeff, xin, yin, ans;
register float fzero = 0.0;
if( pathway_init() ) return(1);
        nc = _receivei( GATE0 );
        _sendi( GATE0, nc-1 );
        nz = _receivei( GATE0 );
        _sendi( GATE0, nz );
        /* capture the first coefficient
         * and pass the rest on.                 */
        coeff = _receivef( GATE0 );
        for ( i = 1; i  nc; i++ ) {
                temp = _receivef( GATE0 );
                _sendf( GATE0, temp );
        }
        /* Horner's rule                         */
        for ( i = 0; i  nz; i++ ) {
                xin = _receivef( GATE0 );
                yin = _receivef( GATE1 );
                _sendf( GATE0, xin );
                ans = coeff + yin * xin;
                _sendf( GATE1, ans );
```

```
        }
        if( pathway_close() ) return(1);
        return 0;
}
```

SUMMARY AND CONCLUSIONS

iWarp is the first of a new class of parallel computer architectures. iWarp integrates both the computation and communication functionalities into a single VLSI component. The communication models support a range from large-grain message passing to fine-grain systolic communication.

The computation agent of an iWarp component contains floating-point units with a peak performance of 20 and 10 MFLOPS for single- and double-precision operations, respectively, as well as an integer unit that performs 20 million integer or logical operations per second. The communication agent operates independently of the computation agent; since both are implemented on a single chip, tight coupling between communication and computation is possible. This permits efficient systolic communication, as well as low-overhead message passing.

iWarp is designed to be a building block for high-performance parallel systems. Not only does iWarp have impressive computational capabilities, it also has exceptional communication capabilities, making iWarp suitable for both scientific computing and high-speed signal processing.

Arrays of thousands of cells are feasible, programmable, and much cheaper than many other supercomputers of comparable power. iWarp systems can have a variety of goals: they can be special or general purpose, and experimental or commercial. The support of well-accepted languages for the cell like Fortran and C, together with parallel program generators to simplify the programming of the array, make it possible to program the diverse parallel machines that can be realized with iWarp components.

REFERENCES

1. **Annaratone, M., Amould, E., Gross, T., Kung, H. T., Lam, M., Menzilcioglu, O., and Webb, J. A.,** The Warp computer: architecture, implementation and performance, *IEEE Trans. Comput. C-36*, 12, 1523, 1987.
2. **Annaratone, M., Bitz, F., Deutch, J., Hamey, L., Kung, H. T., Maulik, P., Ribas, H., Tseng, P., and Webb, J.,** Applications experience on Warp, in Proc. of the 1987 Natl. Computer Conf., AFIPS, 1987, 149.
3. **Dally, W. J.,** *A VLSI Architecture for Concurrent Data Structures,* Kluwer Academic, Dordrecht, 1987.
4. **Dally, W. J. and Seitz, C. L.,** Deadlock-free message routing in multiprocessor interconnection networks, *IEEE Transactions on Computers C-36*, 5, 547, 1987.
5. **Dongarra, J. J., Bunch, J. R., Moler, C. B., and Stewart, G. W.,** *LINPACK User's Guide,* Society for Industrial and Applied Mathematics, Philadelphia, 1979.
6. **Feo, J. T.,** An analysis of the computational and parallel complexity of the Livermore Loops, *Parallel Comput.*, 7, 2, 163, 1988.

7. **Gentleman, W. M. and Kung, H. T.**, Matrix triangularization by systolic arrays, in Proc. of SPIE Symp., Vol. 298, Real-Time Signal Processing IV, Society of Photo-Optical Instrumentation Engineers, August, 1981, 19.

8. **Gross, T. and Lam, M.**, Compilation for a high-performance systolic array, Proc. of the SIGPLAN 86 Symp. on Compiler Construction, ACM SIGPLAN, June, 1986, 27.

9. **Hamey, L. G. C., Webb, J. A., and Wu, I. C.**, Low-level vision on Warp and the Apply programming model, in *Parallel Computation and Computers for Artificial Intelligence*, Kowalik, J., Ed., Kluwer Academic, Dordrecht, 1987, 185.

10. **Kung, H. T.**, Why systolic architectures?, *Comp. Mag.*, 15, 1, 37, 1982.

11. **Kung, H. T., Ruane, L. M., and Yen, D. W. L.**, Two-level pipelined systolic array for multidimensional convolution, *Image Vision Comput.*, 1, 30, 1983. (An improved version appears as a CMU Computer Science Department technical report, November 1982.)

12. **Kung, H. T.**, Systolic communication, in Proc. of the Int. Conf. on Systolic Arrays, May, 1988, 695.

13. **Kung, H. T.**, Computational models for parallel computers, *Philos. Trans. R. Soc.*, 1988.

14. **Kung, H. T.**, Deadlock avoidance for systolic communication, in Conf. Proc. of the 15th Annu. Int. Symp. on Computer Architecture, June, 1988, 252.

15. **Kung. H. T. and Leiserson, C. E.**, Systolic arrays (for VLSI), Sparse Matrix Proc. 1978, Society for Industrial and Applied Mathematics, 1979, 256.

16. **Lam, M. S.**, A Systolic Array Optimizing Compiler, Ph.D. thesis, Carnegie Mellon University, 1987.

17. **Lam, M.**, Software pipelining: an effective scheduling technique for VLIW machines, in ACM Sigplan '88 Conf. on Programming Language Design and Implementation, June 1988.

18. **Pomerleau, D. A., Gusciora, G. L., Touretzky, D. S., and Kung, H. T.**, Neural network simulation at Warp speed: how we got 17 million connections per second, Proc. of 1988 IEEE Int. Conf. on Neural Networks, July 1988, 143.

19. **Raubold, D. and Haenle, J.**, A method of deadlock-free resource allocation and flow control in packet networks, in Proc. of the Third Int. Conf. on Computer Communication, International Council for Computer Communication, August 1976.

20. **Sejnowski, T. J. and Rosenberg, C. R.**, Parallel networks that learn to pronounce English text, *Complex Syst.*, 1, 145, 1987.

21. **Tseng, P. S., Lam, M. and Kung, H. T.**, The domain parallel computation model on Warp, in Proc. of SPIE Symp., Vol. 977, Real-Time Signal Processing XI, Society of Photo-Optical Instrumentation Engineers, August 1988.

22. **Wallace, R. S., Webb, J. A., and Wu, I.-C.**, Architecture independent image processing: performance of Apply on diverse architectures, in Third Int. Conf. on Supercomputing, International Supercomputing Institute, Boston, May 1988, 25.

4   iPSC AND iPSC/2

OVERVIEW

The iPSC and iPSC/2®* are a family of distributed memory concurrent computers. Each processor (node) runs its own program out of its own memory and communicates with other processors by passing messages. Messages are sent from the source node to the destination node by the communication subsystem. The current machines are known as hypercubes because, if the nodes are located at the corners of a multidimensional cube, then the communication channels are along the edges of the cube. Messages sent to distant nodes must traverse several channels to reach their destinations. The Intel machines use a fixed routing scheme to prevent the possibility of deadlock.[1] Messages are explicitly sent and received by subroutine calls in the node programs.

In addition to the nodes of the cube, there is a front-end computer called the System Resource Manager (SRM). The SRM provides the access to a local area network through a TCP/IP Ethernet interface; it boots the cube and runs diagnostic programs; it controls access to the cube; and it hosts the compilers and other tools. It can also run host programs which exchange messages with node programs.

There are four major areas of change in the iPSC/2. The communication subsystem, the node processor, the amount of memory per node, and the user environment have all improved significantly over the first-generation iPSC. The iPSC/2 is also designed so that all parts of the system are modular, enabling future enhancements to be incorporated easily.

The most important advancement in the iPSC/2 is the direct-connect©** routing module (DCM). Each node has a DCM which is both the interface between the node processor and the communication network and a portion of the network itself. When connected together as a hypercube, the DCMs create a circuit-switching network, similar to a telephone network. When node A wants to send a message to node B, the network creates a dedicated circuit from A to

---

*   iPSC® is a registered trademark of Intel Corporation.
** Direct-Connect© is a trademark of the Intel Corporation.

B. The entire message is then sent over the circuit. The channels in the circuit are freed as the end of the message passes through. The two key points here are that the actual switching time is very small, so that the performance of the network is almost constant regardless of the actual number of hops between A and B in the network. Second, the node processors at intermediate hops are not impacted in any way by messages passing through their DCMs. This is in sharp contrast to the store-and-forward implementation of communications in the original iPSC. The only possible effect of a heavily loaded network is contention for the channels. If two messages arrive at a node on different channels and both wish to exit on the same channel, then one message will be forced to wait until the other is finished.

Even for one-hop (nearest neighbor) communications, the DCM is significantly faster than the original iPSC. The latency of a message is less than 350 µs. Each hypercube channel is now a full duplex with a bandwidth of 2.8 Mbytes/s in each direction. Each DCM also has one extra channel for external communication. The SRM is connected to the external channel of node 0. All communication from the SRM to the cube is routed through node 0 and then on through the network. The link from the SRM to node 0 is just as fast as the rest of the network. The external channel of all the other nodes is available for I/O use, as described later. For a more complete description of the DCM, see Reference 2.

The node processor is a four-MIP Intel 80386. It supports a flat 32-bit virtual address space and multiple processes. Three different numeric coprocessors are available. The Intel 80387 provides 250 kFLOPS performance in 64-bit arithmetic. Also available is a Weitek 1167 module which provides 620 kFLOPS on the same loop. Finally, by using pairs of adjacent card slots, a vector processor can be added to each node. The vector board has its own memory (1 Mbyte) and performs vector arithmetic at up to 6.6 MFLOPS in 64-bit arithmetic and 20 MFLOPS in 32-bit arithmetic. Each node board can be configured with 1, 4, 8, or 16 Mbytes per node.

The user environment has improved significantly in the iPSC/2. The SRM is 386 based and runs UNIX System V in the 32-bit native mode. The C and Fortran compilers provided with the original iPSC were designed for PCs and were incapable of handling large user programs. The compilers provided with the SRM are mainframe-quality, 32-bit compilers. They run much faster than the old ones and they also make it very easy to call C from Fortran and vice versa. Concurrent Common LISP and Flat Concurrent Prolog are also available on the iPSC/2. There is also DECON, a symbolic concurrent source-level debugger. DECON allows users to set break points and examine and change message queues simultaneously over all of the nodes of the cube. This makes debugging concurrent programs much easier.

It is also possible to share the cube among several users. Individual nodes belong to one user at a time, but several users can be simultaneously using different parts of the cube. This is particularly convenient for debugging code or for use in teaching situations. The cube-sharing software on the SRM allocates subcubes and assures that users cannot accidentally interact.

Finally, if the SRM is on a local-area network with other workstations, it is possible to access the cube directly from the workstation without explicitly referencing the SRM at all. This allows users to develop and run programs in their typical programming environment instead of the one provided by the SRM.

To use multiple processors to solve problems, it is necessary to partition the work into tasks and then assign the tasks to processors. A good partition minimizes the amount of communication between tasks (low communication overhead) and divides the work equally (good load balancing). On store-and-forward systems there is a third problem: assigning the tasks to processors to minimize the average length of a communication path. For some applications this mapping problem is straightforward and leads to only local (nearest neighbor) communications. For many more applications (both numeric and symbolic) it is impossible to map the partition without significant long-range communication. Such applications cannot be run effectively on large store-and-forward machines. The introduction of direct-connect routing eliminates the mapping problem and allows all applications to be effectively run on large machines.

This result is more noticeable in computing 2D FFTs. The vector processor is quite effective at computing 1D FFTs. If a 2D array is distributed by rows across the processors, then the vector processors can simultaneously compute the FFTs of the rows. However, to finish the 2D FFT, it is necessary to compute the 1D FFTs of the columns. The simplest way to do this is to transpose the matrix. Unfortunately, this is a very communication-intensive task in which each node must exchange a large block of data with every other node. All of these communications are being done simultaneously and most of them are not nearest neighbor. This leads to serious congestion in store-and-forward systems.

For a 1024×1024 single-precision 2D FFT, a 16-node vector system can compute all of the 1D FFTs in about 0.7 s, which is more than 160 MFLOPS, or more than 10 MFLOPS per node. On the original iPSC the transpose takes 5.4 s, which reduces the overall computation rate to 17 MFLOPS. The computation is effectively using less than two of the vector processors. With direct-connect routing the transpose takes less than 0.4 s for a total computation rate of more than 100 MFLOPS, effectively using 10 of the 16 vector processors. On a 64-node iPSC/2, this same computation takes about 0.3 s with an efficiency of about 40/64.

For symbolic applications, the communication tends to be shorter messages with a much more random pattern. To simulate such a pattern, the following program was run. At each phase of the program, each processor sent a message of a specified length to another node. The target nodes were chosen as a random permutation of the node numbers so that each sending node also received a message. If specified, each message was then acknowledged. The program was run for many phases and the average time for each phase was recorded. Various message lengths were used with and without acknowledgments. In 16-node systems, the iPSC/2 is between 13 and 30 times faster than the original iPSC. The improvement in long-distance communication will allow symbolic applications to run effectively on large systems. Further application performance data can be found in Reference 3.

In this section we describe a parallel direct method based on the QR factorization to solve dense linear systems,

$$Ax = b$$

where A is a square dense matrix of size $n \times n$ and b is a vector of size n on the iPSC Intel Hypercube.

This application is decomposed in two steps: first, the QR factorization of A by a series of Givens rotations; second, the resolution of an upper triangular system by the Li and Coleman algorithm.[8] Both algorithms are implemented on the iPSC Intel Hypercube. Each unit of the iPSC system contains 32 micro-computers; every processor is composed of an arithmetic coprocessor and a local memory.

In every iPSC system unit every processor can be connected to up to five neighbors. This topology allows various network structures: rings, grids, torus, binary trees, etc.

We choose to implement these two algorithms on a ring of p processors with $p \leq n$. It is the simplest structure on the hypercube and it is very efficient when using pipelining. Pipelining can be seen as the superposition of the arithmetic operations and the data movements and can be used by alternating computations and communications in every processor. By this technique the communication times can be reduced and even sometimes neglected when the arithmetic times are very large vs. the communication time.

To build a ring of p processors embedded on a hypercube, we use binary-reflected Gray codes.[10] They allow one to define rings where every processor has its two nearest neighbors in the ring, which are its two nearest neighbors in the hypercube. We use then a ring of p processors numbered from 0 to $p - 1$ with a Gray code; all are connected to the cube manager, which functions to distribute the data among the processors and get the results at the end of the computation.

The way the elements of A and b are assigned to the processors affects the communication times, the degree of concurrency and the load balance among the processors.[5] The most efficient way to partition A is the *wrap mapping*: assigning one column to each processor and then wrapping back to the beginning with further columns. This wrap method keeps all processors busy as long as possible, and although it has bigger communications requirements than the *block mapping* (contiguous blocks of columns assigned to the processors), it leads to much better performance because it gives good load balancing and because we can globally overlap the computations and communication among the processors by pipelining them on the ring.

With this wrap mapping every processor, P_i, $i = 1, \ldots, p$, has then m ($m = \lceil \frac{m}{p} \rceil$) columns of A numbered $i, i + p, \ldots, i + (m - 1)p$, that is to say, $i \bmod p$.

The QR factorization of a square dense matrix of order n consists of computing a unitary matrix Q such that $Q^T A = R$ is upper triangular.

We choose to build Q to use Givens rotations rather than Householder transformations because they are easy to implement on distributed memory message passing computers,[4,6,7] they allow one to use pipelining as much as possible,[9] and they are very convenient when solving several linear systems with new right-hand sides, as occurs in least-squares problems.

Our rotations are applied in parallel following a scheme described first by Sameh and Kuck,[11] and then applied to systolic arrays by Gentleman and Kung.[4] This algorithm was written according to the systolic approach: every new step of computation in every processor begins with the reception of a group of rotations, followed by their application to its columns, and ended by sending a new group of rotations. We have then a data flow algorithm as when we are dealing with systolic algorithms. With this approach every processor has m steps of computations to perform on its m columns. Let us now describe the program of the *ith* processor along the ring:

QR factorization algorithm

```
do j = 1,2, . . . , m:
```

1. Receive from P_{i-1} the rotations of the $p - 1$ last columns preceding the *jth* column of P_i, i.e., the $[(j - 1) * p + i]th$ column of A.
2. Simultaneously apply these rotations to the columns $j, j + 1, \ldots, m$ of P_i and send the $p - 2$ last columns of rotations to P_{i+1}.
3. Compute the rotations of the *jth* column of P_i.
4. Simultaneously apply this column of rotations to the columns $j + 1, j + 2, \ldots, m$ of P_i and send this column of rotations to P_{i+1}.

With this algorithm every processor treats one column at a time, beginning with the first column of P_1, then $P_2, \ldots$, up to P_p, then the second column of P_1, then $P_2, \ldots$, up to P_p, and so on, ending with the last column of P_p.

Let us now decompose the tasks during the whole algorithm in order to get the arithmetic and communication times.

We will assume (to simplify our evaluation) that n is divisible by p, that is to say, $n = mp$. In that case, the first column of A is assigned to the processor P_1 and the last column of A, to P_p. If we also assign the right hand side b to P_p, then the last busy processor for the QR factorization of A and the computation of $Q^T b$ will be P_p. The activity of the p processors can then be decomposed in three phases: during the first phase one processor after the other starts working until the p processors are all active, which corresponds to the second phase, ending with the third phase, where one processor after the other gets idle until the processor P_p has finished.

To evaluate the arithmetic and communication times, we have then to compute the time, from the beginning when processor P_1 starts working up to the instant when processor P_p starts working, which we call T_1; and then the time for P_p to perform all its tasks during the second and the third phases, which we will call T_2.

REMARKS

- If β is the start-up time and r the elementary transfer time, then the communication time to send a packet of data of length p from one processor to its nearest neighbor is $\beta + pr$.
- We also set ω the time to perform one floating-point operation. The time to compute a rotation is then 5ω, and to apply a rotation to a couple of reals, 4ω.

Let us start to evaluate T_1: P_1 computes its $n-1$ first rotations and sends them to P_2:

$$5\omega(n-1)+\left[\beta+2(n-1)r\right]$$

P_2 receives them and sends them to P_3:

$$\left[\beta+2(n-1)r\right]\dots$$

P_{p-1} receives them and sends them to P_p:

$$\left[\beta+2(n-1)r\right]$$

So the value of T_1 is

$$5\omega(n-1)+(p-1)\left[\beta+2(n-1)r\right]$$

For T_2, during the first step of computations of P_p, reception of

$$\sum_{i=1}^{p-1}(n-1)$$

rotations and application of them to its m columns:

$$4\omega m\left[\sum_{i=1}^{p-1}(n-1)\right]$$

Computation of its $n-p$ rotations:

$$5\omega(n - p)$$

Sending its

$$\sum_{i=2}^{p}(n - i)$$

rotations to P_1:

$$(p - 1)[\beta + 2(n - 1)r]$$

Application of its $n - p$ rotations to its $m - 1$ last columns:

$$4\omega(n - p)(m - 1)$$

During the second step of computations of P_p, reception of

$$\sum_{i=1}^{p-1}(n - p - i)$$

rotations and application of them to its $m - 1$ last columns:

$$4\omega(m - 1)\left[\sum_{i=1}^{p-1}(n - p - i)\right]$$

Computation of its $n - 2p$ rotations: $5\omega(n - 2p)$.
Sending its

$$\sum_{i=2}^{p}(n - p - i)]$$

rotations to P_1:

$$(p - 1)[\beta + 2(n - 1)r]$$

Application of its $n - 2p$ rotations to its $m - 2$ columns:

$$4\omega(n - 2p)(m - 2)...$$

During the *mth* step of computations of P_p, reception of

$$\sum_{i=1}^{p-1}\left(n-(m-1)p-i\right)$$

rotations and application of them to its last column:

$$4\omega\left[\sum_{i=1}^{p-1}\left(n-(m-1)p-i\right)\right]$$

So the value of T_2 is

$$4\omega\left[\sum_{\ell=0}^{m-1}\sum_{i=1}^{p-1}(m-l)(n-lp-i)\right]$$
$$+(p-1)(m-1)\left[\beta+2(n-1)r\right]+4\omega\left[\sum_{\ell=1}^{m-1}(m-l)(n-lp)\right]$$
$$+5\omega\left[\sum_{\ell=1}^{m-1}(n-lp)\right]$$

And we find for the total computation time

$$T_1+T_2\approx\omega\left[\frac{4n^3}{3p}+\frac{7n^2}{2p}+n^2-\frac{(2p+9)n}{6}\right]+n\left[\beta+2(n-1)r\right]$$

So the arithmetic time is approximately

$$\left[\frac{4n^3}{3p}+\frac{7n^2}{2p}+n^2-\frac{(2p+9)n}{6}\right]\omega$$

and the communication time is $n[\beta+2(n-1)r]$.

If T_1 is the execution time of the serial algorithm, and T_p, the execution time of the same type of algorithm with p processors, then the speed-up is defined by

$$S_p=\frac{T_1}{T_p}\leq p$$

and the efficiency by

$$E_p=\frac{T_1}{pT_p}\leq 1$$

The sequential time for the QR factorization is approximately

$$\left[\frac{4n^3}{3} + \frac{n^2}{2} - \frac{11n}{6}\right]\omega$$

we find a speed-up of

$$S_p = \frac{T_1}{T_p} \approx$$

$$p\left[\frac{1 + 3/8n - 11/8n^2}{1 + 21/8n + 3p/4n - p(2p+9)/8n^2 + 3p\beta/4\omega n^2 + 3pr/2\omega n^2}\right] \approx p$$

and an efficiency of

$$E_p = \frac{S_p}{p} \approx$$

$$\left[\frac{1 + 3/8n - 11/8n^2}{1 + 21/8n + 3p/4n - p(2p+9)/8n^2 + 3p\beta/4\omega n^2 + 3pr/2\omega n^2}\right] \approx 1$$

The parallel code was implemented on a ring of four processors for dense linear systems of orders 32 to 512.

Figure 1 presents the theoretical and experimental speed-ups S_4 for the QR factorization algorithm. As we can see, the theoretical speed-up is larger than the experimental one. The reason for this is that for our complexity evaluation, we made the assumption that during the second phase of the algorithm, when all p processors are active, no one becomes idle awaiting new data.

Unfortunately this hypothesis is not true when the size of the problem is not very big, because in this case the processors do not have enough computations to do between the receives and sends of the data.

Our arithmetic and communication times were approximated under the assumptions that n/p is big and that during most of the time all p processors are active. Under those circumstances our algorithm is nearly optimal, as we get an efficiency very close to 1. Unfortunately the second hypothesis is only true for big systems. That is the reason why, as for the LU factorization algorithm (see Reference 5), these efficiencies are only asymptotic, and we have to run quite large problems to get these good efficiencies.

On the other hand, the wrap mapping is the best assignment of the elements of $[A,b]$ by columns to the processors in matters of load balancing. Computing the computation time of the block mapping with a similar technique to the one

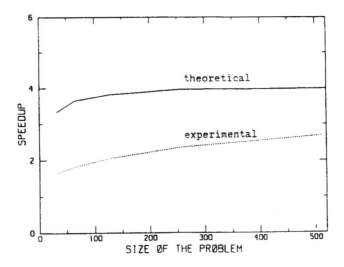

FIGURE 1. Speed-up of the iPSC hypercube with four processors.

described before produced an efficiency of 2/3. From this result we can deduce that all other column mappings, that is to say, all combinations of block and wrap mappings (see Reference 10), for our QR factorization column-oriented algorithm would give theoretical efficiencies between 2/3 and 1.

REFERENCES

1. **Sullivan, H. and Bashkow, T. R.,** A large scale homogeneous machine, in Proc. 4th Annu. Symp. on Computer Architecture, 1977, 105.
2. **Nugent, S. F.,** The iPSC/2 Direct-Connect communications technology, in Proc. of the 3rd Conf. on Hypercube.
3. **Arshi, S., Asbury, R., Brandenburg, J., and Scott, D. S.,** Application performance on the iPSC/2, in *3rd Conf. on Hypercube Concurrent Computers and Applications*, Academic Press, New York, 1988.
4. **Bojanczyk, A., Brent, R. P., and Kung, H. T.,** Numerically stable solution of dense linear systems of linear equations using mesh-connected processors, *SIAM J. Sci. Stat. Comput.*, 5(1), 95, 1984.
5. **Geist, G. A. and Heath, M. T.,** *Matrix Factorization on a Hypercube Multiprocessor*, Proc. of the First Conf. on Hypercube Multiprocessors, Knoxville, TN, SIAM,Philadelphia, 1986.
6. **Gentleman, W. M. and Kung, H. T.,** Matrix triangularization by systolic arrays, *Proc. Soc. Photo-Opt. Instrum. Eng.*, 298, 19, 1981.
7. **Heller, D. E. and Ipsen, I. C. F.,** Systolic networks for orthogonal decompositions, *SIAM J. Sci. Stat. Comput.*, 4(2), 261, 1983.
8. **Li, G. and Coleman, T. F.,** A New Method for Solving Triangular Systems on Distributed Memory Message-Passing Multiprocessors, Technical Report TR 87-812, Department of Computer Science, Cornell University, Ithaca, NY, 1987.
9. **Saad, Y. and Schultz, M. H.,** *Parallel Direct Methods for Solving Banded Linear Systems*, Research Report YALEU/DCS/RR-387, Department of Computer Science, Yale University, 1985.

0. **Saad, Y. and Schultz, M. H.,** *Topological Properties of Hypercubes,* Research Report YALEU/DCS/RR-389, Department of Computer Science, Yale University, 1985.

1. **Sameh, A. H. and Kuck, D. J.,** On stable parallel linear system solvers, *J. Assoc. Comput. Mach.,* 25, 81, 1978.

5 THE PARAGON™ XP/S SYSTEM

OVERVIEW

The Intel Paragon XP/S system offers a high level of performance, flexibility, usability, and programming for demanding supercomputing applications. The Paragon system delivers scalable performance of up to 300 GFLOPS, backed with an architecture that will yield affordable, practical teraFLOPS by the middle of the decade. The system is designed for effective usage in distributed, production supercomputing environments, and includes a comprehensive suite of tools for developing and porting applications that realize high sustained performance.

Paragon XP/S configurations offer peak computational speeds ranging from 5 to 300 GFLOPS. This computational performance is matched by a high-bandwidth, low-latency interconnection network that passes messages between any two nodes in the system at rates of 200 Mbytes/s (full duplex).

Every aspect of the Paragon architecture is scalable. Main storage capacity scales to 128 Gbytes of dynamic RAM and more than a terabyte of high speed internal disk storage. Peripherals and network interfaces are scalable — multiple channels for SCSI-2, VMI, Ethernet, and HIPPI to meet any I/O requirement. The bandwidth of the interconnection network of the system rises with the number of nodes in the system.

The Paragon operating system, a full implementation of UNIX, also scales, ensuring that delivered operating system services match hardware performance as system size increases. Once applications are running on the Paragon system, they too become scalable. For example, a computational fluid dynamics code could be developed and tested on a small number of nodes. When ready for production, the number of nodes used for the production runs can be based on the desired performance — 400 nodes or 1000 nodes could be employed without change of the application.

The system can scale to match the size of the data set without changing the application or the time it takes to get the answer. For instance, often the same algorithm can be used for a larger data set, allowing the user to model an entire structure rather than just a portion of the structure. Using the same application

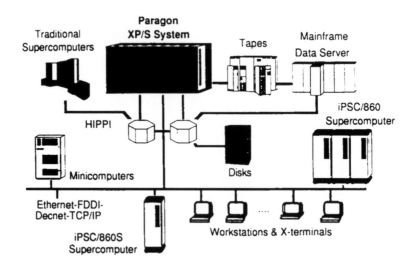

FIGURE 1. The Paragon XP/S system in the technical computing environment.

on a larger number of nodes allows the use of these larger data sets while retaining the quick turnaround time of the answer.

Paragon's transparently distributed UNIX operating system — the first full UNIX implementation for a massively parallel supercomputer — ensures flexibility in managing the Paragon system, developing applications, and integrating the system into heterogeneous distributed computing environments. The operating system is the result of advanced developments in distributed operating system services and microkernel technology and delivers full UNIX services to every node in the machine.

Adding to the system's ease of use and further integrating parallel supercomputing into the technical computing environment (see Figure 1), the Paragon system supports client/server computing via scalable HIPPI, FDDI, and Ethernet networking. For flexible system access, users can submit jobs over a network or can log directly onto any Paragon node. The Paragon system can be used for batch and interactive jobs simultaneously and can be reallocated dynamically to adjust the proportion of interactive and batch services. Comprehensive resource control utilities are available for allocating, tracking, and controlling system resources.

Intel's MIMD architecture supports all programming styles and paradigms. Applications can be developed using the programmer's preferred programming model — object-oriented, single program multiple data (SPMD), single instruction multiple data (SIMD), multiple instruction multiple data (MIMD), shared memory, or vector shared memory.

Tools include optimizing compilers for Fortran, C, C++, Ada, and HPF as proposed by the HPF Consortium; the interactive parallel debugger (IPD);

Intel's highly tuned ProSolver library of equation solvers; Intel's Performance Visualization System (PVS™); and Paragraph.

BEYOND TRADITIONAL SUPERCOMPUTING

The Paragon system's massively parallel architecture provides performance and flexibility beyond the reach of traditional vector supercomputers:

- Transparent software scalability. Once an application is running on the Paragon XP/S system, it can run without change on any number of nodes. Users can move back and forth from a 5 to a 300 GFLOP performance level with no reprogramming or system reconfiguration.
- Simultaneous interactive and batch operations. The Paragon system doesn't force the users to decide between operating in batch mode or interactive mode — all modes of operation are provided simultaneously and all system resources can be dynamically reallocated as the system is running.
- Native or networked development. Developers are given the option to choose software tools directly on the system in native mode or on their own workstations, downloading their applications.

The Paragon™ XP/S System At A Glance

Capacity	5 to 300 GFLOPS peak 64-bit floating-point performance
	2.8 to 172 kMIPS peak integer performance
	Node-to-node message routing at 200 Mbyte/s (full duplex)
	1 to 128 Gbytes main memory, up to 500-Gbyte/sec aggregate bandwidth
	2 to 128 Mbytes processor cache, up to 4.0–Tbyte/s aggregate bandwidth
	9.6 Gbytes to 1 Tbyte internal disk storage, up to 6.4-Gbyte/s aggregate I/O bandwidth
System architecture	Scalable, distributed-memory multicomputer MIMD control model
Node architecture	Nodes based on Intel's 50 MHz i860™ XP processor
	75 double-precision MFLOPS, 42 VAX MIPS peak performance per processor
	16 to 128 Mbytes DRAM per node
	400-Mbyte/s processor-to-memory bandwidth
	800-Mbyte/s processor-to-cache bandwidth
Interconnect architecture	2D mesh topology
	Pipelined, hardware-based internode communications
Operating system	Paragon OSF/1
	Transparent, distributed, scalable services

	Conformance with POSIX, System V.3, 4.3bsd
	Virtual memory
System	Simultaneous batch and interactive operation
access	NQS, MACS utilities for resource management
	Client/server access, direct logins, remote host
Connectivity	Multiple HIPPI channels with 100 Mbyte/s
	bandwidth each
	Multiple Ethernet channels
	Multiple VME* connections
	NFS, TCP/IP, DECnet* protocols, UniTree client
	support
Programming	C, Fortran, Ada, C++, Data-parallel Fortran
environment	Integrated tool suite with a Motif-based GUI
	FORGE and CAST parallelization tools
	Intel ProSolver parallel equation solvers
	BLAS, NAG, SEGlib and other math libraries
	Interactive Parallel Debugger (IPD)
	Hardware-aided Performance Visualization System
	(PVS)
	Operating system support for shared virtual memory
Visualization	X Window System
tools	PEX
	Distributed Graphics Library (DGL) client support
	Connectivity to HIPPI frame buffers

THE PARAGON XP/S ARCHITECTURE

The Paragon system (see Figure 2) employs a scalable, heterogeneous multicomputer architecture. Its various processing nodes populate a two-dimensional interconnection network. Nodes provide a variety of services including computation, input/output services, operating system services, and network connectivity. Main memory is physically distributed among the nodes. The architecture of the Paragon system is reflected in the implementation: flexible, scalable, and modular. All programming models are supported because the topology of the interconnect network allows the programmer to ignore the physical location and function of the nodes. Systems are divided into partitions, which can consist of as little as one node or as much as all of the nodes. In the example in Figure 2, the system is configured with eight nodes dedicated to I/O interfaces, eight nodes to system services, and the remainder to the user applications.

The system makes extensive use of microprocessor technology. Each employs at least two Intel i860 XP processors: one or more *application processors* dedicated to executing a user's application program, and a *message processor* dedicated to the task of internode communication.

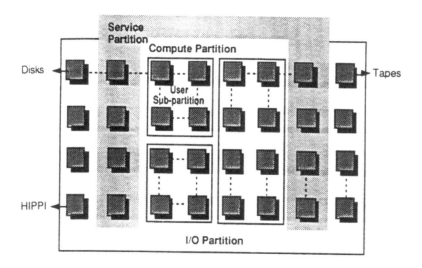

FIGURE 2. Architecture of the Paragon system.

The Paragon interconnection network provides low latency and high bandwidth between every node in the system. Two nodes, anywhere in a Paragon system, achieve process-to-process transfer latency of 25 microseconds — regardless of the size and configuration of the system. Messages are routed automatically, so the application programmer sees a system in which every node is connected to every node, with no difference in performance among nodes.

Within each node, the microprocessor speed is matched by on-chip cache performance and high-bandwidth memory units. Aggregate interconnect bandwidth scales with the number of nodes and supports the efficient operation of systems with many thousands of processors. The bandwidth and latency of the node-to-network interfaces and node-to-node communication channels are matched to the node memory bandwidth and execution speeds. Aggregate I/O performance, like computational performance, scales with the number of I/O nodes. The I/O interfaces are custom engineered to the needs of specific industry-standard bus and channel interfaces. With greater bandwidth than virtually any peripheral device, point-to-point I/O channel, or local area network, the interconnect sustains this I/O performance from or to any node in the system.

Dedicated programmable and fixed-function messaging hardware guarantees that applications achieve the maximum performance from the most time-critical sections of the operating system. Not only is sustained performance increased, but messaging operations, such as broadcasting and other global operations, are reliably and efficiently implemented.

Balance within each Paragon node (see Figure 3) is the result of engineering the memory unit, messaging facilities, and external I/O interfaces to match the performance demands of the Intel i860 XP microprocessor. Capable of 42 MIPS

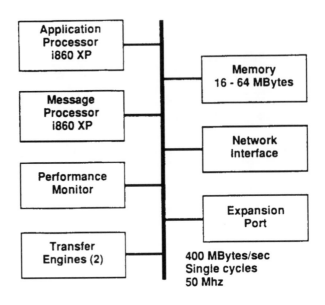

FIGURE 3. Paragon modular node.

and 75 double-precision MFLOPS when operating at 50 MHz (20-ns clock cycle), the 2.55 million-transistor Intel i860 XP provides 4-way set-associative, on-chip, 16-kbyte instruction and data caches. Floating-point-unit-to-cache bandwidth peaks at 800 Mbytes/s. Cache refills to local memory take place at 400 Mbytes/s and are fully supported by an interleaved, dual-bank, single-cycle, 64-bit memory unit design. Local memory is based on 60-ns DRAM arrays with full single-bit error correction and double-bit error detection. The Paragon node is modular in architecture and implementation, with the network interconnect boosted with a dedicated Intel i860 XP separate from the i860 XP processor used for user applications. The design also benefits from a performance monitor with ties to all the major components and a connection to its own (optional) data collection network.

The basic Paragon node design is used within the system in several ways: computation, input/output, and service. Individual compute and service nodes may be configured with various memory options from 16 to 128 Mbytes. Input/output nodes may be configured with from 16 to 64 Mbytes of memory and one of several I/O interfaces, as described below.

A second Paragon node design, currently in development, incorporates multiple processors (MP) per node. This MP node architecture dramatically increases both the volumetric and the peak performance capacity of the Paragon XP/S system. Recent packaging and cache memory technology enables operating speeds well above the nominal 50-MHz clock of the Paragon XP node design. Wide-word memory units maintain the necessary bandwidth to satisfy cache refills and pipelined processor loads and stores.

The Paragon interconnection network is the result of a decade of research and development in multicomputer architecture. Beginning at Caltech with the leadership of Professor Charles Seitz, expanding at MIT under Professor William Dally, and continuing at Intel as part of the joint Intel-DARPA Touchstone project, this line of research has given the Paragon XP/S system a very advanced multicomputer interconnection network.

Interconnect performance is fundamentally determined by the number of wires that cross from one half of the network (the bisection) to the other. Networks with more wires, assuming they are equally short, should be faster than networks with fewer wires. In practice, the number of wires crossing the bisection is limited by electrical factors, such as power dissipation, and mechanical factors, such as trace, connector, and cable density. Designing an optimal network is principally an effort to make the most efficient use of the available wires, given that the wire limit is essentially the same at a given moment in time for all practical designs.

Extensive analytical studies, thousands of hours of simulation, and several prototype systems (most notably the Touchstone DELTA and SIGMA systems) have demonstrated that low-dimensional mesh networks make the most efficient use of available wires. That means, given an equal number of wires at the bisection of the network, a two-dimensional mesh will outperform any toroidal, hypercube, or tree-structured network for uniformly distributed communication traffic in systems containing up to several thousand nodes. The mesh advantage actually *grows* as the communication traffic becomes more localized.

To deliver on the promise of the mesh topology (see Figure 4), the switching elements and electrical design of the Paragon XP/S interconnection network have been carefully considered. The individual switches, located at each vertex in the network, are fabricated in the same 0.75 μm, triple-metal, CMOS technology used to build the Intel i860 XP. Each Paragon mesh routing chip (PMRC) can simultaneously route traffic moving in the four network directions (±X and ±Y), and to and from its attached node, at speeds in excess of 200 Mbytes/s. It takes the PMRC less than 40 ns to make an individual routing decision and close the necessary switches. Internal operation of the router is fully parallel, and all transfers are parity checked. To maximize router bandwidth, it is fully pipelined both internally and externally. Message traffic moving from one router to another is pipelined along the wires so that speed becomes independent of distance for all practical purposes. The routing algorithm is provably deadlock free. The Paragon architecture utilizes a two-dimensional mesh network topology, originally developed under the Intel-DARPA Touchstone project. Every node, regardless of its function in the system, utilizes the interconnection network, permitting every system service and function to scale with the system as it grows to more than a 1000 nodes.

Instead of a passive backplane, as used in most conventional systems, the Paragon system uses an active backplane built from a rectangular arrangement of router chips. Each backplane section connects to its horizontal and vertical neighbors by flexible printed circuits and ultra reliable connectors. Open the back door of a Paragon system and the backplanes are completely clean. Gone

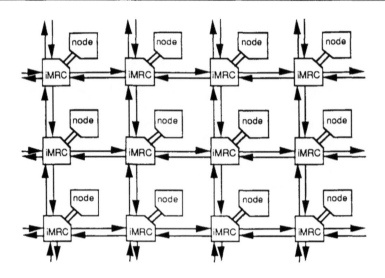

FIGURE 4. Paragon mesh interconnect.

are the fragile cables and wiring mat that have become the trademarks of conventional vector supercomputers and are still found in many highly parallel machines. Also missing are the critically measured and cut clock control lines, as the Paragon interconnect is completely self-timed and clock free. Only the Paragon mesh routers are visible, moving information between the nodes at speeds limited only by their underlying VLSI technology and the speed of light. For systems with as few as 256 nodes, the bisection bandwidth is in excess of 6 Gbytes/s. A Paragon system with 1000 nodes has a bisection bandwidth of over 12 Gbytes/s.

Progress in interconnection architecture and router design continues at a very rapid rate. Anticipating future developments, the Paragon interconnect is open to future technology insertion. Developments are already underway that will increase the bandwidth of the network to over 25 Gbytes/s for 1000-node systems and provide greater routing efficiency for messages traveling long distances.

Studies for the Intel-DARPA Touchstone prototypes revealed that fast nodes and a fast interconnect do not guarantee balanced internode communication. An often-overlooked design element, the *node-network interface*, is key to unlocking overall system performance.

A semicustom VLSI chip, the network interface controller (NIC), provides a full-bandwidth, pipelined electrical interface between a node and its PMRC. Included in this chip is a parity-checked, full-duplex router port and end-to-end error checking for each message transfer. Internal parallel operation permits simultaneous inbound and outbound communication at 200 Mbytes/s. Since the node processor cannot sustain this speed for long messages, two block transfer engines are designed into the memory interface. They can move up to 4096 bytes of data at a time at an aggregate speed just short of the full node memory bandwidth.

To complete the node-network interface, a second full-speed, full-function Intel i860 XP *message processor* is also incorporated into the basic node design. Its purpose is to free the *application processor* from the details of internode communication and to provide a variety of communication services within the Paragon system. Operating in parallel, the message processor handles all communication traffic to and from the node. It autonomously receives messages and prepares them for use by the application processor. Similarly, it handles all the details of sending a message, including protocol and packetization, if required.

The message processor has many subtle benefits that become obvious only after understanding the low-level issues of MIMD multicomputer design. By freeing the application processor of messaging details, the message processor reduces cache turbulence in both processors. The application code and data stay in the caches on board the application processor to sustain performance. The messaging software of the operating system is small enough to remain resident in the code and data caches of the message processor. This guarantees the lowest possible latency, because library code doesn't first have to fill the processor caches before it can run at full speed. The dual-processor design also avoids a costly context switch in the application processor during messaging operations and while handling interrupts from the network interface. This, in turn, helps avoid draining the floating-point and memory pipelines in the application processor.

The message processor at each node also enables the Paragon system to offer a range of global operations that are performance critical in both MIMD and SPMD applications. The global operations that are autonomously implemented by cooperating message processors include broadcasting to and synchronizing a group of compute nodes, as well as mathematical operations, such as global sum and global minimum, for both integer and floating-point operands. Global operations are also available for logical operands, such as global AND and OR, and for aggregates, such as global string concatenation.

It is important to note that while the computational resources of the message processor are utilized by the Paragon system, the performance rating of the message processors are *not* included in the aggregate peak performance ratings of the Paragon XP/S system.

6 ENCORE MULTIMAX

OVERVIEW

Three separate functions were identified in the earliest days of the stored program computer:

- Processor — fetches and executes instructions stored in memory
- Memory — stores instructions and data
- Input/output — for communications with humans and electronic devices.

Computer design can exceed the performance limitations of conventional technology by introducing concurrency into the organization of the computer. The number of operations that are allowed to happen simultaneously determines the degree of concurrency.[6] One approach for introducing concurrency is replication — providing multiple copies of the same resource so that several sections of the computation can proceed concurrently. Replication of processors that share a large main memory yields the multiprocessor of configuration of Figure 1.

Multiprocessors were first developed by mainframe manufacturers as a means for extending the performance range of their product line.[4] These multiprocessors were composed of two to four processors capable of independent instruction execution and able to exchange information through a switching structure.

With the advent of microprocessors, the multiprocessor moved from an interim structure to the basis for a new class of computer.[1,3] The small size, low cost, and high performance of microprocessors allow the design and construction of the Multi(processor) structure that offers significant advantages over traditional computer families.

The Multi is composed of a large number of microprocessors sharing a common memory and input/output devices. Adding more microprocessors increases the performance of Multis in direct proportion to their number and allows Multis to offer a performance range from personal computers to large supercomputers utilizing one technology. In contrast, contemporary computer families are implemented from a range of technologies. The Multi has not only

105

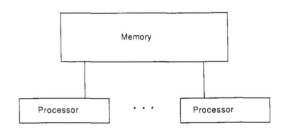

FIGURE 1. Shared memory multiprocessor.

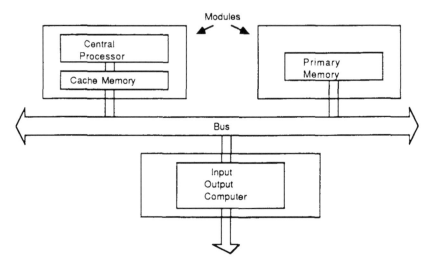

FIGURE 2. Structure of a Multi.

overcome the historical limitations of multiprocessors, but also provides solutions to the limitations suffered by uniprocessor designs.

Figure 2 depicts the generic structure of a Multi. The Multi employs a unified bus over which all communications occur between processors, memories, and input/output devices. This bus structure was pioneered in the Digital Equipment Corporation PDP-11 where it was called a "UNIBUS"™. The small size of microprocessors enables the placement of all the processors in a single backplane, thus eliminating the cable growth problem of earlier multiprocessors. A Multi can be constructed from as few as three module (i.e. card) types: processor, memory, and input/output.

In a Multi, the cache memory associated with a processor services approximately 95% of the processor's requests for memory. Since only 5% of a processor's requests reach the common bus, the requirements for the bus bandwidth are reduced by factor of 20.

While processor caches can significantly improve system performance, they introduce a stale data problem (often termed the multicache coherency problem) due to the presence of multiple cached copies of main memory locations. It is necessary to ensure that changes made to shared memory locations by any one processor are visible to all other processors. Since all processor requests for primary memory appear on the common bus, each cache can independently monitor the bus and delete any data it contains that have been modified in primary memory by another processor. While this simple cache consistency scheme requires a modest increase to the cache hardware, the caches are as effective in multiprocessors as they are in uniprocessors. In addition to reducing the bus bandwidth requirements, the distributed caches also eliminate much of the need for fast main memory that has been historically inherent in other multiprocessor structures. In addition to solving the historical limitations of the multiprocessor, the Multi offers significant advantages in scalability, modular growth, functional specialization, reliability, design, manufacturability, and longevity over the approach of the traditional family of uniprocessors.

Performance Range

System performance in a Multi is linearly proportional to the number of processors until the bus is saturated with requests for access to main memory. Because of this linear proportionality, Multis provide the means for producing a range of compatible computers without a specific processor design and specific technology for each model. With Multis, users can select the computer that meets their need by configuring the appropriate number of processors or memories instead of having to select from different family members.

Modular Growth

The ability to increase processing power, memory size, and I/O throughput by adding modules in an incremental fashion provides a simplified means of upgrading a system. Instead of replacing a uniprocessor by the next higher performance model, a user merely needs to add more of the appropriate card types to the system. However, one should ensure the system remains balanced. In the late 1960s, Gene Amdahl, the designer of the IBM System/360, postulated two rules of thumb for a balanced system. The first rule states that one byte of memory is required for each instruction per second (IPS) executed. The second rule states that one bit of I/O is required to support each IPS. With extensive multiprogramming and the virtual memories used for large programs, a balanced system today would require 2 to 4 bytes of memory and 1 to 2 bytes of I/O per IPS. Since communications between processor, memory, and I/O occur on the single bus in a Multi, the user must also ensure that the total required bandwidth from the bus is adequate not only for average, but also for peak demands.

Functional Specialization

The single bus is attractive for plugging in functionally specialized cards such as vector processors, signal processors, A/I processors, and video cameras. The communications bandwidth requirements of these functionally specialized cards must also be taken into account when calculating required average and peak bandwidth.

Reliability

The Multi has inherent reliability through redundancy because it is built from a small number of card types. Typically, a Multi has a maximum of four card types: processor, memory, and two I/O controller types for disks and terminals. From these components, systems of 10 to 20 cards are formed.

With the appropriate software, various cards can be marked as faulty and taken out of service, allowing the system to function, even in the presence of a failed card. The inherent redundancy provides more than an order of magnitude greater increase over traditional computers of the same size in

- Reliability (probability system can function even though a critical component has failed)
- Availability (probability that the system is available for use)
- Maintainability (increased through simplified logistics of fewer card types, operations in spite of component failure, and tolerance of service delays)

In effect, reliability and availability are added dimensions of configurability that a user can specify at any time by using the appropriate number of each card type.

Design

Multi hardware is inherently simple because of its few (e.g., four) card types, with only one explicitly defined interface standard, the bus. A large Multi consists of a total of 20 cards and can be designed and shipped rapidly. In comparison, a 20-card uniprocessor with one third the performance of a 20-card Multi system requires over 5 times the effort to design; because each card is unique and requires an interface to other cards with which it communicates. A uniprocessor of this type requires over 2 years of design.

A CONTEMPORARY MULTIPROCESSOR SYSTEM: MULTIMAX/UMAX

System Overview

The Encore Multimax incorporates from 2 to 20 National 32032 32-bit microprocessors, each capable of executing 0.75 million instructions per second (MIPS), with resulting performance scalable from 1.5 to 15 MIPS. Memory is scalable from 4 to 32 Mbytes in increments of 4 Mbytes. I/O throughput is scalable from 1.5 to 15 Mbytes/s in increments of 1.5 Mbytes/s. The system backplane can transfer data at a sustained rate of 100 Mbytes/s.[5]

UMAX, the Multimax's operating system, is based upon UNIX. All Multimax processors share a single copy of UMAX — an economy that preserves more memory space for programs and data. Because of the single copy of the operating system, there is minimum memory and bus involvement with data copying and context switching when the system executes multiple processes. Processors are dynamically allocated to the processes having the currently highest priority.

Memory is dynamically allocated to processes, resulting in efficient memory usage and improved interprocess communication. I/O channels can operate on any portion of memory without access speed penalty. UMAX distributes the workload evenly among the resources. UMAX is built to service many simultaneous requests by supporting multiple, simultaneous lines of control — a technique called multithreading. To perform these services efficiently when large numbers of processors and users require system service, UMAX employs techniques such as caching in memory of commonly referenced data structures that would exact large time penalties to reference on disk.

In addition to parity protection and error correction code (ECC) data correction at critical points in most data transfers, Multimax systems perform automatic retries on soft error conditions, thus reducing failures within the system and contributing to greater system availability. The Multimax system control card contains a microprocessor and local memory dedicated to running confidence checks at each start-up, as well as to monitoring normal system operation. It can completely reconfigure the system around failed, redundant cards, without operator intervention, and thus provides significantly higher system availability. Temperature sensors located in critical areas of the enclosure trigger warnings when fans malfunction or airways are blocked. Processors that malfunction can be deactivated automatically; memory banks that accumulate a history of excessive ECC operations — or that manifest any uncorrectable error states — can be taken out of service. Faults are identified on the system console whenever they are discovered; nonfatal faults discovered after the UMAX operating system has been started are made available for logging by the operating system.

Multimax Hardware

The backbone of the Multimax is the Nanobus™. The Nanobus provides 20 slots which can be filled by 4 card types: dual processor cards (DPCs), Ethernet/mass storage cards (EMCs), shared memory cards (SMCs), and the system control card (SSC). Out of the available 20 backplane slots, 11 are dedicated to either DPCs or EMCs (allowing processing power requirements to be traded off against mass storage throughput requirements); 8 slots are reserved for SMCs and 1 slot is reserved for the SCC.

Nanobus™

The ability of the Multimax to support a large number of processor, memory, and I/O cards derives from the extremely high data transfer speed of the main system bus, the Nanobus. The Nanobus is so named because it is 1 ft long — approximately the distance traveled by light in 1 ns.

The Nanobus has separate byte-parity-protected 32-bit address and 64-bit data paths. A separate parity-protected control bus permits control information to be conveyed in each bus cycle, hence requiring fewer cycles to perform complex operations. A total of 64 bits of new information can be carried every 80 ns, even if previous requests are still in progress, resulting in a true data transfer bandwidth of 100 Mbytes/s.

Maximal utilization of Nanobus cycles is provided by supporting multiple transactions in progress simultaneously across the bus. After a "bus requester" asks for data, other, unrelated requests and responses can occur before the "bus responder" returns that data. The bus protocols make this possible by tagging address and data with the requester's identity (using control lines). Bus interfaces can also pipeline multiple bus transaction requests by buffering them at different stages of processing. A memory card, for example, can be simultaneously sending data to a requester, while buffering another request for data. The memory bandwidth matches the Nanobus capacity through eight-way memory interleaving — a technique that spreads memory requests across the banks of memory in the system.

A separate vector bus allows interrupts to be transferred in parallel with address and data transmissions. Vectors can be directed to any processor in the system. Vectors can also be sent to an arbitrated "class" of processors which consists of any number of processors that have been assigned to one of three classes. Special hardware will direct a class vector to the most interruptable processor within the class. Vectors can be delivered at a rate of up to one million per second, and, if sent to an arbitrated class, the vector load will automatically be leveled across the processors within the class.

The Nanobus provides for fast synchronization between processors and I/O devices with negligible performance effects upon other system activities. Synchronization is provided by atomic test-and-set locks. The set-bit-interlocked instruction in the National 32032 microprocessor generates a special Nanobus cycle which reads back that requested byte location. The memory modules in the Multimax recognize this special cycle and atomically write the selected byte to a pattern of all ones. In this fashion, any memory location in the Multimax may be used as a lock. Instructions for setting and testing locks are directly available to the user. Other processors, when testing the state of the lock byte, will first read the contents of the byte from the memory into its cache. Subsequent reads will be from the cache until the value of the lock changes, thereby avoiding a performance load upon the Nanobus or the memory during the time spent waiting for the lock state to change.

The Multimax approach should be contrasted to alternate locking systems or dedicated hard wire buses that physically limit the number of locks available or require operating system intervention, which implies significant time overhead as the number of processors and locks increase.

Since support of multiple bus transactions with request and responses separated in time can diminish, but not eliminate, the likelihood of contention, read/write requests can sometimes be issued to devices that are busy and cannot, for the moment, respond. When this happens, the requester will retry the bus requests. In the unlikely event that a requester has not been serviced within a reasonable time, the arbiter gives priority to all requesters in this condition, while maintaining fair, symmetrical access to the Nanobus. The arbitration scheme of the Nanobus has been optimized for multiprocessing and is implemented in the SCC.

SCC

The system control card, SCC, arbitrates the busses, initializes the system, interfaces to the console terminal and remote diagnosis interface, provides interval timers and time-of-year clock, diagnoses all cards, monitors the environments, and reports status when a failure occurs in another card.

The SCC ensures fair access to system resources for all requesters. Priority among equals is decided on a round-robin basis. When a module is granted access to the bus, it relinquishes priority to the next logical module within its group. When all requesting modules in the group have been granted access, the process repeats.

The vector bus cycle time is 160 ns, allowing for extremely low interrupt latency and an extremely high interrupt burst rate. Vector bus arbitration is mediated by the SCC and allows any requester to post any of three kinds of interrupts:

- To interrupt a specific processor
- To interrupt all processors
- To interrupt the "most interruptable" processor belonging to a given "processor class"

For the last case, processors are assigned to one of three "interrupt classes", and an interrupt is designated for a particular class. Interrupt vector arbitration hardware then load-levels the interrupt among the processors by directing the interrupt to the processor which is most available to take the interrupt. In the case of a light interrupt load when all processors are available, the interrupts are distributed in a round-robin manner across all processors.

The diagnostic processor is the heart of the SCC. The diagnostic processor has access to 64 kbytes of ROM (256 kbytes maximum), 128 kbytes of on-board dynamic RAM (512 kbytes maximum), and 4 kbytes of static RAM (always backed up by an on-board battery containing the system configuration tables). The diagnostic processor performs system tests and initialization after power-up, provides a battery backup time-of-year clock, and supervises the system control panel as well as two serial ports which serve as connection to the system console and remote diagnostic modem or may be used as two local ports. The diagnostic processor also takes control of the Nanobus and all associated modules when a serious system error occurs. If the error is generated by a failed component on a Nanobus card, the SCC can deny that card access to the Nanobus on the next restart. When restart occurs, the SCC can inform the operating system that the card is inactive.

DPC

The duel processor card, DPC, has two National Semiconductor 32032 microprocessors operating at 10 MHz and sharing a single 32-kbyte cache. Each National Semiconductor NS32032 microprocessor on the DPC is a 32-bit central processing unit with a full 32-bit data bus to memory. Executing approximately

0.75 MIPS, the CPU has a compactly encoded instruction set designed to provide efficient and economical support for high-level languages. A fully integrated floating-point instruction set and 13 addressing modes designed for the kinds of accesses compilers generate make the NS32032 comparable to a VAX-11/750 minicomputer.

Memory data are written into the 32-kbyte cache whenever main memory locations are read or written to by either of the two processors; and an index of the addresses of the locations thus stored is kept in the CPU Tag memory array. Thereafter, any effort to access those locations in main memory will result in the access of the same data in the cache. Cache read accesses do not generate the processor wait states incurred by main memory accesses, nor do they result in any load upon the main system bus.

The cache for any given DPC is kept current with relevant changes in main memory (generated by writes from other system devices) by means of the Bus Tag logic. This logic continuously scans the Nanobus for memory writes involving locally cached addresses. When such writes are detected, the Valid Bit for the cache address is switched to its invalid state. The result is that when an on-board processor next needs data from that cache address, it will recognize that the associated cache entry is now invalid and will consequently go to main memory rather than cache for the data. This action will automatically update the entry in cache. Because the Bus Tag logic is independent of the CPU Tag logic and replicates its data, maintaining cache concurrency through bus monitoring can occur without impacting speed of access to the cache by the processors.

Associated with each processor is a memory management unit (MMU) that provides hardware support for demand-paged virtual memory management. The MMU translates the 24-bit virtual addresses generated by the processor into 32-bit physical addresses. Associated with each MMU is a translation look-aside buffer (TLB) and an extended translation look-aside buffer (ETLB), the function of which is to speed up address translation by providing a quickly accessible cache of the most recently used addresses. The National Semiconductor 32081 floating point unit is standard equipment on all Multimax DPCs. These FPU chips provide 32- and 64-bit IEEE floating point operations at approximately 250,000 floating-point operations per second. Each processor has time slice end (TSE) logic associated with it. The TSE provides a way to interrupt a compute-bound process after a program-determined amount of execution time. The TSE logic can provide an interrupt after from 1 to 2^{32} 400-ns clock ticks (that is, 400 ns to approximately 1718 s). This interrupt acts directly on the associated processor of the TSE.

SMC

Each shared memory card, SMC, provides a total of 4 Mbytes of RAM on two independent banks of 256K MOS random access memory (RAM) chips. Each card supports two-way interleaving between banks and two- or four-way interleaving between boards, permitting four-way interleaving on systems having at least two SMCs and eight-way interleaving on systems that have at least four SMCs. (The base address and interleaving characteristics of each memory card are set under software control at system start-up.)

The memory access and cycle time of each SMC is four Nanobus cycles, or 320 ns. During this time, a given SMC can compose 4 or 8 bytes of data for transfer to the Nanobus or accept and store from 1 to 8 bytes received from the Nanobus. Since the bus architecture allows another interleaved board to begin a new 8-byte memory transfer with each successive bus clock cycle, the four boards involved in eight-way interleaving can transfer double-long words of data (64 bits or 8 bytes) at an aggregate rate of 100 Mbytes/s.

All data are stored with an ECC. Single-bit errors in each long word (32 bits) are detected and corrected with each access; double-bit errors are detected and reported. In addition, the SMC "sweeps" the entire memory array during refresh and corrects any single-bit errors found. Since a full refresh sweep on each bank occurs approximately once every 8 s, the likelihood that a double-bit (uncorrectable) error will ever occur is dramatically reduced. Because of ECC, two-memory chips on an SMC (one in each bank) could fail without impacting system operation.

Each SMC carries a diagnostic microprocessor that checks all memory banks at power-up and whenever directed to do so by the system diagnostic processor on the SCC. In addition, the memory bank maintains a control/status register (CSR) through which it reports single- and double-bit errors and bus parity errors to the requesting processor. The Parity logic identifies parity errors in written data so that the requesting processor can retry the write operation.

EMC

The Ethernet mass storage card, EMC, provides interface both to the Ethernet and to the small computer system interface (SCSI) bus. Ethernet is an industry-standard, 10-Mbits/s, local-area network used to connect the Multimax to Annex Terminal Servers and to other computer systems. Processing nodes with Ethernet interfaces are connected by transceiver cables to transceivers, which in turn are connected to an Ethernet coaxial cable that can be as long as 500 m. The Ethernet interface on the EMC connects to an Ethernet transceiver port on the Multimax I/O panel.

The EMC card also supports the SCSI for connecting mass storage devices. Up to four disk controllers (supporting up to eight disk drives) and one tape controller (supporting up to four $^1/_2$-in. tape drives) can be connected via the SCSI bus, either to a single EMC card or, if additional throughput is required, to multiple EMC cards.

Both the Ethernet and SCSI interfaces contain a direct memory access (DMA) engine for transmitting/receiving blocks of data from main memory. When placed in diagnostic mode, the EMC on-board processor disables the Nanobus interface and tests the various components on the card.

Reliability and Maintainability

Since each Nanobus card in the backplane has its own intelligence and local storage, each card can perform fault detection and buffering of error information. In addition, upon command from the SCC during power-up or system reset, each card performs a self-test and reports upon its internal state. If any card fails its self-test, the SCC causes it to be logically removed from the Nanobus and notifies the system software of this action.

During the power-up or system reset sequence, the SCC begins its own self-test while all other Nanobus cards are held inactive. Early in the SCC self-test, the microprocessors on all nonmemory cards begin their own self-test while the bus interface logic on each card is held inactive. These self-tests include a ROM check-sum calculation, a local processor test, a series of RAM tests, and tests specific to the individual card types. Upon finishing its self-tests, the SCC lights the diagnostic processor OK lamp on the Multimax front panel. If the self-test detects minor SCC faults that do not prohibit reliable system operation, the SCC lights the Attention Required lamp. If the self-test detects major faults, the SCC lights the System Fault lamp and the system waits for the SCC to be replaced. If the SCC can communicate with the system console, the operator can request more detailed information. (Internal state is also reflected by diagnostic indicators on the rear edge of each Nanobus card.)

While nonmemory cards are running their individual self-tests, the SCC instructs the SMCs to start the memory self-tests. A quick self-verify test cycles a set of patterns through memory. The test has a high probability of detecting all memory error conditions. As these tests are passed, the SCC builds a map of available memory for later use. Any SMC failing this test is not made part of the memory configuration and is therefore effectively removed from the Nanobus. If sufficient memory remains for system operation, the SCC flags SMC failures by illuminating the Attention Required lamp, printing an error message on the system console, and recording the failure for transmission to the UMAX operating system. If insufficient memory remains, the SCC lights the System Fault lamp, prints an error message on the system console, and proceeds to test the remainder of the system. In battery backup systems returning from power failure, SMCs run no tests that destroy data. Existing data are instead checked and all single-bit errors are corrected. The existence of double-bit errors indicates that the power failure has exceeded the reserves provided by the battery backup system and that a normal destructive test is appropriate.

Proceeding one Nanobus card at a time, the SCC enables the bus interface logic on each nonmemory card in the backplane. Each card reports self-test status to the SCC. As successful completions are reported, the SCC builds a resource table for later use by the operating system.

If any card reports self-test failure, the SCC removes that card from the Nanobus, illuminates the Attention Required or System Fault lamp as appropriate, and records the failure for later use by the operating system. When all cards in the backplane have been dealt with, the SCC initiates and monitors a multitest sequence designed to generate complex bus traffic. Errors, when detected by the SCC or reported to the SCC by the participating cards, are displayed on the system console.

If the system passes the multitest, the SCC searches for a storage device from which it can load its own start-up code or special System Exerciser programs. (All activity up to this point has been executed from ROM.) This device will normally be a system disk drive, although magnetic tape is an option.

The SCC start-up program controls the SCC during normal system operation. The first action performed by this program is to load the software for the mass storage controllers. Then the SCC bootstraps the UMAX operating system. When this is done, the system is fully operational. The SCC has a record in its nonvolatile RAM of the current card configuration, the amount of interleaved and nonleaved memory, and some additional specifications passed to it by the software after start-up. Even if the hardware configuration has changed since the last power-up, no dialogue is required with the operating system.

The System Exerciser programs, on the other hand, present menus on the Multimax console that allow the operator to select from a list of interactive diagnostics that stresses the system in various ways.

Multimax Software

Two multiuser, multiprogramming operating systems are available for the Multimax: UMAX 4.2 and UMAX V. Both ensure that multiple processors can share all the basic physical resources of the machine at extremely high speed. All Multimax physical resources (processors, memory, I/O capacity, and bus bandwidth) are available to every application, with mechanisms for computation and communication that degrade minimally as requirements grow.

Both UMAX 4.2 and UMAX V promote simple and productive software development, as well as high-performance production environments for sequential as well as parallel programs. Both were derived from the AT&T-licensed UNIX operating system. UMAX 4.2 is compatible with UNIX 4.2 bsd (developed at the University of California at Berkeley), currently the standard among scientific, technical, and academic computer users. UMAX V is compatible with UNIX System V (supported by AT&T), the *de facto* standard in office and commercial environments.

Principal UMAX Features

- All of the key user-interfaces, networking capabilities, system services, and utility programs that are found on other UNIX 4.2 bsd and System V systems are found in UMAX 4.2 and UMAX V on Multimax.
- Programs that run successfully on other machines under either version of UNIX generally run without changes under the corresponding version of UMAX on the Multimax.
- The same networking facilities are provided on both UMAX 4.2 and UMAX V. These two operating systems are compatible with one another across local- and wide-area networks and are also compatible with other UNIX-based systems that support the Internet protocol family. In addition, these networking facilities provide communications with annexes for terminal, line-printer, and gateway support.

TABLE 1
Grain Size in Parallel Computing Structures

Grain size	Construct for parallelism	Synchronization interval (instructions)	Encore computer structures to support grain
Infinite	Independent jobs		Multimax
Very coarse	Distributed processing across network nodes to form single computing environment	200–1M ,	Multiple Multimaxes and other vendors' machines on Ethernet
Coarse	Multiprocessing of current processes in a multiprogramming environment	200–2000	Multimax
Medium	Parallel processing or multitasking within a single process	20–200	Multimax
Fine	Parallelism inherent in a single instruction or data stream	<20	Specialized processors added to Multimax

Table 1[2] characterizes parallelism in terms of "grain size", the period between synchronizing events for parallel activities. The subsections that follow examine these types of granularity and describe the corresponding facilities provided by the Multimax computing environment.

Independent Parallelism

UMAX supports the simultaneous execution of several user jobs on the multiple processors of the Multimax in a variation of traditional multiprogramming called "multiprocessing."

Consider a typical multiuser operating system. On a uniprocessor machine, the operating system spends much of its time emulating a multiprocessor, providing a simulation of parallel execution of users' programs and system utilities. This is traditionally called *multiprogramming*, and it means sharing a processor (by rapid, transparent switching of system resources) among several programs. Most modern minicomputers and supercomputers operate in this way.

Both versions of UMAX on the Multimax provide true multiprocessing. That is, both versions support simultaneous execution of user programs on multiple processors. They do so, however, without the overhead incurred by multiprogramming uniprocessors when they switch the process context for each brief session with each program. The advantage of multiprocessing is that, while multiprogramming appears to give simultaneous attention to several programs at once, multiprocessing actually delivers simultaneous attention to those programs. As a result of multiprocessing, overall system throughput (in terms of the number of jobs executed per unit time) increases with the number of processors on the system.

Very Coarse-Grained Parallelism

Very coarse-grained parallelism categorizes those parallel applications in which the period between synchronizing events is separated by several thousand instructions. Distributed processing across the computing nodes in a network is a good example of the application of very coarse-grained parallelism. Since the cost of synchronization of network communications is high, there must be long periods of computation between synchronizations. The UMAX operating systems provide a number of system calls to facilitate network communications:

* Socket — establishes an end point for communications, the type of communications (for example, streams of bytes of datagrams), and the protocol to be used for communications.
* Listen, accept — establish two-way communications over sockets.
* Send, recv, read, write — exchange data between the cooperating processes. Synchronization between the producer of data and the consumer can be automatically provided by UMAX. For example, a process attempting to read from a socket which has no data to be read can be suspended by UMAX and automatically released to synchronize it with the arrival of data.

The UMAX operating systems support the widely used Internet protocol family. Consequently, UMAX provides direct and readily accessible support for very coarse-grained parallelism that invokes both multiple Multimaxes on a network and other systems supporting the same protocols.

Coarse- and Medium-Grained Parallelism

Coarse- and medium-grained parallelism identifies those applications in which the interval between synchronization points is one to two orders of magnitude, respectively, smaller than that of very coarse grained parallelism — that is, on the order of hundreds of machine instructions for coarse- and tens of machine instructions for medium-grained parallelism. With these sizes of synchronization interval, the cost of the synchronization mechanism must be substantially lower than that appropriate to very coarse-grained parallelism.

Coarse- and medium-grained parallelism are used on the construction of the UMAX operating systems. Many of the same techniques used in the development of the operating system are available to the user. By way of example, consider some of the implementation aspects of UMAX.

UMAX is an example of an extremely parallelized application. Since locks in the Multimax hardware are cheap and easy to execute, they have been used liberally to provide medium-grained parallelism.

In an n processor Multimax system, the highest priority n ready jobs are always running. A job can migrate between processors, since all processors are identical.

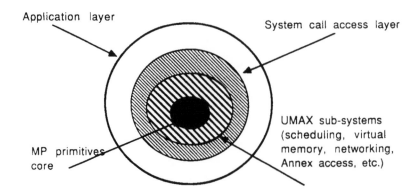

FIGURE 3. UMAX conceptual layering.

The high degree of parallelism in UMAX frees the application programmer from worry about operating system bottlenecks. If multiple processes must make system calls in the execution of a parallel algorithm, the system calls will not interfere, except occasionally at the medium-grained level of access to individual data structures in the kernel.

UMAX allows the programmer to access the shared memory and locking resources without resorting to convoluted device driver accesses. If UMAX were designed to service one user job's operating system requests at a time, multiple requests made simultaneously would have to be queued and then serviced serially, resulting in delay of execution of the requesting programs. Such a potentially large system bottleneck has been avoided in UMAX by allowing all processors running programs to perform operating system services on their own, simultaneously. Such *symmetry* of operation between all processors is critical to system performance.

Symmetrical multiprocessing is achieved through a technique called multithreading. Multithreading allows several simultaneous streams of control through the operating system kernel by multiple processors. Most operating systems for a uniprocessor would not need to contend with multiple streams of control, and special techniques are required.

The principal difficulty in multithreading an operating system is providing *controlled, concurrent* access to shared resources. Multithreading requires locking of shared resources and synchronization between processors that try to gain access to these resources on behalf of processes.

UMAX supports multithreaded operation by a set of core primitives (see Figure 3) that provide synchronization between UMAX processes and Multimax hardware. This is done with three basic mechanisms:

- Spin locks—accomplish synchronization by executing tight instruction loops until the expected condition occurs. Used only for critical, short-duration events. Spin lock primitives are supplied with UMAX as two library routines: *spinlock* and *spinunlock*. These routines do not need system access, since all instructions for setting and testing locks are available at the user level.

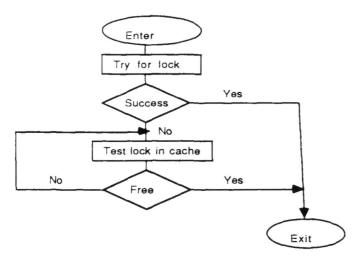

FIGURE 4. Software lock request.

Programmers can access the resulting shared memory (see Figure 4) without resorting to convoluted device drive accesses.

- Semaphores (Dijkstra style) — accomplish synchronization by suspending a process until the resources it needs are available.
- Read/write locks — specialized semaphores that control access to data structures for a single writer or multiple readers. Read/write locks prohibit all pending read access while any write is in progress.

The utility of these mechanisms can be put into perspective by considering the UMAX file system. The file system, in particular, has been rewritten to allow maximal concurrency of access. Multiple processes may desire to read directories concurrently. As long as no process attempts to modify a directory, there is no reason to prevent a simultaneous access. Processes performing directory lookup will therefore acquire a read lock while reading the directories. When a process decides to modify the file system, say by deleting a file, it acquires the write lock.

Access to many UMAX internal resources (in the file system, for example, or in virtual memory management, device drivers, and network protocol code) is controlled by these primitives to achieve multithreaded operation. Locks and semaphores are used *hierarchically* in order to prevent *deadlock* (in which access to data is blocked forever by an event that won't occur) and *indefinite postponement* (or starvation, in which the policy for access to shared resources is unfair). Thus, the operating system achieves a high degree of concurrency in the kernel between processors and processes.

The Multimax architecture will support large user populations making large demands on a very large set of computing resources. For that reason, certain algorithms for manipulating shared data in the operating system have been redesigned to minimize the frequency and duration of accesses (and therefore of mutual exclusion through locking) to these data.

Two techniques are principally used to enhance data manipulation algorithms: *caching* and *fine-grained locking*. Resources that are likely to be reused frequently, such as file and directory entries, are cached by the kernel. For resources that are tables, it is appropriate in some cases to lock individual entries in kernel tables rather than the whole table. In other cases, kernel tables have been divided into subpools of entries, linked together and located by hashing. This minimizes searching times for tables that, in very large systems, can have huge numbers of entries. It also decreases bottlenecks in accessing table entries, since subpools are locked rather than the whole table.

Process Management

UMAX offers "job control", or the ability to control the order, timing, and priority of program execution with simple commands. Program execution can be temporarily suspended with job control to enable a detour to a high-priority activity without losing work in progress. The user also controls whether or not a program or group of programs should be run sequentially, run asynchronously as background processes, or be batched for execution at a later time.

Memory Management Support

Both versions of UMAX take full advantage of the Multimax demand-paged virtual memory to provide up to 16 Mbytes of virtual address space per process.

Software Support for Parallel Programming

Parallel programs require several types of support from system software:

- Facilities for scheduling multiple processors simultaneously to work on a single job.
- Mechanisms for establishing shared data regions between processors.
- Fast accessibility to operating system service from all processors.

UMAX supports the construction of parallel programs in which several processors work simultaneously on shared memory, in order to access and exchange data at extremely high speed to reduce execution time of a job. Sharing data among processors under UMAX typically involves two steps: (1) a program is invoked from the "shell" (resulting in one processor being allocated), and (2) that program requests additional processors, which are given shared access to some or all of the data of the first processor.

To understand the details, it is useful to depict the structure of a UMAX process image, which consists of four parts (see Figure 5).

- Text — this is the pure program code, loaded starting at location 0, read only.
- Data — following text on the next page boundary, the data portion consists of an area containing statically initialized data variables followed by an area allocated to uninitialized data (zeroed when the program is loaded).
- Stack — for temporaries. This region is allocated from the top of virtual memory down.

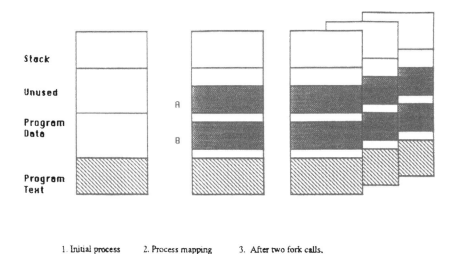

Stack
Unused
Program Data
Program Text

A
B

1. Initial process mapping

2. Process mapping dynamic allocation of shared memory (A) and sharing of previously allocated memory (B)

3. After two fork calls, three processes share text and the two shared segments.

FIGURE 5. Memory allocation for parallel programs.

- Unused — all other virtual locations. Dynamically allocated memory is placed here following the end of the uninitialized data area.

To start a parallel program, a *parent* process is created. The next step is for additional processors to be invoked, running identical copies of the program, and sharing some data. This is done by creating multiple-child processes with the fork call. Then, the application will issue a system call, named *share*, to declare a portion of the process data space shareable by all *child* processes that are created by this program (variations on *share* are described below). After invoking *share*, the parent process can create these child processes, all of which will share the shared memory space. All processes can then start performing work on the problem at hand.

To create these child processes, *fork* is used, which creates processes that get invoked on additional processors.

Note also that when these additional processes are invoked, though they share the designated "shared" data, the stack segments are always private to each processor, in order that every process can maintain its own thread of control without disruption from other sharing processes.

Share can be invoked in two different ways. The first invocation of share is useful for C programs when a data block may be created dynamically:

```
/*Region A below*/
  char*ptr;
  long size;
  ptr = share (0,size)
```

The alternate invocation of *share* may be used by Fortran and C programs to share previously allocated memory:

```
C Region B above
  COMMON/SHARED/A,B,I
  CALL SHARED (SHARED,size)
```

Thus, facilities are provided for constructing parallel applications that utilize multiple processors on a single job. Substantially similar facilities exist for UMAX V.

For medium-grained parallelism, with synchronization intervals measured in tens of instructions, system calls are too expensive for invoking synchronization or to read/write shared data. Thus, the UMAX operating systems provide a multitasking library which implements the notion of a task as a thread of control. A task is a "light-weight" or "cheaper" thread of control than a process, and consequently it has less overhead. A task is analogous to a procedure in a program, except for the important difference that it can execute in parallel with the caller. Like a procedure, a task has arguments passed by its caller. It can have local variables; it has a private stack; and it can access global data that can be shared with other tasks.

The task abstraction is separate from the UMAX process; a process can contain many tasks. A task is a unit of activity that shares an address space with other tasks in the process. The Multitasking library allows a user to create, terminate, synchronize, and share memory between tasks. In general, one can consider creating a very large number of tasks, while it may be inappropriate to create that many processes.

Fine-Grained Parallelism

The classification of fine-grained parallelism encompasses those cases where parallelism can be applied at the level of a single instruction. For example, as provided by special hardware such as systolic or array processors. This granularity of parallelism can only be achieved with special hardware which does not yet exist for the Multimax.

Performance Monitoring Tools

For UMAX, Encore has created a generic tool called *sysmon* that subsumes the functions of such programs as *umstat*, *netstat*, and *iostat*. For UMAX V, the UNIX System V UNIX System Activity Package is supported.

Sysmon can be invoked in either summary or detail mode and allows specification of the rate at which the display is to be updated. Sysmon provides information about the following subsystems:

- SYS — all subsystems
- CPU — the processor management facility (including short-term scheduling and processor statistics)

- PM — the processor management facility (including medium term schedule, UNIX processes, etc.)
- VM — the virtual memory manager (including demand paging)
- FS — the file system
- IO — the I/O systems (including local/remote/pseudo terminals and mass storage subsystems)
- NET — the network software (primarily TCP/IP and the Ethernet software)
- LOCK — the lock management software (including both read/write and multiprocessor locking)

Sysmon, in conjunction with the standard UNIX 4.2 *ps*, *w*, and *uptime* tools, facilitates tuning system software parameters and optimizing user applications.

Terminal Performance on Annex

The UNIX terminal driver has been distributed to make use of the processing power of Annex terminal servers in handling character processing overhead.

Annexes are initialized by down-line loading the necessary terminal and printer drivers from a Multimax or other host on the network. Annexes assume the major terminal support burden for Multimax. This feature has an immense positive impact on overall system performance because it avoids the interrupts and computational overhead for character processing on the Multimax that is present in nondistributed architectures.

The standard UNIX terminal driver mechanisms have been modified to work optimally with the unique capabilities of the Multimax Annex hardware. A proprietary message-based protocol between Multimax systems and Annex terminal servers passes and synchronizes the character data and control information in a way that enables most of the terminal driver functions to be executed in the Annex. The Multimax receives only well-formed packets, rather than individual characters requiring extensive processing. This performance advantage is particularly valuable in large system configurations supporting many terminals.

The Annex also supports multiple simultaneous connections from a single terminal. Users can log into a given Multimax machine, issue an Annex command to switch their terminal connection to another Multimax machine, and thereby be given direct access to the second machine without the need for all transmissions to be routed through the first machine.

Any terminal on any Annex can communicate in this way with any Multimax on the Ethernet. Any terminal can also communicate with non-Encore computers on the network, provided they are running Berkeley 4.2 UNIX or TCP/IP and rlogin or telnet protocols. Similarly, Multimax users can make use of *rsh*, *rcp*, and *rdump* to establish host-to-host contact with any machine on the network that is running UNIX 4.2.

The Annex can function as an editing front end for the Multimax or any UNIX 4.2 host. The purpose of the editing front end is to off-load a host by handling simple changes at the Annex and sending changes to the host editor in batches. This is based on work by Stallman.[7]

The functions performed by the Annex editing front end include

- Performing character and line editing operations for data already on screen, such as character echoing, cursor positioning character insertion, and line insertion.
- Resynchronizing with the host when the user moves to another screen or performs another operation that cannot be performed locally by the Annex.
- At resynchronization time, updating the host file (with edits since the last synchronization) by sending a few well-formed packets from the Annex to the Multimax.

Performance advantages of this architecture are that the Annex effectively makes batch operation for the host out of multiple atomic editing operations. There is no load on the host until a resynchronization operation occurs.

CONCLUSION

Several significant applications have been executed on the Multimax that exhibited a linear speed-up with the number of processors:

- grep — a UMAX utility that searches a set of files for occurrences of a given pattern of characters. The parallelized version of grep executed 9 to 10 times faster on a 14-processor system than the serial version.
- LINPACK — a linear equation-solving benchmark ran 16 times faster on a 20-processor system than the serial version.
- PDE — partial differential equation solver also experienced linear speed-up in the number of processors used.

The importance of parallel programming will increase as the cost of processors decreases, as physical limitations on the speed achieved by a single processor are reached, and as multiprocessor architectures like the Multimax enable the benefits of multiprocessors to be used by programmers. The Multimax hardware and software environment is rich in features to support manageable parallelism at all grain sizes.

PARALLEL BLOCK SOR METHODS FOR
SOLVING POISSON EQUATION

The purpose of this section is to discuss and compare two parallel iterative methods to solve the Poisson equation and their implementations on the Encore Multimax. The solution of the Poisson equation is required in numerous diverse scientific applications. The available architecture, whether it be serial, vector, or parallel type, tends to influence the development of the algorithms for solving the Poisson equation. On a shared memory multiprocessor with p processors, all the computational phases of SOR algorithms except input, termination checking, and the final output, can be performed without any interaction between processors with limited synchronization. Hence, it should be possible to reduce the running time of the algorithms by a factor of p. Since the SOR method for the Poisson equation is typical of many large-scale scientific codes, the performance of the parallel algorithms on the Multimax will be evaluated to give some idea of the possible speed-up on shared memory multiprocessor systems.

Problem Definition

The Poisson equation is defined as

$$\frac{\partial^2 u}{\partial x^2}(x,u) + \frac{\partial^2 u}{\partial y^2}(y,u) = f(x,y) \tag{1}$$

for $(x,y) \in R$ and

$$u(x,y) = g(x,y) \ \ (x,y) \in S$$

where

$$R = \left[(x,y)|a < x < b, c < y < d\right]$$

and the S denotes the boundary of R. We assume that both f and g are continuous on their domains so a unique solution to Equation 1 is ensured.

The method to be used is an adaptation of the finite-difference method for boundary-value problems. The first step is to choose integers n and m and define step sizes h and k by $h = (b - a)/n$ and $k = (d - c)/m$. Partitioning the interval $[a, b]$ into n equal parts of width h, and the interval $[c, d]$ into m equal parts of width k provides a means of placing a grid on the rectangle R by drawing vertical and horizontal lines through the points where for each i,j:

$$x_i = a + ih \qquad i = 0,1,\ldots,n$$
$$y_j = c + jk \qquad j = 0,1,\ldots,m$$

The horizontal lines and vertical lines are called *grid lines* and their intersections are called the *mesh points* of the grid.

We use w to approximate u with local truncation of order $O(k^2 + h^2)$ as follows:

$$4w_{ij} - w_{i+1j} - w_{ij-1} - w_{ij+1} = -h^2 f(x_i, y_i) \tag{2}$$

The linear system of equations which results from the approximation of Equation 2 on $n \times n$ mesh points becomes the form

$$Mw = u \qquad (3)$$

where, M is a block-tridiagonal matrix with n tridiagonal blocks

$$A = (-1, 4, -1)$$

of order n on the main diagonal and with negative identity matrices on the both subdiagonals. u is a vector with n blocks generated from boundary conditions of the problem, and $f(x, y)$, and the SOR parameter ω. Equation 3 is defined in a block linear system of equation form:

$$\begin{bmatrix} A_1 & I & & & \\ I & A_2 & \cdot & & \\ & \cdot & \cdot & \cdot & \\ & & \cdot & \cdot & I \\ & & & I & A_n \end{bmatrix} \begin{bmatrix} w_1 \\ w_2 \\ \cdot \\ \cdot \\ w_n \end{bmatrix} = \begin{bmatrix} u_1 \\ u_2 \\ \cdot \\ \cdot \\ u_n \end{bmatrix} \qquad (4)$$

This problem is time consuming when n is very large. However, the blocks are independently grouped and the vector w can be partially solved. Thus, the system equation can be divided into independently solvable parts and distributed on different processors. It is obvious that the algorithms would yield best results on parallel computers, where the cost of synchronization or communication is a dominant influence on implementation effectivity. On the other hand, the sparse constant matrix M may be stored in only two vectors, (4, 4,...,4) and (−1, −1,...,−1). Thus, memory space is greatly saved.

The color of Red or Black is defined as the odd or even block index of the matrix M. The blocks of the solution vector w are also defined as Red (odd index) or Black (even index) color. Equation 4 may be converted to

$$A_i w_i = u_i + w_{i-1} + w_{i+1} \qquad i = 1, \dots, n \qquad (5)$$

where $w_0 = 0$, and $w_{n+1} = 0$. Equation 5 shows that each of the even-number blocks only depends on its two neighbors, ie., the two odd-number blocks, and vice versa. Thus, parallelism is achieved with this Red and Black ordering since the elements of the same color are not neighbors, which implies all Red and then all Black elements may be updated simultaneously, or combined Red and Black elements, which are neighbors, may be updated simultaneously with special synchronization (or communication) and computing ordering in each processor. These are two key points of our parallel algorithms.

Algorithm 1

Algorithm 1 groups all Red blocks into one set of independent solvable parts and all Black blocks into another set of solvable parts. In each iteration, all Red blocks are computed on different processors simultaneously, then after the barrier synchronization (or the updated value communications), all Black blocks are computed in turn. At the end of each iteration, the solution convergence is checked. This parallel process does not stop until the solution converges in a certain tolerance.

While (*convergence* ≠ *true*) Do
{
   1. Computing all the Red blocks,
$A_i w_i = u_i + w_i + w_{i+1}$ ($i = 1, 3, ..., n - 1$) in parallel.
   2. Barrier synchronization — waiting for all processors to finish the iteration in the shared memory machine.
   (Each processor sends the updated w_i values to its two Black neighbors after the iteration.)
   3. Computing all the Black blocks,
$A_i w_i = u_i + w_{i-1} + w_{i+1}$ ($i = 2, 4, ..., n$) in parallel.
   4. Barrier synchronization — waiting for all processors to finish the iteration in the shared memory machine.
   (Each processor sends the updated w_i values to its two Red neighbors after the iteration in the local memory machine.)
   } (end of *While* loop)
   5. Output the final results,
where n is assumed to be an even number.

Algorithm 2

We have learned from Equation 5 that the solution of any Red or Black block, w_i (except the first block w_1 and the last block w_n), only depends on the solutions of the two Black neighbors or the two Red neighbors, w_{i-1} and w_{i+1}. For example, the dependencies of 12 blocks between the 2 different color groups are as follows:

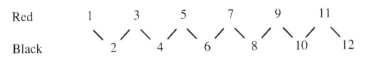

Based on the Red-Black dependencies, algorithm 2 first distributes several groups of blocks, in order, to different processors; in our example, if three processors are used, then we distribute three groups of blocks (1, 2, 3, 4), (5, 6, 7, 8), and (9, 10, 11, 12) to each processor, respectively. The computing order in each processor is Red forward, then Black backward; thus, the computing order in our example is (1, 3, 4, 2), (5, 7, 8, 6), and (9, 11, 12, 10), which are performed in parallel. The number of dependencies to be synchronized is reduced. The only barrier synchronizations or updated solution value communications now are between a single Black block in one processor to a single Red block in another processor, and vice versa. In our example, Block 4 does not start a new iteration until block 5 finishes the computing (or sends the updated solution values); and Block 5 does not start a new iteration until Block 4 finishes the computing (or sends the updated solution values); Block 8 does not start a new iteration until Block 9 finishes the computing (or sends the updated solution values); and Block 9 does not start a new iteration until Block 8 finishes the computing, and vice versa. The new dependencies of the same example in algorithm 2 are as follows:

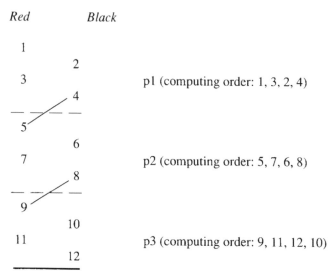

The advantage of algorithm 2 is that it is efficient in synchronization. If the multiprocessing is not out of step, the synchronizations cost almost nothing because the processing order in each processor is already synchronized with itself and the relevant processor. Even if it is out of step, or in imperfect load balancing, the number of barrier synchronization points among the processors is reduced, or the possible waiting time to synchronize the computing among the processors is reduced.

THE EXPERIMENTAL RESULTS

Some results of the present study are summarized in this section. The discussion of the results is focused on the computational speed-up that can be achieved using a parallelized code. In all cases of the tests on the Multimax, 122 × 122 grid points and 128 × 128 grid points were considered.

To evaluate performance of the parallel SOR solvers for the Poisson equation on Multimax, a similar code for a sequential machine was developed. The sequential version of the SOR method was run on the same machine, a Multimax with a single processor, with the same boundary conditions and the same number of grid points. An error norm was defined as the maximum change in the solution between the two successive iterations. The error norm was compared to an acceptable tolerance. When the value of the computed maximum error norm was less than or equal to 10^{-4}, the iteration was stopped. Since the convergence rate of the SOR method is very sensitive to the value of the acceleration parameter ω used, the analytical optimum parameter given by Varga[10] is applied.

$$\omega = \frac{1}{2}\left(1 + \sin\frac{\pi}{n}\right)$$

where n is the system size. The numerical experiments on Multimax showed that the practical optimum ω, which required the minimum number of iterations for a given error tolerance, is identical to the analytical ω.

The speed-up factor we used to evaluate our parallel algorithms is simply defined by

$$S_p = \frac{T_1}{T_p} \; (>= 1)$$

where T_p is the number of time unit steps required by a parallel algorithm running on p processors, and T_1 is the time needed by the corresponding serial algorithm.

Experimental Results on the Multimax

Table 2 gives the computing results of algorithms 1 and 2.

The iteration number is identical in all tests. The CPU times in seconds are the mean of five experimental results. The speed-ups are linear with the number of processors. In experiments, each process distributed to a processor achieves a reasonably good load balancing. Thus, the standard deviation is very small. Algorithm 2, with the efficient *Produce/Consume* method, achieves a linear speed-up line. Since the load balancing in each processor is as good as in algorithm 1, Table 2 does not show any advantage for algorithm 2.

Several other experiments have been done for testing both algorithms with imperfect load balancing situations. Table 3 shows the speed-ups of algorithms 1 and 2. Now the speed-ups of the two algorithms are clearly different when the load balancing is imperfect.

The results demonstrate that both of the two algorithms obtain linear speed-up with the number of processors on Multimax. The multiprocessing system can efficiently schedule the processes. Thus, if the numerical algorithms have balanced structures and reasonably synchronized computing order, the implementation of the algorithms on Multimax can achieve a good speed-up. In addition, algorithm 2 is 10 to 20% faster than algorithm 1 if load balancing is imperfect.

TABLE 2
Experiments of Algorithms 1 and 2 on Multimax

#p	#itns	T 1	sp 1	T 2	sp 2
1	167	1485.15	1.00	1488.07	1.00
2	167	755.26	1.96	753.02	1.97
3	167	508.06	2.92	506.88	2.95
4	167	381.05	3.89	380.04	3.91
5	167	304.68	4.87	305.45	4.90
6	167	253.32	5.86	252.36	5.89

TABLE 3
Imperfect Load Balancing on Multimax

#p	#itns	sp 1	sp 2
1	167	1.00	1.00
2	167	1.76	1.90
3	167	2.58	2.87
4	167	3.47	3.85
5	167	4.34	4.83
6	167	4.86	5.73

REFERENCES

1. **Bell, C. G.,** Multis: a new class of multiprocessor computers, *Science*, 228, 4627, 1985.
2. **Bell, C. G., Burkhardt, H., Emmerich, S., Anzelmo, A., Moore, R., Schanin, D., Nassi, I., and Rupp, C.,** The Encore Continuum: A Complete Distributed Work Station — Multiprocessor Computing Environment, Proc. Natl. Computer Conf., 1985, 147.
3. **Hinden, H.,** Special report on advanced architecture, *Comput. Des.*, August 15, 57, 1985.
4. **Lonergan, W. and King, P.,** *Datamation.* 7, 28, 1961.
5. **Siewiorek, D. P., Anzelmo, A., and Moore, R.,** Multiprocessor computers expand user vistas, *Comput. Des.*, August 15, 70, 1985.
6. **Siewiorek, D. P., Bell, C. G., and Newell, A.,** *Computer Structure: Principles and Examples*, McGraw-Hill, New York, 1982.
7. **Stallman, R.,** The SUPDUP Local Editing Protocol, MIT AI Lab, August 6, 1984.
8. Encore Computer Corporation, Multimax Technical Summary, Marlboro, MA, May 1985.
9. **Kruskal, C. P. and Weiss, A.,** Allocating independent subtasks on parallel processors, in Proc. 1984 Int. Conf. on Parallel Processing, 1984, 236.
10. **Varga, R. S.,** *Matrix Iterative Analysis*, Prentice-Hall, Englewood Cliffs, NJ, 1969.

7   AT&T DSP-3

OVERVIEW

The DARPA-sponsored DSP-3 Multiprocessor Project at AT&T has produced a multi-GFLOP machine applicable to a variety of signal processing and pattern recognition problems. The machine is specifically targeted at those applications whose throughput requirements are so large that they demand multiprocessing and/or specialized topologies. Some of the applications that are well matched to the DSP-3 architecture are

- Beamforming, both adaptive and conventional.
- Neural network simulation.
- Speech recognition and keyword spotting.
- Synthetic aperture radar processing.
- Solving linear systems of equations.
- Image processing and object recognition.

Important features of the DSP-3 multiprocessor are

- The use of a 25-MFLOP, 32-bit floating point processing engine at each node of the multiprocessor.
- The availability of up to 1 Mbyte of high-speed SRAM at each processing node. Bulk (disk) storage is available via a real-time host processor.
- External and internal (between node) I/O rates up to 40 Mbytes/s.
- An interconnection architecture that is capable of:
  - Linking up to 128 processing nodes to achieve a peak throughput rate of up to 3.2 GFLOPS.
  - Providing modularity to permit machine configurations as small as 16 nodes (400 MFLOPS) with higher throughput available in 400-MFLOPs increments.
  - Supporting a machine configuration topology that is most efficient for the specific application (e.g., Linear mesh, linear systolic, binary and nonbinary tree, hybrid).
  - Achieving reconfiguration fault tolerance by isolating and routing around faulty processing nodes.

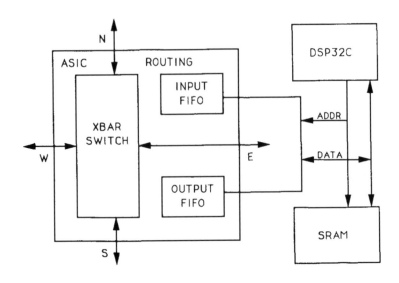

FIGURE 1. DSP-3 node architecture.

- A high-level language software development environment that includes an extensive and growing library of signal processing subroutines.
- A conventionally packaged multiprocessor with a processing density on the order of 1 GFLOPS/ft³ and an advanced multichip 3-D packaging version with an array processing density of 800 MFLOPS in 9 in.³

NODE ARCHITECTURE

The DSP 32C processing element, associated SRAM, bus structure, and routing chip comprise a DSP-3 node, as shown in Figure 1.

At the heart of each node is the AT&T DSP 32C processing element.[1] Key features of the DSP 32C are:

- Low-power CMOS technology.
- 50 MHz operating frequency providing throughput of 25 MFLOPS.
- 32-bit floating-point arithmetic (IEEE standard floating-point format conversion).
- 1024 words of RAM and 2048 words of ROM (or 1536 words of RAM) on chip.
- Up to 16 Mbytes of external memory (addressable as 8, 16, or 32 bits).
- 16-Mbits/s serial I/O ports and 16-bit parallel I/O port with DMA options.
- Internal and external interrupt capabilities and error control logic.

The DSP 32C is upward compatible with its predecessor design — the DSP 32.

A key element of the DSP-3 node is the ASIC routing chip, shown in Figure 1. The routing chip in each node interconnects to copies of itself in four neighboring nodes (north, east, south, and west) to realize the interconnection network. With completely general, program-controlled switching of the communication data paths, information can be routed through a node with no interaction with the processing element and with only a two-clock-cycle latency; this enables efficient implementation of block data transfers between nodes. In fact, a 40-Mbyte/s east-to-west transfer can occur concurrently with a south-to-north transfer at the same rate at the same node.

The data flowing through a routing chip can be conditionally copied into the input first-in-first-out memory, also included in the routing chip, that is the input to the processor. Conditions for copying parts of the data flowing through the communications chip to the FIFO include matching headers on packets with program-loaded match registers in the communication chip, or use of a pair of index counters in the chip to copy every nth datum starting at the pth entry after a data-packet header. A third index defines the length of the entry that is to be copied. These features enable the selection of a channel from a time-division-multiplexed set of inputs, or effecting the transpose of a matrix, as in the SAR application. By placing the input and output FIFOs in the address space of the processor as reserved addresses, it is possible for one processor to store results into its memory space, which after automatic routing defined by program-controlled registers, can then be read by one or more other processors via memory accesses. This low-overhead transfer of data from one PE to another supports very efficient systolic processing almost as if the PEs had shared memory.

NODE INTERCONNECTION TOPOLOGIES AND FAULT TOLERANCE

Since each PE controls its local switch, global coordination among PEs is required to establish a cooperative network. A central command processor, called the real-time host in the DSP-3, distributes global commands to all the processors including instructions defining routing paths. (Since I/O in each PE is achieved via the memory port in the DSP32C, the so-called parallel I/O port on that chip is available for communication with the real-time host.) Software control of the topology in this way results in two useful system attributes:

• Versatile topologies can be formed to satisfy different applications.
• Fault tolerance can be achieved via reconfiguration around faulty nodes.

The physical interconnection of the four-port nodes approximates a rectangular lattice. However, software definition of neighbor relations enables the emulation of a tree, a linear systolic array, and a variety of other topologies. It should be noted that interconnection data packets are routed, in part, by an associative match of header tags. Nodes are not, therefore, limited to an exchange of data with immediate neighbors.

A multiple board interconnection scheme for a linear array is straightforward. Other topologies, such as binary trees, are possible.[2]

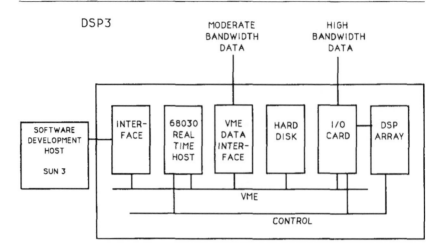

FIGURE 2. DSP-3 multiprocessor system architecture.

Since topologies demanding two- or three-port nodes can be implemented in a combinatorically large variety of ways with the physical four-port lattice, it is possible (under software control) to route around defective nodes in the lattice to achieve fault tolerance. Both linear and tree topologies can continue to be supported in the presence of a failure. If extra processing elements are included in the configuration, multiple failures can be tolerated without resorting to degraded modes.

SYSTEM ARCHITECTURE

The system architecture of the DSP-3 multiprocessor is illustrated in Figure 2.

Two buses are incorporated in the architecture to allow for separation of data and control information. The VME bus is used for data transfer into and out of the DSP array, and the VSB bus is used for transfer of control information. The VME subsystem includes a single-board 68030 processor with a real-time operating system, Ethernet interface to the software development host, bulk RAM, and a hard disk.

There are two primary means of getting external data into and out of the DSP-3. A moderate bandwidth data path uses the VME bus, which can sustain a data transfer rate on the order of 10 Mbytes/s. There are a large number of "off-the-shelf" data acquisition cards that can be plugged directly into the VME bus. Data that is brought in over the VME bus (and destined for the DSP 32C multiprocessor network) is transferred under control of the 68030 host. The 68030 can set up a block transfer or move individual words to the I/O card where a format conversion is performed.

To fully utilize the 40-Mbyte/s data transfer rate of the multiprocessor interconnection network, an alternative I/O capability has been provided for the external interface. This direct I/O path allows a user to bypass the VME bus and input (or output) data directly into one of the multiprocessor board communication ports.

The host processor in the DSP-3 is a VME-based real-time processor. It is a 68030-based single-board computer with 4 Mbytes of RAM and an SCSI interface for a hard disk. The real-time host performs three major functions in the operation of the DSP-3:

- It controls all data flow into and out of the machine.
- It issues all commands to the multiprocessor array needed to initiate and control the execution of signal processing and fault tolerance functions.
- It provides the interface to the software development processor, which, in the illustrated case, is a SUN 3 workstation.

SOFTWARE DEVELOPMENT ENVIRONMENT

Software development for the DSP-3 is done primarily on the SUN 3 workstation and then downloaded to the real-time host and the multiprocessor array. A complete set of tools to develop, compile, load, and debug software for both the 68030 host and DSP 32C array is provided to the user. These items include

- C compilers for the 68030 and DSP 32C.
- VRTX operating system and debugger for the 68030.
- Distributed target linker for handling MIMD programs when a different load image is required for each DSP 32C in the array.
- Configuration routines for synthesizing two topologies (binary tree and linear systolic).
- Kernel functions to communicate with the DSP-3 when configured as a binary tree or linear systolic array.
- A network router tool that allows the user to define interconnection topologies other than linear systolic or binary tree.

PACKAGING TECHNOLOGY (CONVENTIONAL)

The DSP-3 multiprocessor array has been implemented using a conventional triple-height (9U) VME approach. Up to eight of the array boards can be contained in the enclosure. This 6.3-ft^3 enclosure also contains I/O and interface boards, the real time host, a backplane, a hard disk, cooling fans, and the power supply.

PACKAGING TECHNOLOGY (ADVANCED)

There are two fundamental reasons why advanced packaging compares in importance to advanced device technology and architecture. First, in avionic applications such as SAR, volume and weight are clearly critical to the practical use of a system. Second, a compact realization is essential to limit interconnection length in high-performance digital systems if the potential of fast device technology is to be realized. A further by-product is reduced interconnection capacitance, which also reduces the power required, which is another important parameter in avionics.

As a technology demonstration, a version of the DSP-3 is being implemented in a form of wafer-scale packaging. Specifically, the unpackaged dies are assembled on a silicon substrate using a face-down, solder-bump mounting technique. A low-dielectric constant insulation layer is used in realizing the interconnections. The important advantages of the silicon substrate are

- High thermal conductivity.
- Matched temperature coefficient of expansion.
- Low interconnect capacitance.
- An interconnection density that requires at most two layers.

The DSP 32C and eight memory chips are mounted on a silicon substrate 0.9 in. square, and housed in a pin-grid-array package for testing purposes.

To further extend the processing density, the high-density version of the DSP-3 is being realized as a three-dimensional stack, wherein multiple nodes realized as microelectronic interconnect assemblies are mounted on a printed circuit board, and then multiple boards are stacked with a vertical interconnect between boards. Heat is vented primarily via conduction cooling, with the electrical ground planes also servicing as thermal conductors.

The significant consequence of advanced microelectronic packaging and 3-D interconnection is the realization of an 800-MFLOPS multiprocessor array in a volume of 9 in.^{3} Scalability of this processing density to higher throughput levels appears feasible.

BASELINE DEMODULATION ALGORITHM
FOR DSP EVALUATION

To evaluate the performance of the DSP-3, it was programmed using the algorithm illustrated in Figure 3. The algorithm represents the type of demodulation algorithms typically encountered in communication signal processing applications, recognizing that the details will vary from application to application.

The demodulator input is a real data stream, sampled at a data rate of F_s. The Hilbert transform FIR filter transforms the input signal from a real bandpass signal to a complex signal decimated to a sampling frequency of $F_s/2$. Filtering makes this decimation possible, with no loss of information. If a complex multiply of the Hilbert transform output is performed with a digital voltage control oscillator consisting of a phase integrator and a sin/cos look-up table, the filter output is demodulated down to baseband. A real bandpass FIR filter having

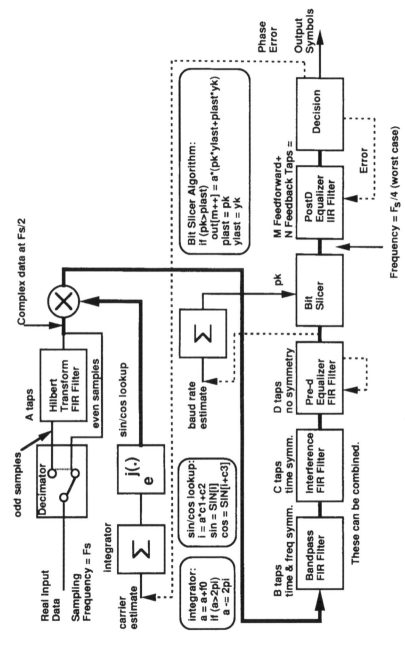

FIGURE 3. Typical demodulation algorithm.

B taps and a complex interference FIR filter having C taps filters the complex baseband signal. The filter output is then passed through a Pre-d equalizer filter, implemented as a D tap complex FIR filter. The bit slicer samples the filtered data stream at the zero crossings of the baud rate estimator, according to the algorithm shown in Figure 3. The soft decisions from the bit slicer are produced at the symbol rate, which is assumed to be $F_s/4$ worst case. These soft decisions then pass through the Post-d equalizer which is implemented as an IIR filter with M feedforward taps and N feedback taps. Finally, the decision function makes the hard decisions on the equalized symbol values.

Phase/frequency tracking, baud rate estimation, and equalizer coefficient adjustment, while important design considerations, are expected to contribute only small performance penalties and were not included in the analysis.

The most important functions to be optimized are the bandpass filter and interference filter. While these filters perform different functions in the demodulator, combining them into one complex filter of B+C taps is possible. A typical B+C value is 256, making a straightforward implementation of these functions impractical. Filtering is therefore performed in the frequency domain using the high-speed overlap and add FFT filtering method that achieves a significant performance advantage for filters of length greater than 32.

Performance of a DSP-3 populated with one processor card containing 16 DSP 32C PEs is capable of achieving an estimated 1.27-MS/s throughput.

REFERENCES

1. **Fuccio, M. L., et al.,** The DSP 32C: AT&T's second-generation floating-point digital signal processor, *IEEE Micro*, December, 30, 1988.
2. **Leiserson, C. E.,** An Area-Efficient VLSI Computation, Ph.D. thesis, Carnegie-Mellon University, 1981.

8 THE MEIKO COMPUTING SURFACE

OVERVIEW

The Meiko Computing Surface provides heterogeneous parallel technology, combining fast processing resources, high-speed configurable point-to-point interconnection, and high-level software that makes a scalable and efficient parallel-processing system.

The Computing Surface architecture fundamentally consists of a collection of heterogeneous processing nodes connected via a high-performance communication network overlaid by a processor-independent parallel-processing environment (CS Tools). Possible processing nodes currently include i860s, transputers, and a range of Sparcs. In addition, the Surface includes a variety of special I/O devices such as high-speed parallel I/O, displays, mass storage drivers, and interfaces to all the expected standard computing resources.

The integral communications network provides the Surface with dynamically configurable, high-speed internode connections which are then used by the software to provide the user with a transparent message-passing system, without loss of performance. Importantly, the software provides the user with a level of insulation from the underlying processor/communications hardware that permits total transportability of code between different processor node types. Thus, applications may be written, executed, and tested in the knowledge that the investment in software development would not be wasted if a change in the processor hardware is needed to meet some subsequent requirement.

Each computing element shown in Figure 1 consists of

- A single 32-bit 10-MIPS processor with a sustained performance of 1.0 MFLOPS in double precision.
- Four high-performance, full-duplex communication links from the computing element supporting single or block DMA transfers at 20 Mbits/s into or out of local memory to other computing elements.

Portions of the text excerpted with permission from "Inside a Heterogeneous Computer," published in the February 1991 issue of BYTE magazine © McGraw-Hill, Inc., New York. All rights reserved

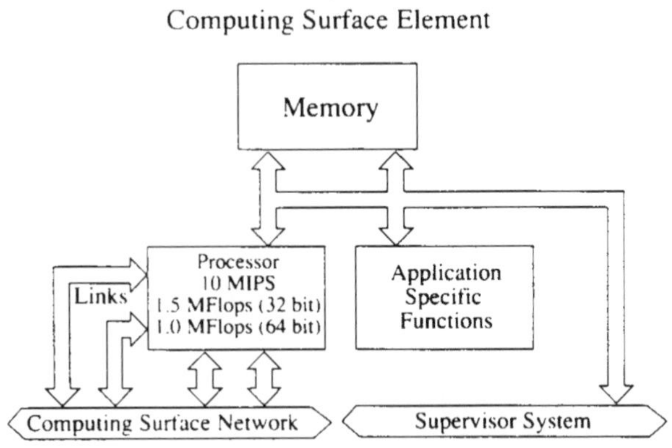

Computing Surface Element

FIGURE 1. Computing surface element.

- Interfaces to the Computing Surface Network (CSN) and the System Supervisor through custom VLSI devices.
- Application-specific hardware providing specialized functions such as graphics or image processing and input/output.

The elements are packaged either one to a board with 8, 12, 16, 24, 32, or 48 Mbytes of memory per processor, or four to a board with 1, 2, 4, or 8 Mbytes of memory per processor. The memory may be upgraded in the field.

Compute power and memory are mixed by the user according to the requirements of the application. The overall system concept is shown in Figure 2.

Graphics output is supported by high-resolution display elements on graphics boards. These boards may be ganged in parallel on a 320-Mbyte/s pixel bus to achieve the desired level of graphics performance by segmenting the image into tiles. Each tile is then updated by a separate graphics element with pixel data flowing over the bus. The graphics elements may be configured with single- or double-frame stores, 8×24 bits per pixel RGB output and resolutions from 512 $\times$ 512 up to 1280×1024.

The overall system management and control of the Computing Surface is provided by the System Supervisor. This is a bidirectional bus, orthogonal to and independent of the links network, communicating with all computing elements via custom VLSI devices and controlled by a module host.

As well as managing the System Supervisor, the module host board provides serial ports to terminals, an Ethernet connection, and an SCSI interface for standalone self-hosted systems. Standard interfaces to IBM PC, DEC VAX, or Sun are available for internally hosted systems.

The major system functions provided by the System Supervisor include monitoring for hardware failures in any of the computing elements, controlling the hardware reset and post-mortem analysis, handling run-time errors and providing debugging support.

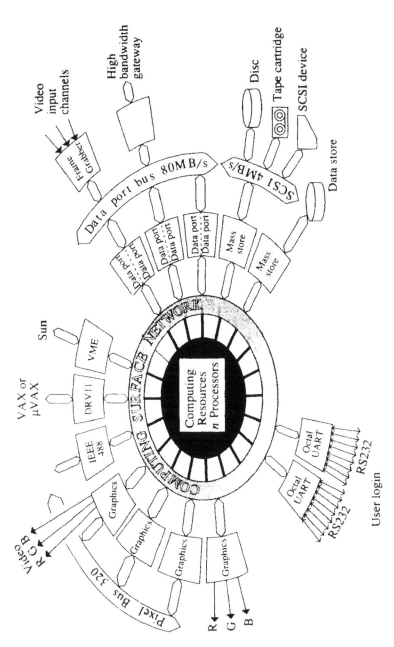

FIGURE 2. Computing surface system architecture.

By developing the CSN as an interface between different processors, Meiko has developed heterogeneous parallel systems that use Intel i860 chips and Sun SPARC processors working alongside Inmos transputers. This made it possible for Meiko to port an industry-standard UNIX operating system to the Computing Surface so that the machine offers an environment that is familiar to scientific users and that lets users program in C.

Perhaps most important of all is the development of system software that makes it easy for Computing Surface users to program this complex parallel system. CS Tools is a software layer that sits above the CSN and lets one configure the processor network topology. Using only high-level software constructs, CS Tools distributes one's parallel program (written in ordinary C, Fortran, or Occam) over the available processors, even when they are a mixture of transputers and i860s.

THREE TYPES OF ELEMENTS

A simple Computing Surface configuration consists of one or more compute elements, a disk I/O element, and a special element, called the module host board, which must be present in all systems. The "vanilla" compute elements all contain a single 32-bit Inmos T800 or T801 transputer with a computing performance of around 10 million instructions per second (integer) and 1.0 million double-precision floating-point operations per second. They also contain varying amounts of working memory; those with smaller memories come four to a board. Single-processor compute elements can contain up to 48 Mbytes of memory per board.

The module host board boots the network, controls basic serial I/O to multiple terminals and Ethernet connection to other machines, and provides an SCSI port to attach disk drives in stand-alone, self-hosted systems. The module host board also controls Meiko's proprietary bidirectional System Supervisor bus to reset the compute elements, to monitor hardware failures and run-time error states, and to provide hardware debugging and post-mortem analysis facilities. The module host board processor maintains a topology map of the transputer link connections for the whole processor network, which it can use to set the VLSI switches (by way of the System Supervisor bus) that allow reconfiguration of the network at run time.

The CSN layer controls this activity in software so that user programs need not see the hardware details of individual switches and link addresses. The effect is that messages can be sent from any processor to any other processor in the network as if there were a direct point-to-point link between them, with the CSN handling all the routing transparently.

Since the CSN layer isolates users from the details of the underlying hardware, it was a natural step for Meiko to use it to incorporate "alien" processors into the Computing Surface. Two new compute elements are now available, one based on the superfast Intel i860 processor and the other on Sun's SPARC RISC chip.

TWO I860S ACHIEVE 80 MFLOPS

The Dual i860 Element, called the MK086, is a board that contains two i860 processors (each with up to 32 Mbytes of memory) and four T800 transputers (two per i860), the job of which is solely to keep the voracious i860s fed with data. Eight of the transputer links provide an external interface into the CSN, shoveling data at 16-Mbytes/s. Internally, the i860s communicate with the transputers via shared memory, using the memory interfaces of the transputers rather than links. Hot-rod chip watchers may find it ironic that although the transputer was once considered the fastest microprocessor in the world, Meiko uses it as mere glue to support the much faster i860. However, this is entirely in keeping with the original design goal for the transputer as a programmable component rather than as a monolithic central processor.

The MK086 element provides a hefty peak performance of 80 MFLOPS, the sort of brute processing power required for supercomputer applications, such as huge matrix calculations. An optimizing, pipelining, vectorizing Fortran 77 compiler is available that enables you to write Cray-type programs for an MK086-equipped Computing Surface. However, the i860 can only achieve full performance on the inherently sequential parts of a parallel algorithm. However, if exploited properly, multiple transputers performing the more finely grained parallel parts can offer synergistic increases in performance.

SPARC PROCESSOR RUNS UNIX

The other new heterogeneous element, the MK083 SPARC Node, while still a high-performance compute element, has a rather different purpose. It allowed Meiko to upgrade its version of UNIX, called MeikOS, into a superset of SunOS 4.0 (Sun's own version of UNIX). This turned the Computing Surface into a full Sun workstation with all the professional software tools that that implies. Sun and Meiko worked closely together on the product.

The MK083 SPARC Node contains a 25-MHz SPARC processor (capable of 15 MIPS) with 64 kbytes of cache and from 8 to 64 Mbytes of main memory. Two communications processors provide a 16-Mbytes/s I/O channel into the CSN.

The SPARC Node acts as a system master. That is, it replaces the module host board and thus is responsible for managing Ethernet, SCSI, and terminal I/O, as well as running the SunOS 4.0 kernel.

A Computing Surface is configured to user requirements using a mix of boards and software, the specifications of which are listed in Table 1.

APPLICATIONS

A typical application for the Computing Surface is at the Charles Stark Draper Laboratory, Cambridge, MA, to process and analyze images created by side-scanning sonar. The side-scanning technique is being explored to find mines in

TABLE 1
Computing Surface Specifications

Processor

Type	Inmos T800 transputer
Data	IEEE-754 compatible, 32 and 64 bit
Performance	10 MIPS peak
	1.5 MFLOPS peak (64 bit)
	1.0 MFLOPS sustained (64 bit)
Number comms links	4 (8 channels)
Comms link speed	20 Mbits/s

Host Board

Number of processors	1
Local memory	8 Mbytes (error checked)
Interfaces	RS232, SCSI, IEEE-488
	Ethernet option
	(access to PC, Sun, VAX)

Compute Boards

Number of processors	1 (Mono boards)
	4 (Quad boards)
Memory per processor	16–48 Mbytes (Mono)
	1–8 Mbytes (Quad)
	Field upgradeable

Graphics Boards

Number of Processors	1
Memory	2 Mbytes dual-ported video RAM
	4 Mbytes program RAM
Resolution	From 512×512 to
	1280×1024 pixels
Bit planes	8 or 24
Pixel bus capacity	200 Mbytes/s

Data Port Element Boards

Number of processors	1
Memory per processor	0.5 Mbytes dual-ported RAM
Data port bus	Max 32 data port elements per bus

Frame Grabber

Number of processors	1
Memory	1 Mbyte dual-ported RAM
Video signals	3 channels
	User configurable (RS170)

TABLE 1 (continued)

SCSI Controller Boards

Number of processors	1
Memory	8 Mbytes
No. attached devices	7

Disk Drives

Drive type	5-$^1/_4$" SCSI
Capacity	140, 300, or 600 Mbytes (formatted)

Cartridge Tape Drives

Drive type	5-$^1/_4$" SCSI
Capacity	120 Mbytes ($^1/_4$" tape)
	2 Gbytes (8-mm tape)

Reel Tape Drives

Drive type	$^1/_2$" tape drive in 19" rack mount unit
Density	1600/3200 or 1600/6250

Enclosures

Dimensions	$67.5 \times 62.5 \times 37$ cm (M10E)
	$110 \times 76 \times 61$cm (M40E)
	$188 \times 97 \times 70$cm (M60E)

Operating System

Type	MeikOS, based on UNIX AT&T System 5.3 and Berkeley 4.3
Interprocessor comms	Meiko Computing Surface Network (NFS protocols)

Programming Environment

Languages	Fortran 77, C, ISO 7158 Pascal, Occam
Parallel programming support	Meiko task farm
	Meiko maths libraries

Networking

Standards supported	Ethernet, TCP/IP, NFS

Graphics Software/User Interface

Window system	MIT $\times$ window system
Graphics libraries	GFX, GKS, GKS 3D, GINO,
	Meiko ray tracing

the ocean, making a high-quality, real-time image very important. The technique involves median filtering to remove noise from each of the sonar images.

The Draper laboratory Computing Surface uses eight high-speed Inmos T800 transputer processors attached to a Sun-4 computer system. The Sun-4 gathers data and sends them to the Computing Surface for median filtering. The data are in turn passed to another computer for neural network processing before being passed back to the Sun-4. From there the data are sent to an image processor and displayed on a high-resolution graphics terminal.

Sonar signals come in rows of data that are 1020 pixels wide. Eight processors did the computing so that one transputer was used as the data manager, sending the entire row of 1020 pixels. The data manager processor divides the row among the seven other processors, each of which then filters only 146 pixels. The median filtering algorithm is written in C, adapted to the Sun station, and ported to the Computing Surface.

The California Institute of Technology (Caltech) is using the Computing Surface to help increase the accuracy of seismic data processing by enabling users to alter graphics representation of the data interactively. The computing system divides the processing tasks between processors from two different manufacturers, matching the task to the type of processor that best performs that task. The system at Caltech uses eight Intel i860 processors and eight Inmos T800 transputers.

The seismic data files at Caltech are between 5 and 7 Gbytes in size. For interactive seismic imaging, the data must be retrieved rapidly from disks and processed at very high speed so that the image on the screen is changed in response to user commands. The T800 processors are used for data management, and the i860 processors are used for the computing. The i860s retrieve the data from the T800s that front-end 16 disks, each with 1 Gbyte of capacity. The T800s receive the data from the disks and pass the data back to the i860s for processing.

The U.S. Army's Natick Research, Development and Engineering Center is using the Computing Surface with its parallel processors to evaluate the performance of military protection equipment and to identify potential ways to reduce future battlefield casualties. The engineer's Computing Surface being used operates with 12 Inmos T800s to help Army equipment designers predict protection of clothing and equipment and how designs will affect performance. The Computing Surface is used to run programs that calculate the path that projectiles would take through the body and their effect. Other programs are used to predict the level of wounds based on experimental and historical data.

The criteria for selecting the computing system requires that users be able to reconfigure topologies dynamically among hypercube, ring, mesh, tree, and pipeline. The system also is required to handle at least four simultaneous events, have a minimum performance capability of 1 MFLOPS, and be capable of expanding to at least 1.280 MFLOPS. The system also must connect to an existing Ethernet network and operate in an office environment.

The Computing Surface has been used to evaluate actual ballistic casualty reduction simulations. The four programs used vary from 29.6 to 153.5

kbytes in size and have one or more data files required for execution. One of the programs took 4 h and 20 min to run on Natick's Univac 1106 main frame computer, and 1 h and 27 min on a 20-MHz Intel 80386/80387-based personal computer. A Meiko single-processor configuration ran the same program in 13 min and 4 s, and it took 4 min and 5 s to run the program using four T800s.

The Edinburgh Concurrent Supercomputer (ECS) at the University of Edinburgh is a 32-processor Meiko Computing Surface. A wide range of neural net models has been developed for a variety of purposes; they differ in structure and details, but share some common features:

- They contain nodes, or units, which are extreme simplifications of neurons, in the sense that the state of each node is usually described by a single real variable, representing its firing activity.
- The nodes are connected, usually in pairs, so that the state of one node affects the potential of all the nodes to which it is connected according to the weight, or strength, of the connection.
- The new state of a node is a nonlinear function of the potential created by the firing activity of the other neurons.
- Input to the network is done by setting the states of a subset of the nodes to specified values; this sets up an image or pattern of activity on these "input" nodes.
- The processing takes place through the evolution of the states of all the nodes on the net, according to the details of the dynamics and the particular connection strengths, until some output activity can be read from another, possibly different, set of nodes.
- The training of the net is the process whereby the values of the connection strengths are modified in order to achieve the desired processing for a set of training data.

For example, for image processing, one can imagine that the array of pixel values from the input image is mapped on to the states of the array of input nodes. The image processing is effected by the dynamics of the net producing the enhanced image or the features identified in the image, etc., at the output nodes. The training data would consist of a set of possibly noisy images with their known, desired output. For an application in medical diagnosis, the input data would be an encoding of the symptoms, and the target output would be the diagnosis, and possibly recommended treatment.

Models based on these ideas are not new. Much of the current effort can be traced to the seminal work of McCulloch and Pitts[3] and Hebb.[4] In the 1960s, Rosenblatt[5] developed the "perceptron" model for pattern classification. An excellent analysis and critique of the perception theory is given by Minsky and Papert.[6] The recent resurgence of interest was in large measure stimulated by the development of networks which go beyond the limitations of the perceptron model; reviews can be found in Reference 7.

Analog neurons were introduced by Hopfield and Tank[8] as a general technique for optimization problems involving boolean variables. It has been used to perform load balancing in parallel computing.[10] We focus here on image restoration, using the algorithm of Geman and Geman[11] for binary images which have been corrupted by noise.

The analog neuron scheme[12] involves representing the array of pixel intensities as a network of neurons, each of which can "fire" on a continuous scale from nonfiring ("black" pixel) to fully firing ("white" pixel). The dynamics of the pixel interactions ("neural activities") are controlled by a cost function determined by the input data and by *a priori* assumptions about the statistical properties of the "clean" images (for example, "edges are rare"). The "best" restored image is that which minimizes this cost function, and the dynamics are designed to achieve at least a good approximation to this. This problem is intrinsically SIMD parallel, with short-range communications, so that it can be expected to run efficiently on most parallel machines; at Edinburgh it was studied both on the DAP[12] and on the MIMD Computing Surface.[27]

The performance of the restoration by this analog neural network method was compared to that of a simple majority-rule scheme (where each neuron continually adopts the intensity of the majority of its four nearest neighbors — or remains unchanged if exactly two of its neighbors are "white" — until the image stabilizes). A third restoration method — performing a gradient descent — was achieved by restricting the neuron firing rates to discrete values ("on"/"off"). The analog neural network method consistently finds lower cost solutions[12] than these schemes.

In a recent paper, Durbin and Willshaw[13] described what they called the "elastic net" method for solving the traveling salesman problem in the plane. A closed loop of elastic "string" is placed in the plane containing the cities which the salesman is to visit, and then slowly deformed into a path which connects all the cities.

The elastic string is modeled as a set of discrete points, each of which is connected to its two neighbors. Attractive forces between each point and its neighbors hold the string together, while each city exerts an attractive force on each point, pulling the point towards it. As the point-to-point forces are relaxed and the city-to-point forces strengthened, the string is gradually deformed into a path. The string dynamics are intrinsically parallel and have been implemented by a straightforward "domain decomposition" on the Computing Surface.[28]

The parallel elastic net solution to the traveling salesman problem is another example which becomes more efficient as it becomes larger. Since the number of points which must be used to model the elastic string grows linearly with the number of cities, the amount of calculation at each step grows as N^2. However, the amount of communication only grows as N, so the ratio of calculation to communication improves as the problems being solved grow larger.

Cellular automata are arrays of cells containing degrees of freedom which take on discrete values (boolean variables 0 or 1 in the simplest case). These variables evolve in discrete time steps according to a transition rule which depends on the states of a cell and its neighbors and which may be deterministic or stochastic (i.e., affected by noise). In one sense, therefore, they can be thought of as primitive molecular dynamics.

The motivation for studying cellular automata is (at least) twofold. Although, from the point of view of physics, their microscopic behavior may not correspond to any specific physical system, their behavior on distance scales is large compared to the cell size and can describe macroscopic continuum phenomena. From the point of view of computation, they are suitable for digital computer simulation.

Microcanonical simulations of the Ising model of a uniaxial ferromagnet can be formulated in this way for example, but the particular case study on which we focus is cellular automaton modeling of fluid flow.[14,15] These models represent an extension of the lattice gas concept from statistical mechanics to hydrodynamics, with "particles" hopping along bonds of the lattice and scattering from each other according to simple local rules. The primary aim is to ensure that the Navier-Stokes equation emerges on the large scale. It turns out that the discrete symmetry of a square-lattice automaton survives in the macroscopic limit. However, a hexagonal lattice has sufficient symmetry to ensure isotropy, which can also be ensured in three dimensions by allowing hopping beyond nearest neighbors.[15-17]

This fluid dynamics type of bit-serial simulation is more ideally suited to the architecture of the DAP than that of the Computing Surface, but the graphical capability of the latter was crucial in motivating the work and in visualizing the results. Whether this approach to turbulent simulation captures hydrodynamic flow effectively, and whether it will emerge with significant advantages over conventional methods using the Navier-Stokes equations, remain matters for research and debate: the existence of parallel computers has certainly been a major factor in stimulating that debate.

Molecular biology has been revolutionized by the development of fast sequencing techniques for nucleic acids. The rate of acquisition of protein sequence data has correspondingly accelerated, and this has led to the urgent need for adequate comparative sequence analysis to promote the efficient use of other research resources.

Proteins and nucleic acids (the genetic material) are linear polymers, the sequences of which may be represented by character strings, with a 20-letter alphabet for proteins and a 4-letter alphabet for nucleic acids. The international database collections of sequences are prime resources for molecular biological research. These databases are currently small; the protein database has approximately one million characters of sequence information, and the genetic base has ten million, but already the task of searching them has led to the development of a number of approximate methods for making comparisons. However, the application of the exhaustive inexact string-matching algorithms reviewed by Sellers[18] has been beyond the capacity of many workstations and mainframe computers. The situation will deteriorate further, as the databases are growing exponentially, doubling in size every 2 years or less.

A suite of programs for exhaustive, inexact string matching has been developed for the DAP by Lyall et al.[19,20] They have been extensively applied in several thousand searches, leading to discoveries of biological significance. These include the relationship of the cystic fibrosis antigen to the bovine s-100a alpha protein chain[21] and the relationship of vitellogenins in *Drosophila melanogaster* to porcine triasyl glycerol lipase.[22] A similarity between prokaryotic and eukaryotic cell cycle proteins has also been discovered.[23] This work is continuing on the DAP and the Meiko Computing Surface, where the Fortran Farm facility of the latter supports parallel comparisons transparently across the database. This is one particular example of the growing application area of parallel databases.

New noninvasive medical techniques such as NMR imaging are being used in bodyscanners to produce three-dimensional (tomographic) images of the body. There have been several attempts over the past few years to produce systems which are capable of manipulating and displaying the vast quantities of data which are required to generate each image, but a satisfactory performance has still not been achieved. Three-dimensional image processing suffers from the same speed problems as in two dimensions, but exaggerated by at least one order of magnitude since there are ten or more two-dimensional sections in one three-dimensional image, and the image-processing operations are inherently more complex.

A key problem in implementing this type of processing on a parallel machine is to achieve a high efficiency of processor usage while minimizing communications overheads. This is typically straightforward for low-level image processing, but less so for intermediate and high-level functions. For example, the common requirement of tracking a surface through the data could very easily result in one processor at a time doing all the work, with all the rest lying idle. In particular, the difficulty of mapping a three-dimensional image onto a two-dimensional array of processors means that data at the boundary may need to be passed not to an immediately adjacent neighbor in the array (as for example in the simulation of fluid flow, where the same local operations can be performed globally right across the entire array of processors), but to some more remote processor, depending on local conditions within the data. While most of the previous examples have been well suited for SIMD parallelism, the higher level processing in this example certainly benefits from the additional flexibility offered by MIMD.

The first phase in handling NMR bodyscan data on the Computing Surface is now complete.[24] The system includes a generalization of the Zucker-Hummel surface detection operator to a nonsquare metric, a three-dimensional surface display, and an implementation of a novel and completely asynchronous arbitrary communications network which carries messages around the processors, ensuring that all are operating as closely as possible to maximum efficiency. The computation runs on 40 T414 transputers at 500 times the speed of a SUN 2, and the display speed is some 50 times greater.

Ray tracing is a way of displaying three-dimensional objects on a two-dimensional screen. The basic idea is to reproduce what happens in a pin-hole camera. The image on the screen is built up from rays of light coming from objects which pass through the pin-hole. Each ray may be identified by starting at the screen and tracing back, through the pin-hole, until it strikes a surface in the three-dimensional world. It is then reflected backwards to determine whether it comes from another surface or from a light source. This backward tracing continues until the ray ends at a light source or passes out of the world. Once the source of a ray has been identified, its path is retraced from the source to determine the color and brightness of the corresponding pixel on the screen. The paths of the reflected rays and the levels of illumination depend on the nature of the light source (e.g., ambient light, which is uniform in all directions, or point sources) and on the type of reflecting surface (e.g., matte, from which illumination is diffuse and in all directions, or mirrored, in which the light is reflected as a single ray); other optical effects can also be included. The complete picture is built up by tracing one ray for each pixel.

The ray-tracing algorithm involves a great deal of computation. For each ray, the first task is to determine the point at which it strikes a surface in the three-dimensional world. This involves solving a system of linear equations, given the equations of all the objects. The properties of the surface determine how the ray is reflected, and the procedure is repeated. This can become very complicated in the case of multiple reflections from smooth surfaces. Although computationally intensive, the algorithm is highly parallel as all of the rays are completely independent and ideally suited to the Computing Surface. The ray-tracing algorithm has been implemented on the ECS, using a load-balancing transputer pipeline. Portions of the screen are distributed by a master transputer down a pipeline of slave transputers. Each slave takes a task from this stream, computes the image corresponding to that portion and sends results back to a graphics processor for display. An example of the high-quality images that may be produced by this technique is found in Reference 25.

Molecular dynamics is another area of physics that is manifestly amenable to parallel computation, because it involves time-stepping the equations of motion for a set of, say, N particles, which is naturally done by the simultaneous calculation of the forces on those N particles. The number of processors available in a realistic simulation is usually less than N, but even when this is not so, they can still be used efficiently by distributing the force calculation for each particle across a number of processors, or perhaps by running independent simulations in parallel under different conditions. A demonstrator program[27] for the ECS has been completed which simulates the dynamics of a system of particles, with a range of masses, moving under the influence of gravitational forces. It provides pseudo-three-dimensional graphics and can be interpreted in terms of the dynamics of a stellar cluster.

The single branch of physics that has probably absorbed the most supercomputer hours is the simulation of quantum field theories, such as quantum electrodynamics (QED), the field theory describing the interaction of charged spin-$\frac{1}{2}$ particles such as electrons with photons, and, more especially, quantum chromodynamics (QCD), the corresponding theory of quarks and gluons, the fundamental constituents of subnuclear particles.

There are two projects under way on the ECS. The first addresses a crucial computational problem in this area, the simulation of spin-$\frac{1}{2}$ particles (fermions). A simulation of fully interacting electrons and photons employing molecular dynamics, Monte Carlo, and sparse matrix techniques has been implemented on a hypercube of transputers. In collaboration with physicists at the Supercomputer Research Institute at Tallahassee, FL, B. J. Pendleton and D. Roweth at Edinburgh developed an algorithm which was demonstrated to be six times more effective than previous approaches for such calculations.[26] A project is now under way in collaboration with the Rutherford Appleton Laboratory to use the same algorithm to study the Hubbard model and its possible role in high-temperature superconductors.

REFERENCES

1. **Bowler, K. C., Bruce, A. D., Kenway, R. D., Pawley, G. S., Wallace, D. J., and McKendrick, A.,** Scientific Computation on the Edinburgh DAPs, University of Edinburgh Report, December 1987.
2. **Bowler, K. C., Kenway, R. D., Pawley, G. S., and Roweth, D.,** *An Introduction to Occam 2 Programming*, Chartwell-Bratt, Bromley, 1987.
3. **McCulloch, W. S. and Pitts, W. A.,** *Bull. Math. Biophys.*, 5, 115, 1943.
4. **Hebb, D. O.,** *The Organisation of Behavior*, John Wiley, New York 1949.
5. **Rosenblatt, F.,** *Principles of Neurodynamics*, Spartan Books, New York, 1962.
6. **Minsky, M. and Papert, S.,** *Perceptrons: An Introduction to Computational Geometry*, MIT Press, Cambridge, 1969.
7. **Hinton, D. G. and Anderson, J. A., Eds.,** *Parallel Models of Associative Geometry*, Lawrence Erlbaum, Hillsdale, NJ, 1981; **Rumelhart, D. E., McClelland, J. L., and the PDP Research Group,** *Parallel Distributed Processing: Explorations in the Micro-Structure of Cognition*, Vols. 1 and 2, Bradford Books, Cambridge, MA, 1986; **Denker, J. S., Ed.,** *Neural Networks for Computing*, AIP Conf. Proc. 151, American Institute of Physics, New York, 1986; **Grossberg, S., Ed.,** *The Adaptive Brain*, Vols. 1 and 2, North-Holland, Amsterdam, 1987.
8. **Hopfield, J. J. and Tank, D. W.,** *Biol. Cypernet.*, 52, 141 1985.
9. **Wilson, G. V. and Pawley, G. S.,** *Biol. Cypernet.*, 1987, in press.
10. **Fox, G. C. and Furmanski, W.,** Caltech preprint C3P 363, 1987.
11. **Geman, S. and Geman, D.,** *IEEE Trans. PAMI*, 5, 721, 1984.
12. **Forrest, B. M.,** Proc. of Parallel Architectures and Computer Vision Workshop, Oxford, 1987, in press.
13. **Durbin, R. and Willshaw, D. J.,** *Nature*, 326, 689, 1987.
14. **Hardy, J., de Pazzis, O., and Pomeau, Y.,** *Phys. Rev. A*, 13, 1949, 1976.
15. **Frisch, U., Hasslacher, B., and Pomeau, Y.,** *Phys. Rev. Lett.*, 56, 1505, 1986.
16. **Wolfram, S.,** *J. Stat. Phys.*, 45, 471, 1986.
17. **Frisch, U., d'Humieres, D., Hasslacher, B., Lallemand, P., Pomeau, Y., and Rivet, J.-P.,** *Complex Syst.*, 4(1), 1987, in press.
18. **Sellers, P. H.,** *J. Algorithms*, 1, 359, 1980.

19. **Lyall, A., Hill, C., Collins, J. F., and Coulson, A. F. W.,** in *Parallel Computing '85*, Feilmeier, M., Joubert, G., and Schendel, U., Eds., Amsterdam, North-Holland, 1986, 235.

20. **Coulson, A. F. W., Collins, J. F., and Lyall, A.,** *Comput. J.*, 30, 420, 1987.

21. **Dorin, J. R., Novak, M., Hill, R. E., Brock, D. J. H., Secher, D. S., and van Heyningen, V.,** *Nature*, 326, 614, 1987.

22. **Bownes, M., Shirras, A., Blair, M., Collins, J., and Coulson, A.,** *Proc. Natl. Acad. Sci.*, 85, 1988, in press.

23. **Robinson, A. C., Collins, J. F., and Donachie, W. D.,** *Nature*, 328, 766, 1987.

24. **Norman, M. G.,** A Three-Dimensional Image Processing Program for a Parallel Computer, M.Sc. thesis, Department of Artificial Intelligence, University of Edinburgh, 1987.

25. **Bowler, K. C., Bruce, A. D., Kenway, R. D., Pawley, G. S., and Wallace, D. J.,** *Phys. Today*, 40, 40, 1987

26. **Duane, S., Kennedy, A. D., Pendleton, B. J., and Roweth, D.,** *Phys. Lett.*, 192B, 1987, 216.

27. **Roweth, D.,** Unpublished data.

28. **Roweth, D. and Wilson, G. V.,** Unpublished data.

9 BB&N BUTTERFLY

OVERVIEW

The Butterfly parallel processor by Bolt, Beranek, and Newman, Inc., is composed of processors with memory and a novel high-performance switch interconnecting the processors. One processor and memory are located on a single board called a processor node. All Butterfly processor nodes are identical. I/O connections can be made to each processor node, making the I/O configuration very flexible. Collectively, the memory of the processor nodes forms the shared memory of the machine. All memory is local to some processor node; however, each processor can access any of the memory in the machine, using the Butterfly switch to make remote references. From the point of view of a program running on one node, the only difference between references to memory on its local processor node and memory on other processor nodes is that remote references take a little longer to complete. Typical memory referencing instructions accessing local memory take about 2 µs to complete, whereas those accessing remote memory take about 6 µs. The speeds of the processors, memories, and switch are balanced to permit the system to work efficiently in a wide range of configurations.

The shared-memory architecture of the Butterfly parallel processor, together with the firmware and software of the Butterfly's operating system, Chrysalis™, provide a program execution environment in which tasks can be distributed among processors with little regard to the physical location of data associated with the tasks. This greatly simplifies programming the machine and permits effective utilization of the multiple processors over a wide variety of applications.

The Butterfly architecture scales in a flexible and cost-effective fashion to meet a wide range of processing and memory requirements. A Butterfly system can be configured with from 1 to 256 processor nodes to achieve computing power matched to the needs of a particular application. Memory, switch, and I/O capacity scale with the number of processors to maintain overall system balance and to prevent potential performance bottlenecks. A large Butterfly configuration achieves its processing power by using many lowcost processors working cooperatively. This is in contrast to large mainframes which use costly technology to achieve their high processing capacity. Each processor node is capable of executing 500,000 instructions per second and is usually configured with 1 Mbyte of memory (expansion memory is available to increase this to 4 Mbytes). The bandwidth through each processor-to-processor path in the Butterfly switch is 32 Mbits/s. Thus, a 256-processor machine has a raw processing power of 128 MIPS, 256 Mbytes of main memory (1024 Mbytes when fully configured with expansion memory), and an interprocessor communication capacity of 8 Gbits/s.

Another benefit of the Butterfly architecture is its relative insensitivity to component failures. The machine can function without one or more of its processor nodes. For example, a 128-processor machine will function with 98% of its capacity even if 3 of its processors have failed or been removed.

Experience with a 256 processor Butterfly system has shown that its raw processing power can be effectively used. Nearly linear speed up is achieved through the entire range of processors.

In summary, the Butterfly parallel processor is modular, powerful, reliable, scalable, and easy to program. These characteristics of the Butterfly system make it attractive for a wide range of applications. It is well suited for computationally intensive applications, such as modeling and simulation. It can undertake the solution of compute-bound problems in the physical sciences, such as those found in molecular mechanics and crystallography, and its capabilities are well matched to the needs of the finite element method widely used in structural mechanics. Relatively small configurations (1 to 16 processors) have been used in data communications applications as Internet gateways, as satellite channel controllers for packet communication systems, and as multiplexing concentrators and demultiplexers for voice terminals. Larger configurations are currently being used for research in image processing and computer vision. The Butterfly computer has been used for circuit simulation to support VLSI design. Symbolic processing and AI applications using the machine have been developed.

The Butterfly parallel processor has several important architectural characteristics.

- It is a MIMD machine. Each processor node independently executes its own sequence of instructions, referencing data as specified by the instructions. The ability of different processors to execute different code on separate data gives programmers freedom to structure programs and data in ways that are natural and efficient for their applications. The MIMD architecture supports a large variety of software structures and permits Butterfly systems to be used in a wide range of applications.
- It is a tightly-coupled, shared-memory machine. Processor nodes are tightly coupled by the Butterfly switch. Tight coupling permits efficient interprocessor communication and allows each processor to access all of system memory efficiently. Program references to remote memory take only slightly longer to complete than references to local memory. This greatly simplifies programming without sacrificing performance. Because there is no significant performance penalty for accessing remote memory, the programmer is free to

organize data without undue concern about which memories store which parts of the data. In fact, for many applications the best performance results are achieved when the application data are scattered uniformly about the machine.

• It is expandable over a wide range of configurations. Each processor node added to a configuration contributes processing power, memory, switch bandwidth, and I/O capacity. As a result, processing, memory, communication, and I/O capacity all grow with the size of the configuration. Furthermore, the complexity and, therefore, the cost of the interconnection switch remain at an acceptable level. The Butterfly architecture overcomes the problems that limit the expandability of bus and crossbar interconnection architectures. Because a bus has a fixed capacity, bus architectures must, in effect, be designed for a given performance level. As a result, bus architectures are relatively expensive in small configurations (since each bus interface must operate at the full bus speed, they must be high to offer high performance for larger configurations); and they cannot be effectively utilized when expanded to configurations that exceed the fixed capacity of the bus. Crossbar architectures have a matrix of interconnection points, called cross points, for connecting system elements. Unlike a bus architecture, the capacity of a crossbar architecture can grow with the number of system elements simply by including enough cross points. Unfortunately, the number of cross points required is proportional to the square of the number of processors. As the size of a configuration grows, the cost of the crossbar comes to dominate the cost of the system, destroying its cost effectiveness.

• It is a homogeneous multiprocessor. Although some nodes may have attached I/O devices, processor nodes are basically identical. As a result, every processor is capable of performing any application task. The uniformity of the Butterfly architecture simplifies programming, since programmers need not concern themselves with allocating certain tasks to specific processors. Furthermore, this uniformity makes it relatively straightforward to write application software that is insensitive to the number of available processors. Such application software is attractive in several ways. It can be developed and tested on small inexpensive configurations and be run on larger operational ones. Should the resource demands of an application grow to exceed the capacity of its current configuration, the application can be moved to a larger, more powerful configuration and run without reprogramming. In addition, application software written in this way contributes to improved overall system reliability and availability since it can run on configurations reduced by failed components.

The Butterfly system has three major components: processor nodes, the Butterfly switch, and I/O hardware. Each is described below. The switch serves to interconnect the processor nodes. Figure 1 illustrates the processor nodes and switch for a 16-processor Butterfly configuration. The figure shows the switch as a cylinder, with each processor node connected to the switch cylinder through an entry port. Data sent by a processor node enters and is routed through the switch, and exits at the destination processor node.

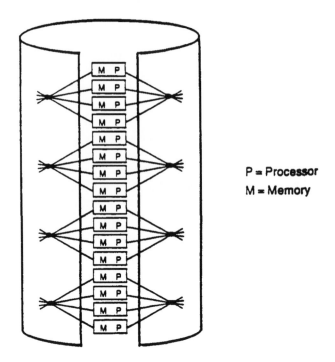

P = Processor

M = Memory

FIGURE 1. Processing nodes and switch for a 16-processor configuration.

The Butterfly processor node contains a Motorola MC68000 microprocessor, at least 1 Mbyte of main memory, a microcoded coprocessor called the processor node controller (PNC), memory management hardware, an I/O bus, and an interface to the Butterfly switch. Figure 2 is a block diagram of the processor node.

All user application software runs on the Motorola MC68000. Although it is not an absolute requirement of the architecture, all codes executed by the MC68000 reside in local memory.

The PNC initiates all messages transmitted over the switch and processes all messages received from the switch. It is involved in every memory reference made by the MC68000. It uses the memory management unit to translate the virtual addresses used by the MC68000 into physical memory addresses. Consequently, all references to memory look the same to an application program. A reference to remote memory simply takes a little longer to complete than one to local memory. As a result, the memory of all processor nodes, taken together, appears as a large single global memory to application software.

Another important function of the PNC is to provide efficient implementations in microcode of a collection of operations that augment the functionality of the MC68000 for parallel processing. These operations include a suite of test and set operations, queuing operations, operations that implement an event mechanism, and a process scheduler that works with the queuing and event mechanisms

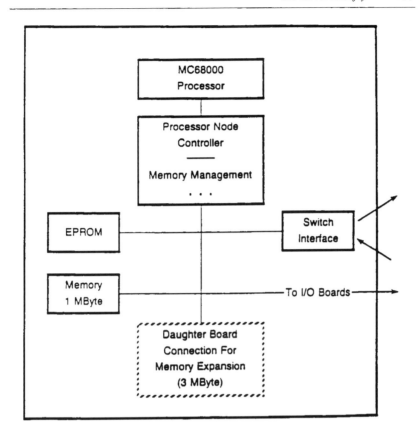

FIGURE 2. Butterfly processing node block diagram.

to provide efficient communication and synchronization between application software modules. Many of these operations manipulate data, and it is important that they be executed in a fashion that guarantees no other processors access the data as it is being manipulated. Because the PNC intercedes in all memory references made by both its MC68000 and remote processor nodes via the switch, it can ensure that these operations are implemented in an "atomic" fashion.

The Butterfly switch uses techniques similar to packet switching to implement high-performance interprocessor communication. The switch is a collection of switching nodes configured as a "serial decision" network. Its topology is similar to that of the fast Fourier transform Butterfly. Hence the name, Butterfly switch.

Figure 3 illustrates the 16-processor Butterfly system from Figure 1 in more detail. The cylinder has been cut down the middle of the processor nodes, flattened, and drawn with half of each processor node on each side. Each Butterfly switching node is a 4 input-4 output switching element implemented by a custom VLSI chip; 8 custom VLSI switch chips are packaged on a single printed circuit board to implement a 16 input-16 output switch.

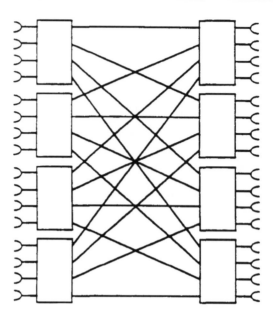

FIGURE 3. A 16 input-16 output Butterfly switch.

There is a path through the switch network from each processor node to every other processor node. Switch operation is similar to that of a packet switching network. The switching nodes use packet address bits to route the packet through the switch from source to destination processor node. Each node uses two bits of the packet address to select one of its four output ports. To illustrate how this works, Figure 4 shows a packet in transit through the 16 input-16 output switch of Figure 3. To send a message to node #14, node #5 builds a packet containing the address of node #14 (= 1110 binary), followed by the message data, and sends the packet into the switch. The first switching node strips the two least significant address bits (10) off of the packet and uses them to switch the remainder of the packet out of its port 2 (10). The next switching node strips off the next two address bits (11) to switch the packet out its port 3 (11) to node #14. Notice that the structure of the switch network ensures that packets with address 1110 will be routed with the same number of steps to node #14 regardless of the processor node sending them.

The Butterfly switch scales well with the number of processors. By adding additional rows and columns of switch nodes, the 16-processor switch can be enlarged to handle more processors. For example, the number of connected processors can be quadrupled to 64, as shown in Figure 5 by interconnecting an additional column of switch nodes and enough rows to accommodate 48 more processors. The communication capacity of the switch grows linearly while the complexity of the switch remains manageable. Each path through the switch supports interprocessor data transfer at 32 Mbits/s. Thus, the switch for a 16-processor system has a capacity of 512 Mbits/s, and the switch for the 64-

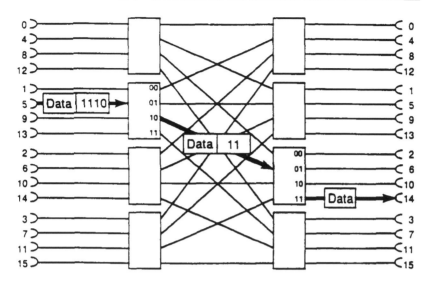

FIGURE 4. A packet in transit through a Butterfly switch.

processor system shown in Figure 5 has a capacity of 2048 Mbits/s. The complexity of the switch (as measured by the number of switch elements or the number of wires) for an N processor system grows as N $\log_4$N. For large configurations this is significantly less than the N^2 elements required for a crossbar switch. For example, the switch for a 100-processor Butterfly requires about 500 wires, whereas a crossbar switch for 100 processors would require about 5000 wires.

Because there is not a dedicated path between each pair of processor nodes, it is possible for two messages simultaneously passing through the switch to reach the same switching node and to require the same switching node output port. When contention of this sort occurs, one of the messages is allowed to proceed to its destination and the other is automatically retransmitted after a short delay.

A switch containing an alternate path between every pair of processor nodes can be constructed by adding extra switching nodes. This is typically done in larger configurations to make the switch resilient to switching node failures. Currently, all systems with more than 16 processors are configured to have redundant paths. Should a switching node on the path between two processors fail, packets are automatically routed along the alternate path. Another benefit of a switch configured with alternate paths is reduction of contention within the switch. Whenever contention within a switch with alternate paths requires retransmission of a message, an alternate path is immediately used.

Machines are configured so that the probability of message collision within the switch is relatively low. As a result, the transmit time through the switch is dominated by the time required for a message to pass through the switch in a bit-serial fashion rather than by contention. The amount of contention is, of course,

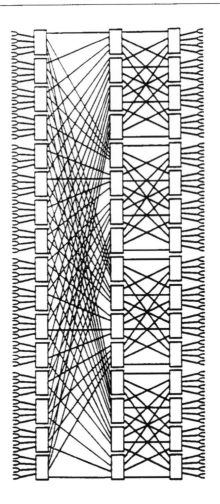

FIGURE 5. Switch for a 64-processor Butterfly system.

application dependent. For applications developed to date, contention overhead amounts to less than a few percent (typically 1 to 5%) of the total application run time.

The way processor nodes use the switch can perhaps best be illustrated by the following brief description of how a read reference-accessing remote memory is handled. When the MC68000 makes a read reference, its local PNC gains control and uses its memory management hardware to transform the supplied virtual address into a physical address, which corresponds to memory on another processor node. To read the referenced location, the PNC sends a packet addressed to the remote processor node through the switch requesting the contents of that physical memory address. The remote PNC receives the packet, reads the referenced memory location, and sends a reply packet containing the value through the switch back to the source processor node. When it receives the reply, the source PNC satisfies the read request of the MC68000 with the value obtained from the reply. The round-trip time for a remote memory reference is about 4 μs. In addition to single-word transfers, the PNC can also transfer blocks of data efficiently between any pair of processor nodes at the full 32-Mbit/s bandwidth of the switch.

The Butterfly I/O system is distributed among the processor nodes. The I/O bus on each processor node supports data transfer at a maximum rate of about 2 Mbytes/s. I/O boards may be attached to any processor node. Presently, two types of I/O boards are available: the Butterfly multibus adapter (BMA) board and the Butterfly serial I/O (BIO) board. The MA board provides efficient data exchange between a processor node and a Multibus to which any Multibus-compatible I/O device, memory, or processor may be connected. The BIO board supports four asynchronous RS-232 interfaces and four synchronous RS-422 lines.

Terminals can be interfaced using either the BIO or the BMA boards. Other devices, such as Ethernet interfaces and disk storage units, can be interfaced using the BMA board.

A 128-processor Butterfly system, including I/O, occupies four standard racks. A 16-processor system occupies a half rack. As noted above, the switch can be configured to provide the level of bandwidth needed for the machine, including alternate switch paths to reduce vulnerability to single point failures and to reduce potential contention. The 16-processor system uses a single board switch, while a 64-processor system uses a switch distributed over 8 switch boards. The 128-processor system uses a switch distributed over 16 switch boards.

The Butterfly memory expansion board (BMEX) increases the capacity of a Butterfly processor node from 1 Mbyte to the full 4 Mbytes allowed by the memory management hardware. Physically, the BMEX board is a daughterboard that sits on a set of stand-offs attached directly to the processor node. It occupies a card slot and mates to a daughterboard connector on the processor node.

The Butterfly floating point platform (BFLPT) provides hardware floating point support. Based on the Motorola MC68020 microprocessor and MC68881 floating point coprocessor, the BFLPT implements the full (64 bit) IEEE 796 standard for floating point arithmetic. In addition, the BFLPT implements microcoded square root, trigonometric, logarithmic, and transcendental functions. Since the MC68881 uses the MC68020 coprocessor facility, hardware floating-point operations are invoked by executing ordinary 68020 instructions.

The 68020 and the 68881 on the BFLPT run at their maximum rated speeds of 16 MHz, and the rest of the processor node runs at 8 MHz. The BFLPT also improves the nonfloating-point performance of the processor node. A fixed point matrix multiplication benchmark runs twice as fast as on a processor node configured with a BFLPT than on one with a 68000. The speed-up is due mainly to the instruction cache and increased operating speed of the 68020.

Physically, the BFLPT is a drop-in replacement for the MC68000. It consists of a small printed circuit board with a set of pins on the underside that mate with the 68000 socket on the Butterfly processor node. Mounted on the printed circuit board are a Motorola MC68020, MC68881, and several smaller chips that take care of minor differences between the 68000 and 68020 bus protocols. The BFLPT is designed so that a processor node with the BFLPT attached takes only a single card slot.

The purpose of the Butterfly VME bus adapter (BVME) is to support I/O devices in which bandwidth requirements exceed the I/O capacity of a single Butterfly processor node.

The VME bus is a 32-bit bus with a specified peak data transfer rate of 40 Mbytes/s. Since the maximum data transfer rate of the Butterfly processor node I/O bus is roughly 2 Mbytes/s, it is necessary to use a higher bandwidth path to connect the VME bus to Butterfly systems. We have chosen to design the BVME so that it transfers data directly to and from the Butterfly switch. This approach has several advantages:

- The 4 Mbyte/s bandwidth of a single switch port is greater than that of the processor node I/O bus.
- By attaching the BVME to more than one switch port, the bandwidth of the connection between the Butterfly system and the VME bus can be scaled to suit the needs of the devices on the VME bus.
- Attaching the BVME directly to the switch means that I/O transfers can use memory from many processor nodes, thereby reducing the load on any given memory.

The Butterfly parallel processor is programmed exclusively in a high-level programming language. To date, most Butterfly software has been written in C. Other languages being installed on the Butterfly system include Lisp and Fortran. Application programs run under the Butterfly Chrysalis Operating System.

Programs for the Butterfly system are written using a cross-compiler and other software development tools on a front-end machine. Currently, these C language development programs run under 4 2BSD UNIX™ on a DEC VAX or Sun Microsystems workstation. This allows the use of the rich set of available UNIX software tools, such as the source code control system and the "make" utility. The Butterfly system is connected to the front end by an Ethernet connection or a serial line. The typical development cycle consists of editing, compiling, and linking a program on the UNIX front end, and downloading, running, and debugging the program on the Butterfly system. A source language debugger for the C language is available that runs on the front end, allowing cross-network debugging of programs running on the Butterfly system.

To run programs on the Butterfly system, an interactive command interpreter, called the bshell (for Butterfly shell), and a terminal window manager are available. The user accesses the system from the front-end computer, and the window manager enables rapid switching between the front-end and Butterfly system environments. The bshell allows loading and execution of applications.

The window manager enables a terminal to be shared for the input and output of multiple processes. It also provides line editing and screen management functions, such as cursor positioning.

The Chrysalis Operating System provides support for applications programs on the Butterfly parallel processor. It is designed to provide a familiar, UNIX-like environment that supports programming on high-level languages, without being restricted to a single language.

The Chrysalis system consists of several components. At the highest level are interactive utilities for accessing, controlling, and debugging programs on the system. These include the interactive command interpreter and window manager, as well as utilities for loading and running applications and for determining their status as well as overall system status.

The lower levels of the Chrysalis Operating System are called as subroutines from the user's program. We describe each level briefly below, but keep in mind that the higher levels provide interfaces more suitable for many applications. The lowest levels are available if needed, and so are described here for completeness.

Application libraries manage many of the details of using system resources, such as allocating memory and setting up processes. They also form the interfaces to the rest of the operating system for application programs. Currently, the most significant application library in the Chrysalis Operating System is the Uniform System library, which supports a methodology for programming the Butterfly parallel processor.

The Uniform System creates an environment where all processors share a common address space. The application is structured so that processes are created on as many processors as are available in a Butterfly system configuration. Each process runs the user's application tasks on a subset of the data. The Uniform System library provides subroutines to efficiently allocate the data structures for the problem, and then calls upon processors to work on the tasks specified by the programmer. Other application libraries provide

- A buffer management package for communications applications.
- A stream-oriented I/O interface, similar to UNIX standard I/O. Its subroutines allow applications to receive terminal input, and to perform formatted output of character strings.
- A library of performance measurement tools oriented toward parallel applications.

Access to system resources, such as the Ethernet, is provided by server processes that run on one or more (but typically not all) processors on the system. Applications programs that need these services communicate with the servers through subroutine calls, such as UNIX-style open, close, read, and write. The interprocess communication facilities of the operating system are used to route these requests to the appropriate server processes. Currently, there are servers that execute functions on remote nodes, translate symbolic resource names to addresses, and access the network. Networking capabilities provide an interface to the Ethernet high-speed local-area network and include the U.S. Department of Defense standard communications protocols, TCP/IP, for reliable data transport. Use of TCP/IP also allows the Butterfly processor to be accessed through gateways from other networks in the DARPA Internet.

Additional servers are being built to support remote debugging, to provide access to remote file servers, and to interface to application programs running on the front-end machines. Remote file access will allow users to read and write files on the front-end machines (Lisp machines, VAX and Sun UNIX systems). A remote procedure call facility will allow applications running on the front ends to communicate over the network with programs on the Butterfly system.

The lowest level of Chrysalis Operating System functions are provided by the kernel, which runs on each of the processors of the Butterfly system. The main functions of the kernel are process support, memory management, primitives for parallel processing and interprocess communication, I/O, and system configuration and initialization. Many Chrysalis kernel services are written in microcode on the PNC for speed; most of the rest are written in C. All Chrysalis kernel functions are called from the user's programs as subroutines or macroinvocations.

The Chrysalis Operating System provides processes and a scheduler that runs on each processor. This allows multiple processes to run on each node, as well as concurrently with processes on other nodes. The process scheduler uses a real-time scheduling algorithm and is mostly microcoded for fast process switching (context switches are typically done in about 120 μs).

Memory in the Butterfly system is accessed through a segmented virtual memory management system. Virtual address references generated by the MC68000 on a Butterfly processor node are 24 bits wide and map to 32-bit physical memory addresses in segments of up to 64 Kbytes. Memory mapping is done through a set of 512 segment attribute registers (SARs), which determine the location, length, and protection attributes of the mapped segments. Through this mapping scheme, programs running on a Butterfly processor node may reference local memory, remote memory through the Butterfly switch, I/O registers, and the microcoded kernel functions. Memory protection is enforced on a segment basis with kernel or user mode read, write, and execute attributes. Memory mapping is managed by the kernel and can be directly controlled by the application program. However, it is typically done by a higher level of system software, such as the application libraries.

Another level of memory management, called object management, is provided by the Chrysalis kernel. This level revolves around objects which are associated with areas of physical memory or special system data structures. The object system provides processor-independent identifiers (object handles) for areas of memory or for these system structures. The object handles can be passed between Butterfly processor nodes through the interprocess communication facilities. The objects can then be mapped into the virtual address space by other Chrysalis kernel calls, which allow the objects to be manipulated by programs.

The special system data structures that are accessible as objects provide many of the primitives for interprocess communication and for parallel process synchronization and locking. Two of the most important of these are the dual queue and the event. A dual queue is an interlocked data queue that can be used as a lock or for passing data, such as object handles, between processes. Events cause processes to be suspended or scheduled to run through wait and post operations. Together, dual queues and events are used to build higher level synchronization and interprocess communication mechanisms. In addition, a set of microcoded atomic memory operations (adds and logical operations) is provided to implement simpler forms of locks and other multiprocessor synchronization facilities.

The I/O functions include support for the Multibus and BIO hardware interfaces, a driver for the Multibus Ethernet controller, and asynchronous terminal support. Standards for writing new device drivers are also provided. The programmer's interface to the I/O system is the same set of I/O calls used to access the server functions (open, close, read, write, etc.).

PROGRAMMING THE BUTTERFLY PARALLEL PROCESSOR

The development of methodologies for programming parallel processors is currently an active research area. The Butterfly hardware and software environment represents an excellent vehicle for pursuing this research.

To date, two distinct approaches to programming the Butterfly parallel processor have seen widespread use. One approach is based on the notion of cooperating sequential processors, as described by Dijkstra, Hoare, and others. The second is the Uniform System approach developed at BBN Laboratories. The Uniform System approach emphasizes the tasks that comprise an application and tends to deemphasize the notion of processes. Combinations of these approaches are possible, as are completely different approaches.

The three following sections provide a brief review of the familiar cooperating sequential processes approach and describe the Uniform System approach in some depth.

To structure an application that uses cooperating sequential processes, the programmer decomposes the application into a moderately sized collection of loosely coupled processes which from time to time exchange control signals or data. The Chrysalis Operating System supports this programming methodology by providing a set of mechanisms for synchronizing and controlling processes and for interprocess communication. Using this method to program an application is similar to programming a multiprocess application for a uniprocessor. The principal difference is that the processes actually run concurrently on the multiple processors of the Butterfly parallel processor, whereas the processes must run in an interleaved fashion on a uniprocessor.

Communications and networking applications developed at BBN for the Butterfly use the cooperating sequential processes approach. The Chrysalis server functions also use this approach. In this case, the cooperating processes are the Chrysalis server processes and the application programs that call upon the servers. Communication between the application programs and the server processes occurs within Chrysalis subroutines. All of this is transparent to the application programmer.

The Butterfly hardware and Chrysalis Operating System comprise a foundation on which to build a variety of software structures. A teachable programming style for using this foundation has evolved from experiments with a wide range of software applications. This style, called the Uniform System approach, has proven to be particularly effective for applications containing a few frequently repeated tasks, e.g., much of scientific computing. It has also been successfully used in applications with less homogeneous task structures.

Beyond the usual concerns of programming, there are two key considerations specific to the Butterfly parallel processor: storage management and processor management. The goal of storage management is to keep all the memories in the machine equally busy, thereby preventing the slow-down that occurs when many processors attempt to access a single memory. The goal of processor management is to keep all the processors equally busy, thereby preventing the inefficiency that occurs when some processors are overloaded and others sit idle without work to do.

The Butterfly switch provides low-delay, high-bandwidth access to all of the memory in the machine. To help the programmer take advantage of this "common memory", the Uniform System implements a large shared memory for application programs and provides the means to spread application data uniformly across the memories of the machine.

The Chrysalis Operating System provides "memory mapping" operations that enable processes to manage their address spaces and, hence, the memory they access. Two or more processes can share memory by mapping the same memory segment.

In practice, memory sharing among processes is typically used in two quite different ways. One approach to programming the machine is to isolate processes from one another by mapping memory so that only a relatively small subset of each process address space is accessible to other processes. This subset can consist of up to 256 separate segments, can be changed at any time, and is often different for different groups of processes. This method facilitates debugging by limiting the number of processes likely to have touched a particular data structure.

The Uniform System uses an alternative approach, which is to share a single large block of memory by mapping the block into the address space of each process. This frees the application programmer from the need to manipulate memory maps and simplifies programming by implementing a large shared address space for application programs. Data that must be shared by two or more processors are allocated without regard to which processors will be using them. Of course the stack and variables local to individual processors are kept locally and, like code, are not fetched across the Butterfly switch.

Collectively, the memories of Butterfly processor nodes form the shared memory of the machine. This means the large shared memory an application program sees is physically implemented by a collection of separate memories. If all the shared data used by an application were to be physically located in a single memory, contention for that memory (as many processors attempt to access the data) would force the processors to proceed serially, thereby slowing program execution. Since the aggregate memory bandwidth of the machine is very large (10 Gbits/s for a 256-processor machine), slowdowns due to memory contention can be reduced by scattering application data uniformly across the physical memories of the machine. When many processors access data that have been scattered, their references tend to be distributed across the memories and can make use of the full memory bandwidth of the machine. The Uniform System library provides a memory allocator that scatters data structures in a way that allows straightforward addressing conventions. It also supports a set of more specialized techniques for cases where that allocator is either inappropriate or ineffective.

To summarize, the approach to memory management used by the Uniform System is based on two principles:

1. Use of a single large address space shared by all processes to simplify programming.
2. Scattering application data uniformly across all memories of the machine to reduce possible memory contention.

This memory management strategy has a cost, due both to the slower access to remote memory and to possible contention in the switch and at the memories. This cost is an increase in execution time, typically from 4 to 8%, and is due less to contention than to the slightly slower access. The benefit of this memory management strategy is that the programmer can treat all processors as identical workers, each able to do any application task since each has access to all application data. This greatly simplifies programming the machine.

The need to make certain operations on memory atomic is another aspect of memory management. This is not unique to parallel systems, it is also necessary in multiprogrammed uniprocessors. The Chrysalis kernel provides an extensive repertoire of primitive atomic operations. When the atomic operations required are more complex than these primitives provide, the primitives can be used to build simple locks that, in turn, can be used to implement arbitrarily complex atomic operations.

The most novel aspect of programming the Butterfly is processor management. This falls naturally into two separate parts: identification of the parallel structure inherent in a chosen algorithm, and controlling the processors to achieve the determined parallelism.

In many applications the parallel structure is both obvious and rich. In others, the structure is less clear and may require reworking the algorithm. Occasionally, an application will be inherently serial and cannot be structured to take advantage of parallel processing. There are, however, a few guidelines.

• Start with the best existing algorithm that implements the application. A Butterfly system with P processors can do no more than speed up an algorithm by a factor of P. Speeding up a poor algorithm may not overcome its inefficiencies. For example, it may take an N^2 parallel sort longer to run on a 128 processor Butterfly than it takes an N log N sort to run on a single 68000.

- Attempt to do the same number and kind of steps as the best algorithm. The order of steps in an algorithm can often be manipulated to achieve parallelism. This may involve adding logic in the form of simple locks to ensure the atomicity of selected operations.

- Look for parallel structure at all levels and in all sizes — the more the better. If necessary, it is usually relatively easy to aggregate small tasks at a later stage into larger more manageable sizes. It is often more difficult to divide a task at a later stage into smaller ones. For example, if an application requires fast Fourier transforms on a number of different channels, the programmer should plan to exploit both the parallelism inherent in an individual FFT and the parallelism due to different channels. The Butterfly parallel processor can work efficiently with individual tasks a few milliseconds in length; if necessary, it can work on tasks in the hundreds of microseconds. For shorter tasks, various overheads begin to interfere with good performance.

There are two strategies for determining the desirable number of concurrent operations to have at any stage in the processing. One strategy recommends a relatively static approach, using exactly P concurrent tasks for P processors. The other strategy recommends using many more than P tasks, typically an order of magnitude times as many or more. Both strategies attempt to deal with end effects — the processor idle time that occurs toward the end of a stage when some processors have finished and others are still working. The first approach minimizes the effect by explicit construction. Here the programmer attempts to manipulate the work so that all processors finish at approximately the same time. The second approach allocates tasks to processors dynamically in an attempt to balance the load. As a processor finishes a task, it is assigned the "next" task ready for execution. This approach relies on having a large number of tasks relative to processors to minimize end effects: some waiting occurs at the end of the problem, but this waiting is generally acceptable since it is small relative to the total program execution time.

The Uniform System encourages the dynamic approach. For many applications the dynamic approach is simpler and more reliable, since it is unnecessary to know in advance how long an individual piece of work will take. Furthermore, it is adaptable to varying numbers of processors and sizes of problems.

After the programmer has determined the processing that is to occur in parallel, he or she must then control the Butterfly parallel processor to make this happen. There are several ways to do this. The Chrysalis kernel provides a collection of relatively low-level operations for starting processes on various processors and for communicating among them. The Uniform System provides a higher level abstraction for managing the processors, one that is natural for a large class of applications.

The Uniform System treats processors as a group of identical workers, each able to do any task. To use the Uniform System, the programmer is required to structure an application into two parts:

1. A set of subroutines that perform various application tasks.
2. One or more "generators" that identify the "next" task for execution.

To illustrate this, consider matrix multiplication as an example. One way to structure a matrix multiplication program would be to write a routine that computes the dot product of a row and a column and to ensure that the routine for the dot product task gets called once for each element of the result matrix, using the appropriate row and column of the operand matrices as parameters.

Usually a well-designed program will be structured as a set of subroutines to improve program modularity, whether or not it is intended for parallel execution. Normally, there will be a subroutine per task type, each subroutine taking arguments that define individual tasks in terms of subsets of the program data to be operated on. To use the Uniform system, the programmer simply ensures that these subroutines correspond to the tasks he/she wants to do in parallel. In the case of the matrix multiplication example, there is a single task type, computing dot products, and corresponding to that task type, the dot product routine, whose row and column parameters specify particular tasks.

The second part of the application code comprises one or more subroutines able to identify the "next" task for execution. Such a subroutine is called a "generator", since its function is to generate tasks. In a serial program the generator function is usually embedded in the control structure of the program (e.g., do this, do that, then do ten of these). For parallel processing via the Uniform System, the programmer is expected to make generation of the next task explicit. For the matrix multiplication example, the task generator would be responsible for generating a call on the dot product routine for each element in the result matrix.

It is helpful to think of the generator concept in terms of three procedures and a task descriptor data structure. A generator activator procedure (GA) takes as parameters a worker procedure (W), a description of data (D) upon which work is to be done, and a task generation procedure (TG).

GA (W, D, TG)

The generator activator procedure (GA) first builds a task descriptor data structure (TD) that specifies the task generator in terms of the worker (W) procedure, the data (D), and the task generation (TG) procedure. It then "activates" the generator by making the task descriptor (TD) available to other processors. The processor that invoked the generator activator along with other available processors then use the task descriptor (TD) and the task generation procedure (TG) to make repeated calls on the worker procedure (W), specifying subsets of the data to work upon. Each call of the worker procedure (W) is a task. When the last task is done, the processor that called the generator activator procedure (GA) continues execution of its program, while the other processors that worked on the tasks look for something else to do. In the matrix multiplication example, the worker procedure is the dot product routine, and the data are the operand and result matrices. The dot product worker routine is called once for each combination of row and column index; these indices are stored in the task descriptor and are incremented indivisibly each time the task generation procedure is executed by a processor.

Conceptually, the generator notion is similar to the various "map" functions in the Lisp language. The unique thing about the Uniform System is that it achieves parallel operation by using processors as they are available to execute the various calls upon the worker procedure. Task generation and the processor management associated with it are implemented in a distributed fashion in the sense that each processor performing tasks participates in their generation.

Often the required generator is quite simple. In the matrix multiplication example, where a dot product is computed for every element in the result matrix, the generator can find the next task by incrementing row and column counters that identify the element in the result matrix to be computed next. Occasionally a generator must be more complex. A generator that selects the next node to process in an alpha-beta tree walk, for example, would rely heavily on using the most up to date information about the state of processing of the tree. Occasionally a generator will involve a simple queue, in which case it would operate much like a process scheduler found in many time sharing systems. The next task for execution would be the one at the front of the queue. In general, though, a large number of applications can be constructed from a small set of generators. The Uniform System library includes a collection of commonly used generators.

The Uniform System library provides a way to bind task generation procedures to worker procedures. The basis for this binding mechanism is a "universal" generator activator procedure. To use this universal generator activator procedure directly, application programs specify both a worker procedure and a task generation procedure. The library also includes a set of generator activator procedures that embody many commonly used task generation procedures. When an application program calls one of these "specific" generator activator procedures, it specifies only the worker procedure. The generator activator passes its associated task generation procedure and a task descriptor to the universal generator activator along with the worker procedure supplied by the application program.

Often an algorithm will require multiple, perhaps nested, instances of generators. As long as the algorithm does not depend upon the order of task generation between different generators, the programmer is free to make multiple calls to task generators to start the system working on all of them at once. If the algorithm does depend upon the order, the programmer must either provide a task generation procedure to properly answer the question about what to do next, or carefully manage the use of existing generator activator procedures to ensure the ordering requirements of the algorithm are met.

The Uniform System approach to processor management offers three important benefits.

1. The generator mechanism is very efficient. It is implemented using one process per processor in a way that ensures no unnecessary context swaps

occur. Each processor executes a tight loop consisting of "generate next task — execute next task." The programmer supplies both the task generation and worker procedures, usually finding an appropriate generator activator procedure in the library. Both the task generation and the worker procedures execute at the application level. As a result, once a generator gains control of a processor, the Chrysalis kernel need not be involved until the generator has exhausted all the work it knows how to find.

2. Programs that use the Uniform System task generation mechanism to exploit parallelism are insensitive to the number of processors. It is possible to debug programs on small configurations and run them on larger ones. Should an application grow to exceed the capacity of its current configuration, it can be moved without modification to a larger one. Perhaps more important, programs are able to run on "reduced" configurations, for example, one where processors have been removed for repair.

3. The load can be balanced dynamically. Whenever a processor becomes free, a generator identifies the next task to be executed. Since the task generation procedures are supplied by the application programmer, the task choice can be based on the current state of the computation and the requirements of the application.

PARALLEL VISION APPLICATIONS WITH THE BUTTERFLY COMPUTER AT THE UNIVERSITY OF ROCHESTER

We program the BPP in the C language augmented with system calls to Chrysalis,[9] the BPP operating system. Chrysalis is a protected subroutine library that provides operations for process management, interprocess communication, and memory management. Many of the most common operations are implemented in microcode in the PNC.

The user-level view of the memory architecture is defined by several Chrysalis system calls. These calls allow a programmer to establish a limit on the number of segments a process will use, allocate memory and assign it to a newly created memory object, initialize a SAR (to add a memory object to a process address space), and remove a memory object from an address space.

The interface provided by Chrysalis is too low level for convenient use by application programmers. We have found, however, that its primitive operations constitute a very general framework upon which to build efficient, higher level communication protocols and programming environments. We have built a number of the software packages on top of Chrysalis. Their success has depended on the fact that Chrysalis allows the user to manage processes, memory, and address spaces (explicitly), and provides low-level mechanisms for synchronization and communication.

As for larger operating system issues, Chrysalis has no support for paging, does not contain a file system, has a poor user interface, and is essentially a single-user system. Recent releases of the operating system allow the machine to be partitioned into multiple virtual machines, but do not support multiple users in a partition. Protection loopholes in both the hardware and in Chrysalis allow processes, with a little effort, to inflict almost unlimited damage on each other and on the operating system. It is inappropriate as a programming environment for most users because it requires too much expertise. It is also not a development environment; programs are developed on host VAXen or Suns and downloaded for execution on the BPP.

The BBN Uniform System library package[8] implements light-weight tasks that execute within a single global address space. The Uniform System interface consists of calls to create a globally shared memory, scatter data throughout the shared memory, and create tasks that operate on the shared memory. During initialization, the Uniform System creates a manager process for each processor which is responsible for executing tasks. A task is some procedure to be applied to a subset of the shared memory and is usually represented as an index into the shared memory and an operation to be performed on that memory. A global work queue (accessed via microcode operations) is used to allocate tasks to processors. Since each task inherits the globally shared memory upon creation, the Uniform System supports a very small task granularity.

The Uniform System is the programming environment of choice for most applications, primarily because it is easy to use. All communication is based on shared memory, and the mapping of tasks to processors is accomplished automatically. Moreover, the light weight of tasks provides a very cheap form of parallelism. Nevertheless, there are some disadvantages to using the Uniform System. The work queue model of task dispatching has led to an implementation in which tasks must run to completion. Spin locks must be used for synchronization. Waiting processors accomplish no useful work, implementation-dependent deadlock becomes a serious possibility, and programs can be highly sensitive to the amount of time spent between attempts to set a lock.[47] In addition, spin locks steal remote cycles, exacerbating the problem of memory contention.

The Uniform System also limits the amount of memory that can be shared. Like any Chrysalis process, a Uniform System manager can have at most 256 segments in its virtual address space. Since all managers have identical memory maps, only 16 Mbytes (out of a possible 1 Gbyte of physical memory) can actually be used by a computation under the Uniform System. Similarly, the Uniform System limits how the data are structured. One of the main advantages of a segmented address space is that memory segments can be allocated to logical quantities regardless of their size, since each segment is of arbitrary size. This is not a reasonable approach under the Uniform System because the number of available SARs, and hence memory segments, is severely limited. In order to be able to access large amount of memory, each segment must be large. Data must be structured on the basis of this architectural limit, rather than logical relationships. Large amounts of data, irregular in structure, must be allocated in regular patterns to economize on SARs.

Finally, the Uniform System model does not encourage the programmer to exploit locality. The Uniform System creates the illusion of a global shared memory, where all data are accessed using the same mechanisms. However, the illusion is not supported by the hardware, since access to individual words of remote memory is undesirable. Thus, in many applications, each task must work around the shared memory illusion by copying data from the "shared memory" into local memory, where they are processed and then returned to the shared memory.

As a result of its 3 years of research with the BPP, the University of Rochester, NY, has developed several tools, languages, and libraries that extend the functionality of the BPP and the model of computation that it supports. These include the Lynx language (comparable to ADA), a Modula-2 code generator, Chrysalis ++ (a C ++ encapsulation of Chrysalis calls), Structured Message Passing (a message-passing and process family library), Instant Replay (a debugging tool that allows nondeterministic computations to be repeated, even when debugging statements are added), the Elmwood operating system kernel, the BIFF image-processing library, and other packages. These are described in the literature and also in the *Butterfly Project Report* series available from the University of Rochester Computer Science Department. Many of the software packages are available through BBN-ACI at 10 Fawcett St., Cambridge, MA.

We use the Image File Format (IFF) as an internal standard for vision programs. The file format, access functions, and several utilities were written at the University of British Columbia by W. S. Havens and others, and are under license by UBC. Most IFF utilities are available in two forms: an interactive program that operates on image files, and a kernel library function used by the interactive version that can also be called by user programs. An important feature of standard IFF programs is that they can be used as filters, that is, they will read or write UNIX pipes if input and output file names are not supplied.

We have built BIFF,[35] an IFF-like facility on the BPP that uses extensive parallelism to realize substantial speed-up over the sequential UNIX implementation for low-level operations. The interactive user interface of the BIFF system resembles its UNIX counterpart. We decided to get maximal performance out of the Uniform System by running pipelined programs in sequence, giving each the full resources of the BPP while it was running. Our solution was to use part of the BPP's huge main memory as a virtual file system, storing BIFF images in named memory objects. (The operating system uses a similar trick to store executable programs.) This solution yields the simple programmer interface and pipe simulation we sought. Each stage of a pipeline must reinitialize the Uniform System and pay the associated time penalty. The image pseudofiles are unpacked into arrays upon transfer from the UNIX host to the BPP, so that successive pipeline stages need not perform this inherently sequential task. The arrays are stored "scattered", that is, an equal number of scan lines are stored in the local memory of each processor. They can thus be copied into and out of Uniform System memory in parallel without excessive memory contention.

As an example routine, consider a 3 x 3 Laplacian operator. The interactive driver program is called ifflapla(), and by convention its kernel program is called lapla(). ifflapla() interprets the command line, initializes the Uniform System, reads in the image header, annotation record, and image array to shared memory, calls lapla( ), updates the annotation record to say what was done, and outputs the resulting image pseudofile. lapla( ) may be called from any Uniform System application and is basically a FOR loop that runs in parallel to create each output scan line. The tasks of computing output scanlines are distributed over all available processors. For each output scanline, the three input scanlines contributing to it are block copied into local memory, the output scan line is computed and then copied into shared memory space. Such copying of shared information into local memory is usually necessary to avoid memory contention and is standard practice in Uniform System applications.

To date we have implemented the following utilities[37] iffload — load an IFF image onto the BPP from a UNIX host; iffstore — write an IFF image from the BPP to a UNIX host; iffhistogram — compute the greyscale histogram of an image; ifflapla — approximate the laplacian of an image; iffconvolve — convolve an image with an arbitrary integer mask; iffexpand — perform the so-called expand-white or expand-black operation for binary image smoothing; iffslice — given an image and a greyscale interval, prepare a binary image in which ones correspond to pixels that were in that interval in the original image; iffzerocr — given a signed image, prepare a binary image in which ones correspond to positive pixels with negative neighbors; and iffpyramid — given a greyscale image, produce a pair of scalespace pyramids.[14]

Timing of the computationally intensive programs indicates that on a 120-node BPP, all utilities get reasonably close to linear speed-up (processor utilization better than 80%) and are between 50 and 100 times faster than the UNIX versions of the same utilities running on a VAX 11/750.

Our implementation of the Hough transform illustrates some of the typical steps in removing bottlenecks. The first implementation was a control, without local copies of row pointers, for two-dimensional arrays. Speed-up becomes sublinear as soon as the input contains a significant number of votable features. This presumably reflects contention for the row pointer arrays for the look-up tables. Version Two is identical to Version One except that row pointers are copied into local memory. For a 10000-vote image and a worst-case image there is a dramatic improvement both in linearity and in the raw times (from 70 to 19 s), due to nearly halving the number of remote memory references. Version Three spawns 512 logical processes (one for each input scan line) to do the magnitude and phase look-up for each feature in the input once and for all. Their output is an array indicating the number of features in each line and an array of arrays of records containing the stored magnitudes and phases for each feature. The effect is to cut the time to 2.5 s for the 10000-vote image. Inner-loop improvements to Version Three (register variables, simplified expressions, and a faster but less safe Chrysalis macro for block copy) yielded the final Version Four . Speeding the loop was indicated by the linear speed-up of Version Three. If communication was the dominating factor, speed-up would fall off as the number of features increased. Simulations showed that a local cosine table (or fast transcendental function hardware) would speed the 10000-vote program

25% to 1.19 s. Finally, we simulated using much faster (68020-class) processors with the same switch hardware. This communication-limited result was 0.5 s for the 10000-vote image.

One natural structure of a program using the Uniform System on image analysis problems is to compute (in parallel) on each of many subimages, examine (in parallel) each of the boundary areas along adjoining subimages, and merge the data structures associated with features spanning the boundary. The key idea of this section is to cast the problem in terms of set manipulations, and then to take advantage of the properties of the set operations to coordinate the processing. Often the data structures to be merged may not be amenable to simultaneous update by many processors. This work follows an approach using the UNION-FIND algorithm that provides a straightforward method for organizing a computation to perform its data structure merging without excessive serialization.[13,37]

Rephrased in terms of set operations, the program structure above resembles

1. label: Compute on each of many subimages, labeling crucial features with set names obtained with NEWSET operation.
2. merge: Examine each of the boundary areas along adjoining subimages, and perform UNION operations on the sets that represent features that span the boundary. As a side effect of the UNION operation, update the feature data structures based on the structures being merged.
3. relabel: Use the FIND operation to find the root set of each of the original set names. Each UNION of distinct sets will have left one set representing the merged feature, and the other set subordinate to (and pointing to) it; relabeling can be done to eliminate references to the subordinate set names.
4. tally: Examine the root sets (either serially or in parallel) to do whatever per-feature processing is to be done.

A suite of BPP routines was built to embody these ideas — it implements the UNION-FIND algorithm, as described in References 3 and 46. This algorithm has the desirable property that the data structures it builds (inverted trees) are shallow and balanced, so tree traversals will be brief, reducing the likelihood of unnecessary contention.

A typical connected component problem is as follows. The input is a 1-bit digital image of size 512×512 pixels. The output is a 512×512 array of nonnegative integers in which pixels that were 0s in the input image have value 0; pixels that were 1s in the input image have positive values; two such pixels have the same value if and only if they belong to the same connected component of 1s in the input image.

A typical edge detection and border-following task is as follows. The input is an 8-bit digital image of size 512×512 pixels. Convolve the image with an 11 $\times$ 11 sampled Laplacian operator. (Results within 5 pixels of the image border are ignored.) Detect zero-crossings of the output of the operation, i.e., pixels at which the output is positive, but with neighbors where the output is negative. Such pixels lie on the borders of regions where the Laplacian is positive. Output sequences of the coordinates of these pixels that lie along the borders.

To construct a chain code for the borders in the image, the program performs a local operation for each edge-pixel to choose at most two adjacent edge-pixels as neighbors. Each pixel's neighbor selections must agree with the selections of its neighbors; selections that do not meet this criterion are also excluded. This thins adjacent edge operator responses and guarantees that multiple intersecting borders are treated as separate borders when they are tracked.

In this thinned edge image, border tracking reduces to connected component labeling, where the connections are dictated by the neighbor relation. This amounts to the use of single-pixel subimages, with merging of adjacent subimages where the neighbor relation holds. The update code executed during a UNION operation keeps track of the end points of assembled borders, taking advantage of the fact that the two-neighbor constraint causes merges to occur only at the end points of a partial border. After all the merging, the program has end point and neighbor information for each border. This is sufficient to trace out the border coordinates or construct a chain code. Writing out the border coordinates is a serial operation that was not timed in the results below. Correctness was verified by running the output through a drawing program to watch the borders get traced out one at a time.

The programs were tested on a natural 512×512 binary image. The programs are untuned and straightforward, lacking some rudimentary Uniform System optimizations (even the usual local caching of remote data structures), which should yield significant improvement in performance. Certain phases of processing exhibit serious serialization problems; some phases even execute more slowly with additional processors. All these effects can be traced to the use of the Uniform System memory allocation system, which uses an internal lock to serialize access to its data structures. The performance results for reading the input image, allocating temporary storage, initializing the UNION-FIND package, writing the output image, and freeing memory were affected. Subsequent work at Rochester has parallelized the allocator.[12,22] The remaining phases of execution exhibit improved performance with additional processors. The best results are seen in local operations: convolution, zero-crossing detection, labeling, and neighbor selection.

Only moderately successful performance is obtained in the merge phase, which is a global process in which its performance depends on the processors avoiding contention by working on different features as much as possible. Experiments showed that data structure contention was the major potential pitfall during the merge phase, which is the only phase in these programs when locking and busy-waiting occurs. Early versions of the program obtained little improvement in run time, because they set the worker tasks loose on adjacent parts of the image. This caused many workers to contend for the feature data structures related to one portion of the image, while other portions of the image were still completely unprocessed. The solution to this problem was to scramble the task assignments so that tasks generated close together in time would work on pieces of the image distant in space. A simple bit-reversal function applied to the row index sufficed to make performance begin to improve. There is still some room for improvement in the order of task processing, but a 512×512 image may not have enough features to use 120 processors efficiently.

All of the current applications of the UNION-FIND package have used either the row or the point as the basic subimage shape. This has had the advantage of making the computation/labeling step quite simple, since it need only compute over a one-dimensional image or a single pixel, and then merge along easy-to-describe boundaries. In general, the boundary merging task for a subimage only examines half of the boundaries, because other tasks examine the other half.

Processing (and locking) of data structures in this utility is on a per-feature basis. Through time, the features get larger and fewer, and efficient parallel operation is best achieved by maximizing the number of active structures. The order of performing unions in this algorithm can be determined by the user. An alternative is to work with regions organized in a fixed tree, such as a quad-tree. In both cases, smaller structures are coalesced into larger through time. In the quad-tree, data structure locking takes place at fixed boundaries, and merging happens in a fixed order that could also result in end effective task imbalance (when some quadrants are more active than others).

Active and Reactive Control Hierarchy (ARCH) is an evolving software system to control an observer moving through the world, along with (initially) simulations of the observer's transportation capability and the visual world. The project has three goals.

1. Build an example system that interacts with the world and that is made up of asynchronously operating processing and communicating, both through messages and shared memory.
2. Evaluate the Lynx language[41] as a system-building language for MIMD intermediate- and high-level computer vision on the BPP. Develop and suggest language and operating system extensions (e.g., the Psyche system[43]) to support enhanced models of computation.
3. Serve as a front-end (i.e., a data acquisition and processing facility) and back-end (i.e., a producer of effects in a real or simulated world) for vision and manipulation cognitive tasks, such as object recognition, planning (of motions, active vision tactics, and sensor positioning), and acting (real-time reactions to events beyond the power of the planner to anticipate or deal with).

ARCH is organized in a straightforward way,[45] reflecting the obvious functionality and communications between modules in a parallel, distributed system for sensing, navigation, planning, and action.

Active vision, or the direct manipulation of sensor parameters to aid perception, will be an important part of the ARCH testbed.[5,6] There are many scientific advantages to active vision,[4,7] and there is still work to be done in the low-level systems that implement such basic capabilities as fixation. Some relatively unexplored opportunities exist with foveal and peripheral vision.

Besides vision, ARCH raises several high-level issues; for instance,

1. Strategic planning: cognitive decision making, using metric representations of the world. Some standard problems, such as trajectory planning, fit this category. So also does planning viewpoints for efficient scene mapping or searching, and generally the use of quantitative representations of the world, the goals, and the resources in generating behavior.
2. Reactive planning: reflex- or skill-level reactions to real-time situations. This idea has undergone a renaissance, and we believe it to be important;[2,11,24] especially interesting issues are the learning of skills and the smooth knitting together of actions through time.
3. Representation of environment: development of a coherent and usable world representation using multiple active sensors and multiple viewpoints.[19,21]

ARCH is meant to evolve from a pure BPP implementation, featuring simulated world and sensing, to one that interacts with the world through camera control, ultimately achieved with a six-degree-of-freedom robot "body" as well as a three-degree-of-freedom robot "head". The sensors and effectors will bring other computers (Suns) and parallel pipelined video-rate image processing hardware into the picture. The initial implementation simplifies the problem to that of a mobile observer on a two-dimensional world. This phase of the project is called ART (Automated Road Trip).

ART is meant to include the modules and communications needed for a performing real-time system like ARCH, but to be simplified and simulated. Initially ART will simply be an observer that drives around on roads in flat landscape, using simulated proximity and vision sensors for navigation. ART will be gradually expanded to incorporate real sensors, effectors, and other computers. ART's world is planar, consisting of labeled areas. There are five types of labels for areas: ground, road, obstacle, sign, and landmark. A landmark or sign label has an associated identifier (a string or digit) that supports or allows look-up of its characteristics. For instance, a sign could have a road name, and a landmark a unique ID.

ART's transporter is a Newtonian vehicle with acceleration, deceleration, and direction controls. The world surface has friction. The camera is mounted above the transporter and has pan, tilt, and zoom controls. A rasterized version of the image is sent to low-level vision routines (to find boundaries between road and non-road, for example, or to "read" a road-sign). Since each pixel has its semantics built in, high-level visions such as road-sign reading and landmark recognition is automatic.

ART's navigator uses a road map that is a labeled digraph showing distances between road intersections, landmark locations, etc. A typical assignment is to get from the current location to another location.

ART's pilot is responsible for real-time decision-making and interpretation of sensor output. It makes requests of navigation and vision and responds to information volunteered by other modules. "Reflex"-level actions at the lowest level can take place before the pilot is aware, but it seems reasonable that the pilot be able to override certain reflexive behavior if it is predictable and wrong. In a real-time system, the pilot is the "reactive planner", who must arbitrate the (possibly conflicting) control directives from other modules. Resource allocation is a basic problem the pilot must confront. The advent of pipelined parallelism means that low-level vision no longer shuts down all cognitive activity, and the emphasis in active vision is on where to look, at what resolution, etc., in order to survive and achieve goals.

ART avoids most of the vision problems that ARCH will confront. Ultimately ARCH will have several visual outputs from different vision channels available for higher-level vision and reflexes in real time. Such a capability seems possible and is necessary to capitalize on the available parallelism. It raises several issues on the cooperation of multiple sensors[15] and will let us investigate interesting biologically motivated ideas, such as different channels for color, motion, and shape, or even different channels for "what/where" distinctions. The earlier levels of processing will be provided by the MaxVideo pipelined video-rate processor, which we hope to be able to configure flexibly through software libraries on the Sun-3, as well as to adjust incrementally by direct calls from simpler software on the BPP. Vision software that has already been written for the BPP may not be directly usable since it was built with the Uniform System. However it may well be possible to modify it for use in an enhanced computing environment that supports both message passing and shared memory.

The integration and dissemination (I&D) module collects sensory information and makes it available (in its variously digested forms) to other modules. It collects various directives, recommendations, and transporter control signals from the planner and navigator. It must be aware of reflex actions as well. The collected action-related information must be made available to the pilot for decision-making.

I&D is a practical necessity, functionally related to the Blackboards or action schedulers of other systems. It is also used for efficiency. Multiplexing data in Lynx (getting it to more than one process) involves sending a message (i.e., doing a Remote Procedure Call) to each process needing the data. A trade-off must be made between parallelism and message overhead, and the I&D facility is a flexible way to do the job. For instance, a scheme whereby the visual data is directly read from the world by parallel early vision processes is attractive, but perhaps impractical in a pure Lynx system. Thus, this use of the I&D is a symptom of a potential extension to the Lynx model.

We see special-purpose video-rate processors as the best current choice for front ends in image analysis applications. The MaxVideo system[32] is a flexibly configurable set of boards performing general image processing functions. Digitized video data flows through the boards generally at video rates, with the possibility of being stored or processed asynchronously along the way. We have used MaxVideo to track moving objects, to do temporal image differencing, and to perform arbitrarily large convolutions, and we are working to integrate the system with our Sun and Butterfly processors. Pipelined processors seem to offer the best cost/benefit ratio for low-level vision processing. We are designing an integrated parallel system with pipelined parallelism at the lower levels and MIMD parallelism at the higher levels.

Like other pipelined systems, such as the Warp processor,[25] MaxVideo is a complex system. The Apply programming model was a major step in making Warp usable for nonexperts. We perceive a need for a similar set of abstractions and uniform accessing methods for MaxVideo for the same reasons. This section outlines the hardware components and then considers a case study, the computation of optic flow, to see how the system could be used to implement a relatively complex low-level computation.

At Rochester we have nine MaxVideo boards; the DIGIMAX, ROISTORE-512, ROISTORE-2048, two VFIR-MKIIs, MAX-MUX, MAX-SP, FEATUREMAX, and EUCLID. Each board has a straightforward functionality but is made of subsystems that can be useful in themselves. Broadly, DIGIMAX is a frame grabber, with D-to-A and A-to-D capabilities and look-up tables. ROISTOREs are smart raster memories, which allow Regions of Interest (windows of the stored image) to be shipped through the system. VFIR performs up to 8×8 convolutions at video rates, MAX-MUX is a multiplexing board for manipulating data streams, MAX-SP is for pixelwise signal processing, FEATUREMAX does histogramming and reports the spatial location of feature tokens, and EUCLID is a general-purpose, signal-processing computer with local memory for data and programs, the operations of which are in general not at video rates.

Data flows through the system in pipelined streams. The MAXBUS is for digitized data, and the VME bus is used for communication with host computers. The source for a stream of data is either the original camera data from the DIGIMAX or is from one of the ROISTORE boards. The data can then split and go through many of the boards in sequence, eventually ending either at the DIGIMAX video outputs, the ROISTORE boards, or as extracted features in the FEATUREMAX.

It is possible to combine components from the various boards to expand the system's functionality. For instance, VFIR-MKII has an adder. The original purpose of the adder is to make cascading 8×8 convolutions easy so that larger convolutions can be done. However, this adder can be used to add a stream of 8-bit numbers to 16-bit ones by setting the convolution part of the circuit to do nothing (i.e., a mask with all 0s and one 1). Looking at the boards this way, our boards have two adders, two multipliers, three delay lines, three frame buffers, and eight look-up tables (LUTs).

There are a limited number of I/O ports to the MAXBUS on the MaxVideo boards, connected together by patchcord cabling. The ports impose a constraint on the generality of data flow, the impact of which we try to minimize. It is in fact possible to connect each board to many others by using a MAXBUS feature in a slightly nonstandard way.

Our goal is to reach a stable enough design using the MaxVideo so that one wiring system can be set up and all the common applications can run on that system without rewiring. We could then build a single general-purpose MaxVideo access program, which would simplify use of the hardware considerably. This program would allow the user to design a system simply by specifying the flow paths of data for each pass of a computation, with the sizes of ROIs and the widths of the data paths specified as needed. The program would automatically reconfigure the busses and route the signals along the desired paths, in parallel if enough boards are available and the data allows it, compensating for all board delays. It could also ease the problems of arithmetic precision and could precompute LUT contents to perform desired functions. Without stabilized MaxVideo wiring, a more complex program could also design the connections.

The derivation of the optic flow equations is familiar.[26] They require computing the gradient in x, y, and t, as well as the local averages of the flow velocities and some fairly complex equations:

$$u^{(k+1)} = u_{av}^k - f_x(P/D) \qquad\qquad v^{(k+1)} = v_{av}^k - f_y(P/D)$$

$$P = f_x u_{av} + f_y v_{av} + f_t \qquad\qquad D = K^2 + f_x^2 + f_y^2$$

These equations show that the estimate of optical flow at time $t = k + 1$ consists of velocity in the x direction (u) and in the y direction (v). These depend on the local averages of previous estimates of u and v as well as the x, y, and t gradients f_x, f_y, and f_t.

An important consideration is the precision needed in the computations. We shall consider 8-bit integer log values to be sufficient, but a careful study is required to determine what is really required. The limitation of the MaxVideo (and PIPE processor) to 8- and 16-bit integers is a decided drawback, compared, say, to the Warp processor, which uses floating point exclusively. The user must always be conscious of scaling factors.

The incoming data are a 512×512 array of pixels at a rate of 30 frames per second. The most straightforward method is to run the data through the MaxVideo hardware in pipelined mode. With the nine boards of our configuration, we shall see it take about six passes. At a rate of 30 ms per pass, this is $^6/_{30} = ^1/_5$ second per iteration.

Another approach is to send the raw data to the Butterfly microprocessor. We could expect a 68020 to process the points at a rate of 100 μs each at best (an informal estimate derived from extensive experience with 68020s), which means a single processor would take $512 \times 512 \times 100$ μs $= 25$ s per frame, which means that 128 processors allow a rate of $^1/_5$ second per iteration. This solution isn't feasible because of the transfer rate of the VME bus, but it indicates that for low-level vision, a large general-purpose processor configuration has approximately the same speed as a special-purpose, pipelined image processing system.

A good estimate for the speed of processing on the EUCLID boards is that it is three to four times faster than a 68020 for signal processing applications. Thus, it would require 32 EUCLID boards to get comparable performance with straight EUCLID processing, ignoring data transfer rates.

Thus the MaxVideo hardware is the fastest solution, and even that is six times slower than real time. Another tradeoff is to sacrifice resolution and speed-up the processing closer to real time. Using one of the VFIR-II boards to reduce the image to 128×128 pixels (i.e., $^1/_{16}$ the pixels of the full frame size) allows 6 passes over the data in a frame time (6 passes $\times$ $^1/_{16}$ frame time per pass = $^6/_{16}$ frame time per iteration). In fact, the system could almost handle 256×256 images in real time.

A multi-resolution solution is also possible, using a high-resolution "fovea" and a lower resolution "periphery". A system with a peripheral matrix of 128×128 and a foveal matrix of 64×64 has a data rate that can be processed in real time. The use of mixed resolution input is anticipated for the four-camera setup.

The following discussion assumes that the input signal is reduced to 128×128 before further processing takes place.

Pass 1 — Once this 128×128 image is ready, the actual processing can start. On the first pass the VFIRs are used to find f_x and f_y, if they are stored in one of the ROISTOREs, and MAX-SP is used to find the sum of their squares. Its multiplier is used to square f_x, 1 of the 16 banks of the LUT for squaring f_y, and the ALU to add them. This result is stored in the second ROI. Since VFIRs are used for the differentiation, a "gradient" template of up to 8×8 can be used.

Pass 2 — On the second pass, one VFIR is used to compute u_{av} from u. This is done with the Horn-Schunck averaging template:

$$\begin{bmatrix} 1/12 & 1/6 & 1/12 \\ 1/6 & 0 & 1/6 \\ 1/12 & 1/6 & 1/12 \end{bmatrix}$$

It is multiplied by f_x using MAX-SP and the result stored. At the same time, $f_x^2 + f_y^2$ is fed into the second VFIR with a unity convolution mask and added to a constant Λ^2 coming from one of the ROIs. This goes through the 16-bit LUT in MAX-MUX and is turned into a log value. This log value is only eight bits of log (appearing in the least significant bit); the most significant bit (MSB) is either 1 or -1, indicating the sign of the input (in this case the input is always positive, so it will be a 1). The 8-bit result D is stored in the ROISTORE in the MSB area corresponding to where the newer image is stored, offset by the correct delay so that Pass 4 can get both at once.

Pass 3 — One VFIR is used to compute v_{av} from v, the result is multiplied by f_y using MAX-SP, and that is added to $u_{av} f_x$ from the last pass in the ALU. This then goes to the second VFIR, where it is added to the first half of f_t, which is an average over a few pixels of the newest image. This result is stored.

Pass 4 — The result from Pass 3 has the second half of f_t, the average of the older image in this local area, subtracted from it in one of the VFIRs. This completes the P term. It is fed to the 16-bit MUX table to be converted to log, and the sign word is stored. D from Pass 2 is subtracted from the log value using the second VFIR board, and the antilog of the result determined using a bank of the MUX in MAX-SP and stored.

Pass 5 — The derivative f_x is multiplied by P/D and either added to or subtracted from u_{av}, depending on the sign bit from P in Pass 4, using MAX-SP. The result is the new u value and is stored.

Pass 6 — The derivative f_y is multiplied by P/D and either added to or subtracted from v_{av}, depending on the sign bit from P, using MAX-SP. The result is the new v value and is stored.

The last two passes don't use the VFIR boards and only require that results be stored in one of the ROISTOREs. This means they can be overlapped with the DOWN-SAMPLING stage. Thus, only four passes have to occur in alternation with grabbing the video frames, which means that more pixels or processing could be fit in without slowing the algorithm in pipelined mode.

Tables — This solution uses 2 of the 16 banks of the MAX-SP LUT for squaring and antilog tables, the 16-bit LUT of MAX-MUX to compute the logarithm, and 6 templates for the VFIRs (unit y, x gradient, y gradient, t averager, Horn-Schunck averager, and Gaussian averager).

Any complex algorithm on a small number of MaxVideo boards will require multiple passes to perform the required operations. Through clever resource allocation, some of these passes can occur in parallel on MaxVideo hardware. However, any algorithm too complex to run in one pass will accordingly be slower than real time if working with full-frame image data. Therefore, the image data must be down-sampled in order to achieve real-time processing.

From a programmer's point of view, there are as yet few abstractions of the data or processing in the MaxVideo system. A high priority capability, if the system is to be useful as a general tool and not just a custom-tailored, special-purpose device for each single application, is software that will help configure the boards to achieve their potential functionality. Despite their difficulties, pipelined image processors are cost-effective front ends to vision systems.

ART is specified so that parallel processing is necessary if the system is to operate in anything near real time. Since the paths of communication between the various components as described above are used relatively infrequently and carry small amounts of data (aside from the lower levels, which deal in data like velocity profiles and image data), it seems appropriate to accept the overhead of message passing (MP) over shared memory to gain the significant advantages in ease of synchronization afforded by MP systems. Furthermore, many of the individual pieces have several different more or less independent things occurring disjointly (the pilot can be alert to the navigator while controlling the transporter, for instance). Thus, it seems appropriate to use multiple threads of control within the components. The Lynx-distributed programming language is the obvious local choice for implementation, since it was designed with these capabilities in mind[41] and is available on the BPP.

Lynx was designed for use on a more loosely-coupled (no remote memory access [NORMA]) multiprocessor than the BPP. The significantly lower communication latency available in the BPP run time implementation of Lynx (using the switch for message passing) is thus an important aspect of the language for real-time applications. Lynx is able to pass messages over links on the BPP much faster than is possible with traditional local area nets and communication protocols.[42]

Each of the independent pieces of ART (e.g., the Navigator, Pilot, I&D, reflexes, vision processes) is implemented as one or more Lynx processes and has its own dedicated set of Butterfly processor nodes (one node per Lynx process). Each interconnection between components is implemented by a Lynx link.

The Lynx thread concept (which is similar to traditional co-routines, including semantics ensuring that only one thread within a process runs at a time and which permits thread switches only at well-known, predefined points) was convenient for building components involved in multiple conversations with other components simultaneously. When a conversation starts, one Lynx thread is created to represent it in each of the conversing components. When one of the components is performing some operation demanded by the conversation, the corresponding thread in the other component is blocked, awaiting the reply. Thus, we are able to encapsulate the state of the conversation in the blocked partner simply by the program counter and stack of the blocked thread, and we are saved from the (sometimes quite painful) task of explicitly restoring the state of the conversation on each end of every exchange.

Furthermore, Lynx provides for the implicit creation of threads. That is, one is able to specify that when some entity makes a certain request on a certain link, a thread should be created to service the request (which may be a simple query, in which case the thread will die quickly, or could be the beginning of a long conversation). This is unlike many MP systems (like Mach), in which each message must be received explicitly, and if a separate thread is desired, it must also be created explicitly.

Lynx presents a distributed model of computation, opposed to a parallel model, which means the design of ART had to be limited in ways that have suggested extensions to the language or to its available libraries. Processes may only communicate via remote procedure call (RPC), and not shared memory; the RPC latency in the best case is too high to do some things that shared memory would provide, such as allowing reflexes to examine vehicle state information effectively instantaneously. Doing this in Lynx would require a reflex to block waiting on an RPC in the middle of its action. The reply could take many tens of milliseconds, due to latency and serialization at the other end.

Threads within a process may not run concurrently. Thus, an RPC to a process (either to an extant thread or to an implicitly created one) at best must wait for whatever thread is running in that process to block before it is serviced. In the average case, the call must wait for all of the other currently waiting calls to block. This model of FIFO "fairness" is not unusual, but real-time systems need different schedulers, such as deadline scheduling and priorities.[40]

SMP is an example of a locally written research system to prove the concept of message passing as a viable model of parallelism on the BPP. Locally developed software packages such as Chrysalis++, SMP, the X library, and Lynx have enormous potential to make the programmer's life easier. However, in a research institution it is not always economical to bring such tools to production standard, and often the developer has moved on to other things, meaning no manpower is available for systems support. Rochester has quite a good staff/ researcher ratio, but the number of systems is larger. Thus, the luxury of using the BPP's inherent flexibility to provide varying models of computation is often bought at some cost in programmer frustration.

We have been experimenting with different programming abstractions[28] and visual applications for MIMD computers. We easily implemented the seven problems in the DARPA Benchmark[39] (besides a few BPP utilities, they included subgraph isomorphism, visibility calculations, geometrical constructions, and minimal cost path[12]). This work involved four different programming environments. Some of our general conclusions follow.

Many low- and intermediate-level vision operations are best performed on special-purpose architectures. Among the most cost effective today are the parallel pipelined computers for video-rate image analysis. The BPP seems to offer a promising alternative to a more loosely coupled system as the high-level component in a hierarchical parallel system.

Several easy-to-learn and complementary models of computation and debugging tools exist for the BPP. A programmer starting with only a knowledge of standard sequential programming can now produce good parallel programs in a day or two. Achieving excellent performance is still something of a black art.

While programming environments that emphasize one or another parallel model are available now, a single environment that gives the programmer access to a mix of computational models is not. For vision it would be useful. At Rochester, the Psyche project[43] has the goal of providing unified support for a variety of parallel computation models, including both shared memory and MP.

Two features of the Uniform System are responsible for its success as an environment for low-level image processing. First, the process model handles load balancing automatically and dynamically, so that the user does not need to know how many processors are working on the problem. Second, shared memory makes it possible for very light-weight processes to be used, even though each process requires a huge amount of state information (typically one or more images).

For serious work in the area of scientific computation and image analysis, hardware floating point capability is indicated.

Large-scale shared-memory multiprocessors are practical. We have achieved significant speed-ups (often almost linear) using over 100 processors on a range of applications including connectionist network simulation,[23] game-playing, Gaussian elimination,[28,29] parallel data structure management,[22] and various computer vision and graph algorithms.[12,13,17,35,36] We have also discovered that many interesting effects become obvious only when 100 processors or more are in use.

Locality of reference is important, even with shared memory. Although the BPP allows each processor to access the memory of others, remote references are five times slower than local references. This disparity is not so great as that found in local-area networks, where two or three orders of magnitude are common, but it cannot be ignored without paying a substantial performance penalty. Any measurable difference between local and remote access time requires the programmer to treat the two differently; caching of frequently accessed data is essential.

An efficient implementation of a shared name space is valuable even in the absence of uniform access time. The primary advantage of shared memory is that it provides the programmer with a familiar computational model. Even when nonuniform access times warp this model by forcing the programmer to deal explicitly with caching, shared memory continues to provide a form of global name space. Programs can pass pointers and data structures containing pointers without explicit translation. The Uniform System programmer uses the shared memory as a global name space through which data can be passed from one local memory to another.

Contention has the potential to impact performance seriously. Remote references on the BPP can encounter both memory and switch contention. The potential for switch contention was clearly anticipated in the design of the BPP hardware and has been rendered almost negligible.[38] On the other hand, the potential for memory contention appears to have been underestimated, since remote references steal memory cycles from the processor containing the memory. Only 1 processor can issue local references to a given memory, but over 100 processors can issue simultaneous remote references, leading to performance degradation far beyond the nominal factor of 5 delay. The careful programmer must organize data not only to maximize locality, but also to minimize contention.

It is difficult to exercise low-level control over parallelism without accepting explicit control of other resources as well. Programmers use a multiprocessor for performance gains and therefore must maximize the (true) parallelism in an application program. Since it is difficult to anticipate the needs of each application, a parallel programming environment will usually provide low-level mechanisms for mapping processes to processors. Unfortunately, in allowing the programmer to control parallelism (and the corresponding processes), the environment will often force the programmer also to manage other resources. For example, the programmer may be required to manage address spaces explicitly in order to colocate a process and its data. All of the parallel programming environments on the BPP couple the ability (or inability) to manage parallelism with the ability (or inability) to manage memory. Chrysalis allows the programmer to create a process on any BPP node, but it also requires the programmer to explicitly manage shared memory. SMP does not require the user to explicitly manage the address space of a process, but it does allocate processes to processors using a fixed-allocation algorithm, which can be inappropriate for some applications. A better balance between flexibility and ease of use must be found.

The programming environment must support multiple programming models. We have implemented many different applications using an assortment of operating systems, library packages, and languages. Empirical measurements demonstrate that nonuniform memory access time (NUMA) machines like the BPP can support many different programming models efficiently and that no single model can provide optimal performance for all applications. Moreover, subjective experience indicates that conceptual clarity and ease of programming are maximized by different models for different kinds of applications. Some large applications may even require different programming models for different components; therefore, it is also important that mechanisms be in place for communication across programming models. These concerns form the motivation behind the Psyche operating system.

Better monitoring and debugging tools are essential. The lack of such tools contributes dramatically to program development time and is probably the most frequently cited cause of frustration with parallel programming environments. Performance is paramount in multiprocessors, yet few general tools exist for measuring performance. Bottlenecks such as memory or switch contention are difficult to discover and must usually be measured indirectly. Single-process debuggers cannot capture parallel behavior, and performance monitoring and debugging tools for distributed systems[27,33,34] have yet to inspire similar tools for multiprocessors. The problem is particularly acute in NUMA machines, since they lack a shared communication medium that could facilitate monitoring. Only recently has some progress been made in monitoring and debugging tools for shared-memory multiprocessors.[31,44] Architectural support for program monitoring should be considered in designing future machines.

Amdahl's law is extremely important in large-scale multiprocessors. Serial program components that have little impact on performance when 30 processors are in use can lead to serious bottlenecks when 100 processors are in use. Serial access to a large file, for example, is unacceptable when 100 processes are available to process the data. Serial access to other resources (such as process templates in Chrysalis) limits the utility of those resources in highly parallel programs. We are currently exploring issues in the design of a parallel file system that scatters data across multiple disks on multiple processing nodes.[18]

Architectural variety inhibits the development of systems software that will port to new machines. There are many new systems problems that are introduced by large-scale parallelism, problems that have not been completely solved in any general-purpose operating system. Substantial investments in software development are currently required for every new machine. In many cases it may even be difficult to develop a production-quality operating system fast enough to make truly effective use of a machine before it becomes obsolete. The problem is less severe in the sequential computer world, since uniprocessors tend to resemble one another more than multiprocessors do. While an operating system like UNIX can make effective use of an impressive variety of conventional sequential computers, simply porting UNIX to the BPP would not provide fine-grain parallelism, cope effectively with NUMAs (the so-called "NUMA problem"), or address a host of other issues. The emergence of Mach may improve matters significantly, but its effectiveness for NUMA architecture has yet to be demonstrated.

Programming environments are often more important than processing speed. Many application programmers in our department who could exploit the parallelism offered by the BPP continue to use Sun workstations and VAXen. These programmers have weighed the potential speed-up of the BPP against the programming environment of their workstation and found the BPP wanting. The only applications truly committed to the BPP are those that simply could not be run otherwise. New processors, switching networks, or memory organizations will not change this fact. The most important task ahead for the parallel programming community is not the development of newer and bigger multiprocessors, but rather the development of programming environments comparable to those available on sequential computers.

REFERENCES

1. **Accetta, M., Baron, R., Bolosky, W., Golub, D., Rashid, R., Tevanian, A., and Young, M.,** Mach: a new kernel foundation for UNIX development, in Proc. USENIX Technical Conf. and Exhib., June 1986.
2. **Agre, P. E. and Chapman, C.,** Pengi: an implementation of a theory of activity, in Proc. AAAU-87, 1987, 268.
3. **Aho, A. V., Hopcroft, J. E., and Ullman, J. D.,** *The Design and Analysis of Computer Algorithms*, Addison-Wesley, Reading, MA, 1974.
4. **Aloimonos, J., Weiss, I., and Bandopadhay, A.,** Active vision, in Proc. 1st Int. Conf. on Computer Vision, London, June 1987, 35.
5. **Ballard, D. H.,** Eye Movements and Spatial Cognition, TR 216, Computer Science Department, University of Rochester, NY, November 1987.
6. **Ballard, D. H., Brown, C. M., Coombs, D. J., and Marsh, B. D.,** Eye movements and computer vision, *Computer Science and Engineering Research Review*, Computer Science Department, University of Rochester, NY, October 1987.
7. **Bandopadhay, A.,** A Computational Study of Rigid Motion Perception, TR 221, Ph.D. thesis, Computer Science Department, University of Rochester, NY, December 1986.
8. BBN Laboratories, The Uniform System Approach to Programming the Butterfly® Parallel Processor, Version 2, BBN Report 614, BBN, Cambridge, MA, June 1986.
9. BBN Advanced Computers Incorporated, Chrysalis® Programming Manual, Version 3.0, BBN, Cambridge, MA, April 1987.
10. BBN Advanced Computers Incorporated, Inside the Butterfly Plus, BNN, Cambridge, MA, October 1987.
11. **Brooks, R. A.,** A robust layered control system for a mobile robot, *IEEE J. Robot. Automat.*, 2, 14, 1986.
12. **Brown, C. M., Fowler, R. J., LeBlanc, T. J., Scott, M. L., Srinivas, M., Bukys, L., Costanzo, J., Crowl, L., Dibble, P., Gafter, N., Marsh, B., Olson, T., and Sanchis, L.,** DARPA Parallel Architecture Benchmark Study, BPR 13, Computer Science Department, University of Rochester, NY, October 1986.

13. **Bukys, L.,** Connected Component Labeling and Border Following on the BBN Butterfly Parallel Processor, BPR 11, Computer Science Department, University of Rochester, NY, October 1986.
14. **Burt, P. J.,** Fast filter transforms for image processing, *Comput. Graph. Image Process.*, 16, 20, 1981.
15. **Chou, P. B. and Brown, C. M.,** Probabilistic information fusion for multi-model image segmentation, in Proc., Int. Joint Conf. on Artificial Intelligence, Milan, Italy, August 1987.
16. **Chou, P. B., Brown, C. M., and Raman, R.,** A confidence-based approach to the labeling problem, in Proc., IEEE Workshop on Computer Vision, November 1987.
17. **Costanzo, J., Crowl, L., Sanchis, L., and Srivinvas, M.,** Subgraph Isomorphism on the BBN Butterfly Microprocessor, BPR 14, Computer Science Department, University of Rochester, NY, October 1986.
18. **Dibble, P. C., Scott, M. L., and Ellis, C. S.,** Bridge: a high-performance file system for multiprocessor systems, in 8th Int. Conf. on Distributed Computing Systems, November 1987, submitted.
19. **Durrant-Whyte, H. F.,** Consistent integration and propagation of distributed sensor observations, *Int. J. Robot. Res.* 6, 3, 1987.
20. **Durrant-Whyte, H. F.,** Integration, Coordination and Control of Multi-Sensor Robot Systems, MS-CIS-86-67 and Ph.D. thesis, Department of Computer and Information Science, University of Pennsylvania, Philidelphia, August 1986.
21. **Durrant-Whyte, H. F.,** Integration of disparate sensor observations, in Proc. Int. Conf. on Robotics and Automation, 1986, 1464.
22. **Ellis, C. S. and Olson, T. J.,** Parallel first fit memory allocation, in Proc. IEEE Int. Conf. on Parallel Processing, August 1987, 502..
23. **Fanty, M.,** A Connectionist Simulator for the BBN Butterfly Multiprocessor, TR 164: BPR 2, Computer Science Department, University of Rochester, NY, January 1986.
24. **Firby, R. J.,** An investigation into reactive planning in complex domains, Proc. AAAI-87, 1987, 202.
25. **Hamey, L., Webb, J., and Wu, I.,** Low-level vision on Warp and the Apply programming model, in *Parallel Computation and Computers for Artificial Intelligence*, Kowalik, J., Ed., Kluwer Academic, Dordrecht, 1988, in press.
26. **Horn, B. K. P., and Schunck, B.,** Determining optical flow, Artif. Intell., 17, 1–3,185.
27. **Joyce, J., Lomow, G., Slind, K., and Unger, B.,** Monitoring distributed systems, *ACM TOCS*, 5(2), 121, 1987.
28. **LeBlanc, T. J.,** Problem decomposition and communication trade-offs in a shared-memory multiprocessor, in *Proc. Numerical Algorithms for Parallel Computer Architectures Workshop*, Springer-Verlag, New York, 1987.
29. **LeBlanc, T. J.,** Shared memory versus message-passing in a tightly-coupled multiprocessor: a case study, in *Proc. Int. Conf. on Parallel Processing*, August 1986, 463. (Expanded version available as BPR 3, Computer Science Department, University of Rochester, NY, January 1986.)
30. **LeBlanc, T. J., Gafter, N. M., and Ohkami, T.,** SMP: a Message-Based Programming Environment for the BBN Butterfly, BPR 8, Computer Science Department, University of Rochester, NY, 1986.
31. **LeBlanc, T. J. and Mellor-Crummey, J. M.,** Debugging parallel programs with instant replay, *IEEE Trans.*, C-36, 4, 471, 1987.
32. *MaxVideo Software Primitives*, Datacube Doc. #UM00002, Datacube, Inc., Peabody, MA, 1986.
33. **Miller, B. P., Macrander, C., and Sechrest, S.,** A distributed programs monitor for Berkeley Unix, *Software Practice Experience*, 16(2), 183, 1986.
34. **Miller, B. P., and Yang, C. Q.,** IPS: an interactive and automatic performance measurement tool for parallel and distributed programs, in *Proc., 7th Int. Conf. on Distributed Computing Systems*, September 1987, 482.

35. **Olson, T. J.,** An Image Processing Package for the BBN Butterfly Parallel Processor, BPR 9, Computer Science Department, University of Rochester, NY, August 1986.

36. **Olson, T. J.,** Finding Lines with the Hough Transform on the BBN Butterfly Multiprocessor, BPR 10, Computer Science Department, University of Rochester, NY, August 1986.

37. **Olson, T. J., Bukys, L., and Brown, C. M.,** Low-level image analysis on an MIMD architecture, in Proc., First IEEE Int.. Conf. on Computer Vision, London, June 1987, 468..

38. **Rettberg, R., and Thomas, R.,** Contention is no obstacle to shared-memory multiprocessing" *CACM,* 29 (12), 1202, 1986.

39. **Rosenfeld, A.,** A report on the DARPA Image Understanding Architectures Workshop,in Proc., DARPA Image Understanding Workshop, Morgan Kaufmann, February 1987, 298.

40. **Schwan, K., Bihari, T., Weide, B. W., and Taulbee, G.,** High-performance operating system primitives for robotics and real-time control systems, *ACM Trans. Comput. Syst.,* August 1987.

41. **Scott, M. L.,** Language support for loosely-coupled distributed programs, *IEEE Trans. Software Eng.,* SE-13 (Special Issue on Distributed Systems), 1, 88, 1987; TR 183, Computer Science Department, University of Rochester, NY, September 1986.

42. **Scott, M. L. and Cox, A. L.,** An empirical study of message-passing overhead, in *Proc., 7th Int. Conf. on Distributed Computing Systems,* Berlin, September 1987, 536; BPR 17, Computer Science Department, University of Rochester, NY, December 1986.

43. **Scott, M. L. and LeBlanc, T. J.,** Psyche: A General-Purpose Operating System for Shared-Memory Multiprocessors, BPR 19: TR 223, Computer Science Department, University of Rochester, NY, August 1987.

44. **Segall, Z. and Rudolph, L.,** PIE: a programming and instrumentation environment for parallel processing, *IEEE Software,* 2, 6, 1985.

45. **Stentz, A. and Goto, Y.,** The CMU navigational architecture Proc. DARPA Image Understanding Workshop, February 1987, 440.

46. **Tarjan, R. E. and van Leeuwen, J.,** Worst-case analysis of set union algorithms, *J. ACM,* 31 (2), 245, 1984.

47. **Thomas, R.,** Using the Butterfly to Solve Simultaneous Linear Equations, Butterfly Working Group Note 4, BBN Laboratories, Cambridge, MA, March 1985.

10 SEQUENT

OVERVIEW

To provide high performance, reliability, and extensibility, Sequent has developed a shared-memory, multiprocessing architecture that employs the following elements:

- A parallel architecture that uses multiple industry-standard microprocessors. This architecture affords a wide range of computing power in a single architecture and a potential to exploit advances in microprocessor design.
- The UNIX operating system, for its power, popularity, flexibility, and extensibility. The UNIX operating system is becoming a standard for academic and commercial computing worldwide.
- Standard interfaces such as VMEbus, MULTIBUS, SCSI, and Ethernet, because of their widespread acceptance and potential to support high-performance applications.

To combine these elements, Sequent developed a high-bandwidth bus; an extensible, high-performance memory system; a dual-channel disk controller; custom VLSI support circuitry; diagnostic software; operating system enhancements; and parallel programming tools.

In the Symmetry system architecture, all CPUs operate on a peer basis, executing a single copy of the operating system executive, or "kernel." There is no designated "master" CPU except during system start-up and diagnostic operations. Any process (program) in any state can execute on any CPU.

The following characteristics distinguish the Symmetry system architecture from other types of multiprocessors or "multimicro" architectures:

- *Tightly coupled.* All CPUs share a single pool of memory to enhance resource sharing and communication among different processors.
- *Common bus.* All CPUs, memory subsystems, and I/O interfaces plug into a single high-speed bus. This feature simplifies the addition of CPUs, memory, and I/O bandwidth.
- *Symmetric.* Within a system, all CPUs are identical, and any CPU can execute both user code and kernel code.

- *Transparent.* Programs written for a uniprocessor system can run on a Symmetry system without modifications for multiprocessing support. CPUs can be added or removed without modifying the operating system or user applications.
- *Dynamic load balancing.* CPUs automatically schedule themselves to ensure that all CPUs are kept busy as long as there are executable processes available.
- *Shared memory.* An application can consist of multiple instruction streams, all accessing shared data structures in memory.
- *Hardware support for mutual exclusion.* To support exclusive access to shared data structures, the system includes a user-accessible hardware locking mechanism.

Symmetry multiprocessor systems improve performance of ordinary sequential applications because the availability of multiple CPUs to handle work in the multiuser environment increases system throughput. In a system with only one CPU, but with multiple active processes, the processes wait in line for a slice of the CPU time. When the processing load is light, the system performance may be adequate. However, as the processing load increases, system performance can degrade quickly as each process receives a smaller portion of the CPU time. The CPU becomes burdened while constantly switching from one process to the next.

With the Symmetry system architecture's symmetric multiprocessing, adding CPUs is like adding clerks at an airline check-in counter — the waiting line moves faster the more you add. (See Figure 1) When there are more CPUs than executable processes, the waiting line is empty. As the processing load increases, several CPUs are available to service the increasing number of processes. Each process is given more time with a CPU, each process spends less time waiting, and the system spends less time switching from one process to the next (*task switching*). The result is dramatically increased performance.

The parallel nature of the Symmetry system architecture is transparent to sequential programs, and most programs can be ported from other systems with little or no change to the source code.

Programs can be automatically or manually tailored to use multiple CPUs simultaneously and thus obtain a dramatic increase in *execution speed.* Parallel applications can be structured to adapt themselves to the number of CPUs present in the system. Thus, the number of CPUs is simply a tuning parameter; the same parallel application can run on systems of different configurations without software modifications.

In Sequent systems, a single application can consist of several closely cooperating processes. Each process can run on a separate CPU, decreasing the run time.

A surprisingly large number of applications can be converted from sequential to parallel algorithms and run in this parallel fashion. The result is a linear or near-linear performance improvement as more CPUs are dedicated to the task. Additionally, a sequential program can often be "parallelized" automatically by

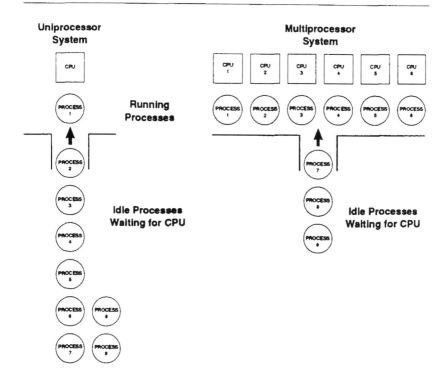

FIGURE 1.   Uniprocessor and multiprocessor system.

using a code optimizer, or manually by modifying only a small fraction of the program code.

Note that the Symmetry system architecture is designed to support both general-purpose, multiuser environments and dedicated, compute-intensive parallel applications. Parallel programs and ordinary sequential programs can run simultaneously on the same Symmetry system.

Computer system requirements are subject to rapid change under the pressure of competitive and technological trends and increasing workloads. Consequently, Sequent computer systems are configurable and adaptable to changing requirements; they use standard I/O interfaces and can be expanded through the addition of software, CPUs, memory, or peripheral devices, and through incorporation in local- or wide-area computer networks.

To provide the high bus bandwidth needed to support multiple CPUs, Sequent uses a high-performance data bus tailored to multiprocessing. The 64-bit system bus carries data among the system's CPUs, memory subsystems and peripheral subsystems, and supports pipelined I/O and memory operations. The bus can transfer variable-size data packets and can achieve an actual data transfer rate of 53.3 Mbytes/s.

To allow up to 30 Intel 80486 or 80386 CPUs to run at full speed while sharing the system bus, the system must minimize bus traffic. To accomplish this, each

CPU uses its own cache memory. These 512-kbyte (128 kbyte for 80386 CPUs), two-way, set-associative caches allow each processor to operate most of the time within its own memory environment, reducing the frequency of bus access to system memory. VLSI logic maintains consistency between copies of data in CPU caches without requiring CPUs to write system memory when shared data is modified.

The System Link and Interrupt Controller (SLIC) is a VLSI component that manages the control of multiple processors. Each CPU and each board connected to the system bus has its own SLIC chip. SLICs communicate with each other by means of the SLIC bus. Together, the SLICs manage system initialization, interprocessor communication, distribution of interrupts among CPUs, and diagnostics and configuration control.

To provide a disk architecture that can keep up with a high-performance multiprocessor architecture, these computers use a dual-channel disk controller, or DCC. The DCC can control up to eight storage module device-extended (SMD-E) disk drives. Each DCC provides two independent data channels to the system bus, with up to four disks on each channel. The two channels can transfer data simultaneously. The DCC manages overlapped seeks, so that all disks connected to the DCC can seek simultaneously.

Sequent's Symmetry family of multiprocessor computers includes six models: the S16, S81, S2000/200, S2000/400, and S2000/700. Architecturally, all models within the Symmetry 2000 line are equivalent, as are all models within the Symmetry line: succeeding models simply accommodate larger numbers of CPUs, disks, terminal multiplexors, and other peripheral devices. Table 1 lists some of the features and expansion capabilities of the family.

THE MEASURED PERFORMANCE OF PARALLEL DYNAMIC PROGRAMMING IMPLEMENTATIONS — A SEQUENT USER REPORT

One focus of research is to understand in detail the performance of algorithms on shared memory parallel machines. By means of careful measurements we are exploring how well certain classes of algorithms can be parallelized, and we are identifying the bottlenecks that stand in the way of achieving the goal of linear speed-up.

Our approach is to take an algorithm and develop several different parallel implementations of it. We then take detailed timings of each implementation to attempt to gain a complete understanding of performance. We currently have a 20-processor Sequent Symmetry.

In this report we consider dynamic programming algorithms. Dynamic programming is an important technique with a wide range of applications. We consider the class of dynamic programming algorithms that involve the computation of the entries of a k-dimensional array where the array locations can be ordered in such a way that the value at each location is a function of locations that have already been computed.

TABLE 1
Symmetry Family Configuration Options

System resource	Configuration options[a]
CPUs	From 2 to 30 Intel 80486 or 80386 32-bit CPUs
System memory	From 16 to 384 Mbytes
Disk storage	From 150 Mbytes to 85.4 Gbytes
Tape drives	Up to four $1/_2$-in., reel-to-reel tape drives
MULTIBUS interface	Up to 4 MULTIBUS subsystems, 44 slots (for tape drives, terminals, printers, communication devices, custom devices)
VMEbus interface	Five VMEbus slots on S16 and S2000/200 systems or peripheral expansion (for terminals, communication devices, custom devices)
Single-ended SCSI connections	Up to four single-ended SCSI ports for disk and tape drives
Differential SCSI connections	Up to six differential SCSI ports for disk and tape drives
Local-area networks	Up to four Ethernet interfaces supporting Ethernet 1 and Ethernet 2 of IEEE 802.3
Synchronous data communications	Up to eight RS-232-C synchronous ommunication lines (which can support X.25)
SNA- and BSC-compatible ports	Up to four ports that support SNA and binary synchronous communications (BSC) protocols
Asynchronous ports	Up to 256 directly connected asynchronous RS-232-C lines (additional terminal connections are possible using the Ethernet and X.25 facilities)

[a] Note that not all system configurations support all options.

We consider two problems that can be solved by dynamic programming and give several parallel implementations for each problem. Our implementation techniques are applicable to other dynamic programming problems of two or more dimensions. The two problems that we consider are to find the longest common substring of a pair of strings, and to compute the solution of a multiple-class queueing network performance model.

In our study we have several very different implementations of parallel dynamic programming. We have measured their performance in detail and have an understanding of their relative performance and their sources of slowdown.

The two problems that we consider are the longest common substring (LCS) problem and the solution of a queueing network model. These problems both involve computations over grids, but they differ in the dimensionality of the grid and the complexity of the computation performed at grid points.

The LCS problem: given a pair of strings A and B, find a string C of maximum length that is a substring of both A and B. The standard sequential algorithm[6] for finding the LCS of strings $a_1a_2...a_n$ and $b_1b_2...b_m$ constructs an $n \times m$ matrix A where $A_{i,j}$ gives the length of the LCS of the strings $a_1a_2...a_i$ and $b_1b_2...b_j$. The value of $A_{i,j}$ is a simple function of $A_{i,j-1}, A_{i-1,j}$, and $A_{i-1,j-1}$. It is straightforward to implement the computation of the matrix A with a pair of nested loops, yielding an $O\ (nm)$ algorithm.

The QNM problem is to simulate a queueing network model that contains a set of service centers and a set of customers which travel around the network obtaining service at the various centers. The problem is to compute performance measures such as utilizations and throughput. The specific case we consider has a set of customers divided into several classes. The solution algorithm that we have parallelized reduces the a problem for a given number of customers and classes to a set of problems where there is one fewer member of each class. Thus, the problem for four classes of twelve customers each (represented [12, 12, 12, 12]) can be solved by first solving the problems (11, 12, 12, 12), (12, 11, 12, 12), (12, 12, 11, 12), and (12, 12, 12, 11). The test cases used in this paper come from this case. It is a 4-dimensional dynamic programming problem with $12^4 = 20,736$ entries computed in its solution. (For further information on queueing network models, see Reference 4.)

The QNM problem and the LCS problem give us two different types of grid-based dynamic programming algorithms. Both are solved sequentially by a computation that has several nested loops. A big difference between the two problems is that the computation of each grid point for the QNM problem involves a moderate amount of work, while the computation done in the LCS problem is very simple.

In order to implement dynamic programming in parallel, we need units of work that can be performed independently. A unit of work is the computation of a value at one of the grid points. The value at a grid point depends on the values at other grid points. It is often convenient to view this dependency structure as a directed, acyclic graph, which is referred to as a task graph. A task graph can aid in the construction of a parallel algorithm if it is possible to decompose the graph into sets of tasks that can be performed simultaneously.

Diagonal — In the two-dimensional case of the LCS problem, all of the entries on a diagonal of the grid can be computed simultaneously. For an $n \times n$ matrix, there are $2n - 2$ phases of computing entries along a diagonal. The values along a diagonal are computed by dividing them equally amongst the processors. A processor waits at a barrier until all processors have computed their values for a given diagonal.

Pipeline — The pipeline method can be viewed as an attempt to directly parallelize the sequential algorithm, which computes the rows of the resulting matrix in order. If we have p processors available, we can perform the computation on the first p rows by staggering the starting times on each row. If the i-th processor starts on row i at time i, the values it needs from the preceding row will have been computed by processor $i - 1$. When a processor finishes a row, it can go on to start the next untouched row. This method turns out to be exactly the same as the Diagonal method when there are n processors.

One important issue in implementing this method is the synchronization along the rows: it is essential that the processor on row i does not catch up to the processor on row $i - 1$. This can be achieved by setting locks. Our implementation of the pipeline method differed between the LCS problem and the QNM problem. For the LCS problem, processors were given a number of row values to compute at a time so as to reduce the number of locks. For the QNM problem, since the individual tasks were larger each entry had a flag, and that one processor would spin on if another had not released it yet.

Task Graph — The previous two methods use a static assignment of work to processors; the assignment is determined in advance and does not depend upon actual execution rates. A different style is to dynamically assign work to processors. One way to do this is to have a central controller which passes out pieces of work to the processors. When a processor finishes a job, it makes a call to the controller to request a new task. The controller keeps track of the precedence of the jobs, so that a job is not scheduled until after its predecessors have been completed. In our implementations, the controller was a subroutine with a single lock on a shared list data structure.

In the QNM problem, two variations of the approach were tried, one in which the controller scheduled the tasks in FIFO order, and one in which the controller scheduled tasks in last-in-first-out (LIFO) order. For the LCS problem, only a FIFO order was used.

In the LCS problem, the individual tasks are very small (they consist of a few array references, an increment, and some comparisons), so it would not make sense to schedule each task with the scheduling routine, since the cost of the scheduling would far exceed the cost of the computation. The solution is to make each task a larger unit of work. This can be done by partitioning the original set of tasks into larger groups and making each group a *supertask*. The supertasks also have a precedence structure, so they were assigned to processor with a dynamic scheduling routine.

Synchronization-Free — The three preceding methods, while different in many respects, are similar in that each identifies concurrently executable components of the sequential algorithm and uses some form of synchronization to coordinate the processors. Our final implementation represents an entirely different approach: we avoid all synchronization. The cost of this is that certain values may be computed multiple times. The benefit is that synchronization overhead is eliminated.

The method is based upon a recursive definition of the problem. To solve a particular node N, the method recursively solves any unsolved predecessors, and then solves N. The key observation is that if several processors solve the same node, correctness isn't compromised.

Three variations of this method were implemented, all for the QNM problem. They differ in how the processors select nodes for evaluation. (The objective, of course, is to minimize redundant computation without resorting to explicit synchronization.) In the first variation the processors used a pseudorandom number generator to decide which predecessor of the current node to evaluate first. (The remaining predecessors were evaluated in order following the first.) In the second variation, random choice was again used to select the initial predecessor, but nodes that were being worked on by other processors (detected

TABLE 2
"Honest" Speed-Ups for the LCS Implementations

No. processors	String length 1000				
	1	2	4	8	16
Diagonal	0.84	1.69	3.28	5.98	9.38
Pipeline	0.97	1.80	3.47	6.72	12.18
Task	0.99	1.95	3.79	7.16	12.71

by means of a flag) were skipped and revisited in a second pass if their (recursive) solution had not yet been completed at that time. In the final approach, a precomputed table, indexed by the processor number and the index of the diagonal of the graph that contains the node, was used to select the predecessor to work on first; the table was computed using a greedy heuristic, which attempted to maximize the distance between the processors on a given diagonal of the graph the first time they reached that diagonal.

We performed a large collection of timings on each of the implementations. Timings were done to both understand relative performance and identify sources of slow-down. Only a fraction of the results are reported here. We placed an emphasis on having a good experimental methodology and in performing accurate timings. The Sequent Symmetry provides an excellent timing facility, which greatly aided our experiments.

The speed-up of a parallel algorithm is the ratio of the parallel algorithm with a sequential algorithm for the same problem. In the ideal case, a p processor algorithm will achieve a speed-up of p. In reporting speed-ups, a distinction should be made between an "honest" speed-up and a "relative" speed-up. An honest speed-up is the comparison with respect to a good sequential algorithm for the problem. The honest speed-up is the appropriate measure for determining the benefit achieved by parallelization. The relative speed-up is the comparison with the one processor version of the parallel algorithm. The relative speed-up gives a better understanding of the scalability of an algorithm.

Table 2 gives "honest" speed-ups for the LCS problem. The diagonal method performed substantially poorer than the other two algorithms. The diagonal algorithm was slower than the one-processor algorithm and also suffered from performance roll-off as the number of processors increased. The pipeline and task algorithms were comparable, with the task method being slightly superior. The single-processor versions of these two algorithms were just about as good as the sequential algorithm.

The performance of the Task Graph method depends upon how many tasks are used. If too few tasks are used, then there is insufficient parallelism, while if too many tasks are used, the overhead of the task scheduling gets too large. The figures we report are for the number of tasks that gave the best running time. For example, for strings of size 1000, the number of tasks for 2 processors was 324 and the number of tasks for 16 processors was 4096.

TABLE 3
"Honest" Speed-Ups for the QNM Implementations

No. processors	1	2	4	8	12	16
Task (FIFO)	0.80	1.56	3.04	5.88	8.30	9.08
Task (LIFO)	0.80	1.57	3.08	6.03	8.77	10.69
Pipeline	0.89	1.78	3.46	6.84	10.11	12.36
Synch.-free	0.90	1.79	3.54	6.97	10.38	13.34

The speed-ups for the QNM problem are shown in Table 3. The results are for a problem with 4 classes, each of which has a population of 12. From the sequential timings, we see that the Task Graph method has the highest cost in the single processor case. The pipeline and the synchronization-free methods have approximately the same cost in the single processor case. The best speed-ups were obtained by the synchronization-free method and the pipeline method.

There are many different sources of slow-down that parallel algorithms may face. A study such as this in which a collection of implementations are examined allows us to see quite a few of these effects and to assess their impact in practice. We divided the observed slowdown into the following seven categories: parallel overhead, synchronization and control locking, critical sections, starvation, cache locality, memory contention, and extra work over the sequential algorithm.

The diagonal method was the poorest approach for the LCS problem. It exhibited a significant amount of parallel overhead. The diagonal method also failed to achieve linear speed-up. The major problem was the high cost of barrier synchronization.

The pipeline method was implemented for both the LCS problem and the QNM problem. It has some overhead that was introduced in the parallelization, but less than the diagonal method. The pipeline method traverses the matrix in row order. However, each time a processor starts on a new row, it makes a procedure call to find out which row to use, which introduces some overhead. Additional overhead is introduced by the code to maintain synchronization.

The pipeline method did not produce linear speed-up because of synchronization, bus traffic, starvation, and the cost of advancing to the next row. Since the processors did not make progress at the same rate, some processors would block at synchronization points. Synchronization accounted for about half of the slow-down observed in the QNM problem. A second cause of slow-down was contention for the memory bus. In the pipeline method, different processors work on adjacent rows, so that processors will always be fetching values which are in the caches of other processors, producing bus traffic. Bus traffic accounted for most of the slow-down not caused by synchronization. The pipeline method suffers from some starvation when waiting for the pipeline to fill up initially. In the LCS problem, the processing of each row is staggered by a large amount to reduce the synchronization costs, but this increases the amount of starvation. A final cause of slow-down with more than two dimensions is the cost of selecting the next row. The program must compute the population of the first element of the row. With a larger number of processors the average number of classes whose population differs between successive rows selected by a processor will increase slightly. The effect of this speed-up turns out to be negligible.

The task graph method for the LCS problem performed roughly as well as the pipeline method, but the sources of slow-down were somewhat different. The actual tasks processed were $k \times k$ submatrices, so there was little overhead associated with the approach. However, there was an overhead introduced in scheduling and manipulating the tasks. The advantage of the task graph method is that the units of work can be larger, so that synchronization does not play as major a role in the slowdown. The disadvantage of the task graph method is that there is starvation both at the start and the end of execution. The choice of the number of tasks to use is based on a trade-off between the work of processing tasks and the amount of starvation. When the number of tasks is too large, the method fails miserably. This is because the task scheduling routine is a critical section.

The synchronization-free method performed very well in those situations where the amount of duplicate work that was performed was held to a small proportion of the total processing. Duplicate work was a major problem if the processors acted randomly. Precomputing the initial paths worked well with the $12 \times 12 \times 12 \times 12$ test case. We conclude that the synchronization-free method requires that the potential parallelism be much greater than the number of processors, since otherwise the cost of duplication is going to be too high.

This report has had two goals: to demonstrate through implementation that there are a number of viable parallel implementations of dynamic programming, and to understand the sources of degradation in these programs.

The diagonal method was clearly the worst of the methods because of the parallel overhead and because of its use of barrier synchronization, which caused its performance to deteriorate rapidly as the number of processors increased and the task size became small. The pipeline and task graph methods both performed well. On the LCS problem the task graph method performed somewhat better when the problem size was large. We believe that this is because the task graph method worked with square supertasks, while the pipeline method was limited to rectangles of width one in this implementation. For problems with more than two dimensions, the pipeline method is less general than the task graph method because the pipeline method deteriorates when given more processors than the length of a row in the matrix. The pipeline method also deteriorated badly when dedicated processors were not provided.

The synchronization-free method performed well when there was enough excess parallelism to hold the amount of duplicate work down. Our results provide a demonstration that this method outperforms approaches which use synchronization calls to avoid duplicate work when there is sufficient excess parallelism to support this approach. An area for future research is to evaluate synchronization-free parallelizations of other algorithms.

REFERENCES

1. **Almquist, K., Anderson, R. J., and Lazowska, E. D.,** The Measured Performance of Parallel Dynamic Programming Implementations, Technical Report #89-01-03, Computer Science Department, University of Washington, Pullman, January 1989.
2. **Edmiston, E. and Wagner, R. A.,** Parallelization of the dynamic programming algorithm for comparison of sequences, in *Int. Conf. on Parallel Processing,* 1987, 78.
3. **Li, G. and Wah, B. W.,** Systolic processing for dynamic programming problems, in *Int. Conf. on Parallel Processing,* 1985, 434..
4. **Lazowska, E. D., Zahorjan, J., Graham, G. S., and Sevcik, K. C.,** *Quantitative System Performance,* Prentice Hall, Englewood Cliffs, NJ, 1984.
5. **Papadimitriou, C. H. and Ullman, J. D.,** communication-time tradeoff, in 25th Symp. on Foundations of Computer Science, 1984, 84.
6. **Wagner, R. A. and Fischer, M. J.,** The string-to-string correction problem, *J. ACM,* 21(1), 168, 1974.

1 1   TERADATA

OVERVIEW

Relational database capabilities serve a variety of industries including airlines, research laboratories, insurance companies, financial institutions, and utilities like KCP&L. With these capabilities, service providers can solve complex problems; airlines can analyze the effect of frequent flyer programs on coach seating during peak fly times; and telephone companies can determine the percentage of the communications network downtime which exceeded a minute in the last 2 hours.

For industries such as these, however, databases can present a problem. Huge databases can drain the mainframe of computing power and storage capacity. Moving the database off the main system would seem a logical way to ease the computing burden, but few system administrators want to invest in a mainframe just to contain the database.

Teradata's approach is the DBC/1012, a fault-tolerant computer designed for relational database management. It supports a variety of mainframes.

The minimum configuration provides processing capacity of 24 MIPS and 5 Gbytes of storage capacity. The system can be expanded to more than three BIPS (billions of instructions per second) and nearly 5 Tbytes of data storage. The disk storage system is composed of separate Winchester drives or disk storage units (DSUs), each containing 1.2 Gbytes of storage.

The unit is composed of the DSUs, a number of processors, and the Ynet, a high-speed intelligent interconnection.

The three types of processors include interface processors (IFPs), access module processors (AMPs), and communications processors (COPs). IFPs connect to mainframe hosts; the AMPs manipulate the database, access the disk storage units, and prepare data; and the COPs serve as the link to minicomputers, workstations, and PCs on a local-area network.

When the user requests information stored on the database, the Structured Query Language (SQL) is captured by the Teradata Director Program and dispatched to the database machine where it is received by an interface processor and translated. The Ynet broadcasts the request to the AMPs, which in turn process the request with the disk storage units.

All data tables are distributed across the DSUs. Working in parallel, each AMP selects rows within the DSUs that qualify for the request and posts the necessary columns (fields) to the Ynet.

The Ynet performs a hardware sort/merge, collating each AMP response set into one master answer set. The sort sequence is determined by the order clause of the original SQL request. The answer set (file) is then sent back to the host by an interface processor and posted to the requesting user's application area. In a recent demonstration of the performance of Teradata Corporation's DBC/1012 Data Base Computer, the system achieved 1176 transactions per second with a subsecond response time. This result was compared to IBM DB2 Version 2's reported capability of 438 transactions per second.

The DBC/1012 performance measurements utilized what Teradata believes to be the same benchmark parameters — very simple transactions and credit authorization (the lost/stolen credit card check). The tests were run on a system being prepared for shipment utilizing currently available software release 3.0 and standard 80286-based processors.

Teradata's test configuration system consisted of 191 parallel processors (191 MIPS) and 286 disk storage units. This configuration represents approximately one fifth the size of a maximum Teradata DBC/1012.

In an additional test, one of the results was the DBC/1012 performance in evaluating query processing. A query which sorted eight million data rows ran 11 min, 49 s on the DBC/1012 with a sort time of only 6 min, 47 s, compared to IBM's reported query time of 116 min for DB2 Version 2.

The DBC/1012 Data Base Computer is a dedicated relational data base management system. Except for the host system communication interface software (provided by Teradata), all hardware, firmware, and software reside in an environment separate from the host computer. The DBC/1012 system consists of the following subsystems:

- Host system communication interface (HSCI)
- Interface processor (IFP)
- Communication processor (COP)
- Ynet (interprocessor bus)
- Access module processor (AMP)
- Disk storage unit (DSU)
- Systems console with printer

The interrelationship of these subsystems in a basic system is shown in Figure 1.

The HSCI consists of a library of service routines and an interface program. This software resides in the host computer and enables users to conduct sessions with the DBC/1012 Data Base Computer. Sessions may be conducted in interactive, batch, or on-line transaction processing environments.

Several software components located in DBC/1012 processors supplement the host-resident software and hardware subsystems. These software packages include

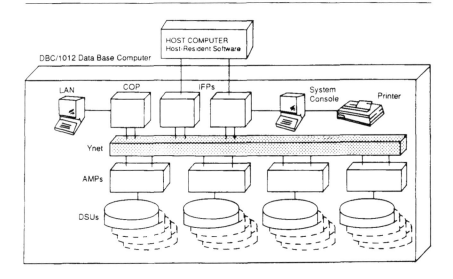

FIGURE 1. Basic DBC/1012 configuration.

- The Teradata Operating System (TOS), a virtual-memory, multiuser, real-time operating system.
- A set of services that supports DBC/SQL.
- A data base system that supports DBC/1012 data definition and manipulation functions.

IFPs manage channel traffic between the host computer and the DBC/1012 Data Base Computer. The IFP translates requests from a host into internal commands, forwards the commands over the Ynet to the AMPs, and coordinates the responses as they return over the Ynet from the AMPs. As this request traffic increases over time, it may be necessary to add IFPs to a DBC/1012 system.

The COP allows access to the DBC/1012 from a local area network (LAN) and manages network traffic between the host and the DBC/1012 Data Base Computer. The COP translates requests from the host into internal commands, forwards the commands over the Ynet to the AMPs, then coordinates the responses as they return.

The Ynet is an independent bus that interconnects all IFPs, COPs, and AMPs and performs — in hardware — many of the tasks associated with multiprocessor system management. To enhance system operation and provide a fail-safe backup, two fully operational Ynets are designed into each DBC/1012 system. In contrast to an ordinary, passive bus, the Ynet is an array of active logic. The name "Ynet" is derived from the pictorial representation of this subsystem, a network of high-speed circuits that resemble upside-down Ys arranged in a lattice.

AMPs receive requests forwarded over the Ynet by IFPs and COPs and perform the required data manipulations. The AMPs then send appropriate responses back over the Ynet to the IFPs and COPs, which return the responses

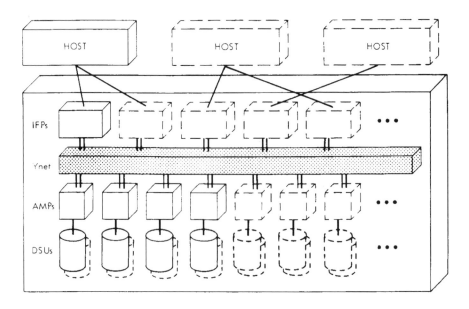

FIGURE 2. Expandability of the DBC/1012 system.

to the original point of request. Data are distributed evenly across all AMPs. As a result, the workload in all AMPs is balanced and data can be processed in parallel. When AMPs are added to a DBC/1012 system to increase storage capacity (or to increase performance), the data are automatically redistributed.

The DSUs serve as the data storage medium for the system. Each AMP can support as many as four 500-Mbyte DSUs, for a total of 2 Gbytes of data storage capacity per AMP.

The system console (an AT&T or IBM-compatible personal computer) allows the system operator to communicate with the DBC/1012 Data Base Computer in order to get reports of system status, current configuration, and performance. Through the console, the operator also controls system and diagnostic operations. A printer allows the operator to make a hard copy of anything that is displayed at the console.

The DBC/1012 system can be expanded to accommodate a number of mainframes, IFPs, COPs, and AMPs, as shown in Figure 2.

The data on the DBC/1012 Data Base Computer are stored in one or more relational tables. The data in a table can be thought of as being arranged in vertical columns and horizontal rows. The sample table in Figure 3 has five columns and nine rows.

A row is an entry in the table. Each row is characterized by a number of attributes, called columns. For example, each row in the employee table is characterized by an employee number (EmpNo), name (Name), department number (DeptNo), position (Position), and years of experience (YrsExp).

The intersection of a column and a row is called a "field". The term "field value" refers to the specific data value in a field. In Figure 3, the field value for row 3 of the Position column is "Engineer".

EMPLOYEE TABLE

```
Columns  |            |            |            |            |
         |            |            |            |            |
         V            V            V            V            V
      EmpNo         Name        DeptNo     Position      YrsExp
      -----------------------------------------------------------
      | 10001  |  Peterson J | 100  |  Bookkeeper  |  5  |
      -----------------------------------------------------------
      | 10002  |  Moffit H   | 100  |  Recruiter   |  3  |
      -----------------------------------------------------------
      | 10003  |  Smith T    | 500  |  Engineer    | 10  |
      -----------------------------------------------------------
      | 10004  |  Jones M    | 100  |  Vice Pres   | 13  |
      -----------------------------------------------------------
      | 10005  |  Kemper R   | 600  |  Assembler   |  7  |
      -----------------------------------------------------------
      | 10006  |  Marston A  | 500  |  Secretary   |  8  |
      -----------------------------------------------------------
Field ---> 10007 |  Reed C     | 500  |  Technician  |  4  |
      -----------------------------------------------------------
      | 10008  |  Watson A   | 500  |  Vice Pres   |  8  |
      -----------------------------------------------------------
      | 10009  |  Regan R    | 600  |  Manager     | 10  |
      -----------------------------------------------------------
```

FIGURE 3. Relational data base sample table.

Data are defined in terms of tables and are selected through operations on tables. A table is defined by giving it a name and identifying its columns by their names and characteristics. All fields in a column have the same characteristics. For example, all values in the YrsExp column in Figure 3 are integers.

A row is the smallest unit that can be inserted into or deleted from a table. An insert operation adds one or more rows to a table. A delete operation removes one or more rows from a table. In general, rows have no inherent order and can be selected in any order a user specifies. Similarly, columns have no inherent order and can be selected in any order the user specifies.

A field is the smallest unit of data that can be modified. An update operation replaces one or more field values for one or more rows. Updating never replaces a part of a field value.

Basic operations on tables are

- Creating or deleting tables.
- Adding or deleting columns.
- Inserting or deleting rows.
- Updating existing rows.
- Updating existing columns.
- Selecting data from existing rows.

A user can select individual columns (or all columns) from specific rows (or all rows) from one or more tables. A select operation that involves data from more than one table is called a *join*.

A view is a *window* through which a user looks at selected portions of a table. The view is an *apparent* table derived from one or more real tables. Views can also be derived from other views, or combinations of views and tables.

A view looks like a stored table. It has rows and columns and, in general, can be used as if it were a space. A view is typically used to simplify query requests, to limit a user's access to or manipulation of data, and to allow different users to look at the same data from different perspectives.

A relational data base management system has many practical advantages. In particular, a data base is easy to understand because there is only one simple data structure, the table. As a result, data is easy for a user to select, manipulate, and control using the nonprocedural DBC/SQL language. Using DBC/SQL, the user simply states the results desired, rather than a procedure for deriving the results.

For example, given the employee table in Figure 3, the DBC/SQL statement

```
SELECT Name
FROM Employee
WHERE EmpNo = 10006;
```

produces the name of the employee whose employee number is 10006: Marston A.

In contrast to DBC/SQL, the data manipulation language of most software data base management systems is procedural. To select a single record, an application programmer must first know an access path to the desired record and then traverse the path, step by step, to reach it. Each step requires an explicit statement to describe it, and the navigation process must be repeated for each record to be selected. This navigation requires a high level of user expertise, and productivity is relatively low. Also, as the data base grows in size and complexity, the navigation process originally selected may become cumbersome or ineffective.

Businesses continually store new data, generating new requirements for using and relating data in the process. Often these requirements necessitate structural changes in a data base. Structural changes in a conventional, nonrelational data base management system require changing the application program to reflect the new access paths. Thus, these structural changes are difficult to make.

By contrast, the relational data base management system of the DBC/1012 Data Base Computer facilitates changes in data structures and contents. This flexibility provides two long-term benefits to the DBC/1012 user: individual programs are safeguarded against obsolescence, and the system as a whole has a much longer, more productive life cycle.

The HSCI supports user sessions with the DBC/1012 Data Base Computer in the following host system environments:

- MVS/XA (all releases); MVS/SP (release 1.3 and above)
- VM/SP (release 3 and above)

The HSCI consists of the following:

- Call-level interface (CLI)
- User-to-TDP communication (UTC) routines

- MVS: SVC and cross memory services (XMS)
- VM: interuser communication vehicle (IUCV)
- Teradata director program (TDP)

User sessions can be conducted with the DBC/1012 in either an interactive or a batch environment. The MVS host operating environment is shown in Figure 4. In the figure, one end user has direct, interactive access to data via the Interactive Teradata Query facility (ITEQ). ITEQ is an application program that operates under MVS/TSO or VM/CMS. Other "users" of data in the DBC/1012 data bases include an application program running in batch mode and an on-line transaction program operating under CICS (Customer Information Control System). The TDP operates as a separate "server" undedicated address space.

Figure 5 shows the VM/SP host operating environment. In this illustration, one user has access to data via an on-line transaction program; a second user has interactive access to data via ITEQ, while a third is operating in a batch environment. The TDP operates as a separate "server" virtual machine.

Under VM, all users access data through the conversational monitor system (CMS), the VM time-sharing facility. For users in either the MVS or VM host environment, call-level interface (CLI) service routines provide a protocol for communicating requests to the TDP and the DBC/1012, and for receiving responses. CLI routines are used in different ways:

- They may be called by Teradata-supplied software, such as ITEQ or BTEQ, as a result of DBC/SQL statements entered directly by a user.
- They may be called indirectly by user-prepared application programs that are written in high-level languages such as COBOL and PL/I.
- They may be called directly by programs written in Assembly language or any other language that has a "call" facility.

The Teradata Director Program (TDP) manages communication between users and the DBC/1012 Data Base Computer. The TDP runs in dedicated address space in an MVS environment, or as a virtual machine in the VM environment.

CLI routines and the TDP communicate via a number of special routines. These routines manage requests between Teradata-supplied applications or user-supplied applications and the TDP.

The application program invokes CLI routines to build a DBC/SQL request to the TDP. The TDP, in turn, creates a request message and communicates it over a block multiplexer channel to an IFP for processing.

The application invokes CLI routines to provide responses from the DBC/1012 Data Base Computer via the TDP, then returns them to the originating application. If the effect of a DBC/SQL request produces a single result, the routines return that one row of data. If a request produces a number of rows, the results are "deblocked" to serve the requestor one row at a time.

DBC/SQL statements may be embedded in a COBOL program or a PL/I program. The COBOL Processor or the PL/I Preprocessor, both provided by

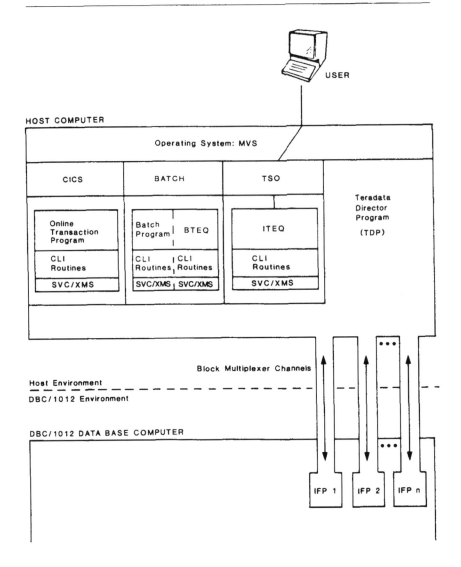

FIGURE 4. MVS host operating environment.

Teradata, convert embedded statements into calls to preprocessor support routines. After the program is compiled, it may be link-edited at run time. Applications written in high-level languages that have a CALL statement (including COBOL and PL/I) may also call CLI routines directly.

The IFP manages the dialogue between a user session in a mainframe and the DBC/1012 Data Base Computer. As shown in Figure 6, at least two IFPs are normally assigned to each mainframe channel in order to achieve redundancy and fail-safe operation.

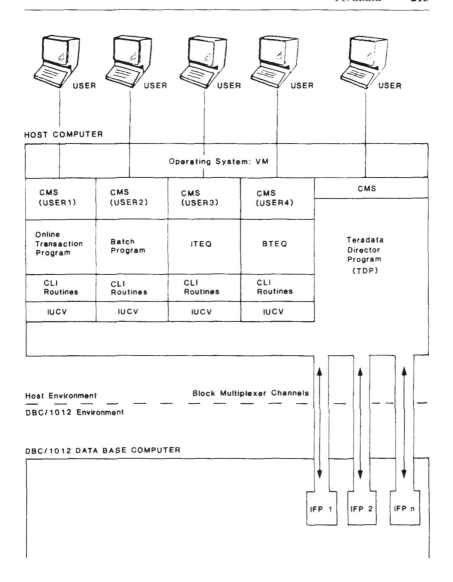

FIGURE 5. VM/SP host operating environment.

Each IFP consists of the following elements:

- Session control
- Host Iinterface
- Parser
- Dispatcher
- Ynet interface

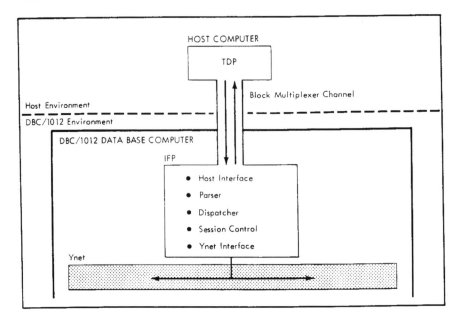

FIGURE 6. IFP configuration.

Session control processes logon and logoff requests from a host and establishes sessions.

The host interface controls the exchange of messages between the TDP in a mainframe and the DBC/1012 Data Base Computer.

When an IFP receives a DBC/SQL request message from a TDP, the parser interprets it, checks for syntax, then evaluates it semantically. In processing the request message, the parser refers to the data dictionary/directory (system tables containing information about DBC/1012 data bases) to resolve symbolic data names and make integrity checks. Finally, the parser decodes the request into a series of work steps necessary for routing and resolving the request. It then passes them to the dispatcher.

The dispatcher controls the sequence in which steps are executed and passes the steps to the Ynet interface.

The IFP Ynet interface controls the transmission of messages over the Ynet to and from the AMPs. As noted below, when rows are inserted in a table, the data are automatically distributed evenly across all AMPs in the system. This distribution increases throughput by allowing table data to be accessed simultaneously on all AMPs. This principle is illustrated in Figure 7, where 12 rows of a table are distributed among DSUs attached to 4 AMPs. Depending on whether a request is for data in a single row (a "primary index" request), the IFP either transmits steps to a specific AMP (see IFP 1 in Figure 7) or causes the Ynet to "broadcast" the steps to all AMPs (see IFP 2 in the figure). Also, in order to minimize system overhead, the IFP can send a step to a subset of AMPs when the request asks for data that reside on more than one (but not all) of the AMPs.

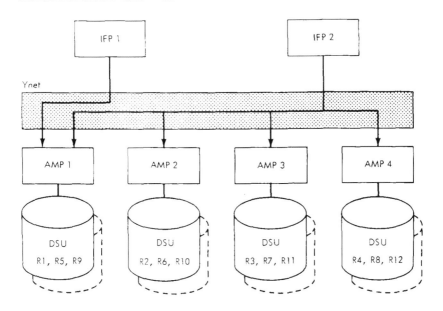

FIGURE 7.   IFP routing of DBC/SQL request messages.

As an example, consider the following two DBC/SQL statements from a table of checking account information:

1.   SELECT * FROM table_01 WHERE AcctNo = 129317;
2.   SELECT * FROM Table_01 WHERE AcctBal > 1000;

For purposes of illustration, assume that IFPs 1 and 2 receive requests 1 and 2, respectively, that data for account 129317 is contained in table row R9 stored on AMP 1, and that information about all account balances is distributed evenly among the DSUs of all four AMPs. The IFP 1 parser determines that its request is a primary-index retrieval, which calls for access and return of one specific row (record). IFP 1 then issues a message to the Ynet containing an appropriate read step and R9/AMP 1 routing information. Once the desired record is received from AMP 1, IFP 1 transmits the data back to the TDP over the block multiplexer channel.

The IFP 2 parser determines that its request is an all-AMP data retrieval, then issues a message to the Ynet containing the appropriate read step to be broadcast to all four AMPs. Once results are received from the AMPs, IFP 2 transmits the data back to the TDP over the block multiplexer channel.

To enhance system performance, the DBC/1012 executes steps in parallel whenever possible. Parallel steps can work with multistatement requests, macros, and single statements and can provide a significant improvement in response time. For example, the response time of a multistatement request consisting of four statements that can be independently executed may be cut in half.

Dual interprocessor buses, or Ynets, connect the IFPs, COPs, and AMPs of the DBC/1012 system, as shown in Figure 8.

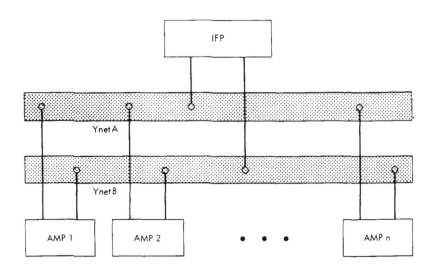

FIGURE 8. Ynet.

Both Ynets operate concurrently to share the communication load. In contrast to an ordinary bus, which contains no logic, each Ynet is an array of active logic that provides routing and sorting functions. Each of the Ynets is independent of the other. This independence extends to physical partitioning, electrical power, Ynet clocks, and interface logic. Each Ynet thus serves as a backup for the other, ensuring continuous system operation in the event of failure.

Each Ynet provides communication

- Between single processors (e.g., between an IFP and an AMP).
- From a single processor to a group of processors (e.g., from one COP to all AMPs).
- From a group of processors to a specific processor (e.g., from all AMPs to one IFP).
- Between various processors to synchronize operations on data.

As data returning from the AMP are merged back across the Ynet to the originating IFP or COP, they are automatically sorted into the sequence specified by the user's request. In this way, the Ynet also acts as an ultrahigh-speed sort/merge processor.

The AMP handles the actual data base functions of the DBC/1012 Data Base Computer. An AMP responds to IFP steps transmitted to it across the Ynet by selecting data from or storing data on its DSU(s). Figure 9 shows the components of the AMP:

- Ynet interface
- Data base manager
- Disk interface

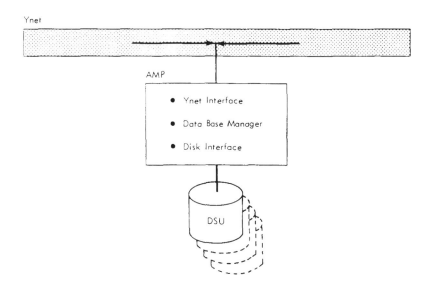

FIGURE 9. AMP configuration.

The Ynet interface accepts messages from an IFP or COP and returns responses. The interface also participates in system-wide synchronization of operations that involve many AMPS.

The data base manager performs select, insert, delete, and update operations on existing data tables. Other functions of the data base manager include data definition, rollback and recovery, data base reorganization, and logging.

The disk interface controls reading and writing of data to and from the DSUs attached to the AMP.

In general, as rows are inserted into a table, the data are evenly distributed among all AMPs. Processing a request for data from a table can thus be performed in parallel. That is, the data can be accessed simultaneously by all AMPs. This simultaneous access greatly increases system throughput. This parallel processing capability means that as AMPs and DSUs are added, storage capacity is increased and system performance improves.

The DSU is the information storage medium for the DBC/1012 Data Base Computer. Figure 10 shows the relationship between the Ynet, the AMP, and the DSUs. Up to four DSUs can be attached to an AMP.

The space on a DSU can be divided conceptually into the following areas:

1. System area
2. Primary copy area
3. Fallback copy area (optional)

The system area stores programs and tables. The primary copy area stores the primary copy of the data. The primary copy area is accessed under normal conditions.

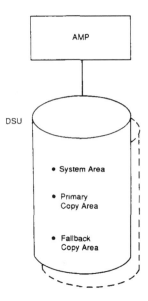

FIGURE 10. DSU.

The fallback area contains a duplicate copy of the data stored on other AMPs within the system. The fallback copy is accessed whenever the primary copy becomes unavailable, thus providing data redundancy as well as hardware redundancy. When a data base is created, or any time after a table has been created, a user can specify that the DBC/1012 keep a fallback copy of the data in the table. Like the primary table data, this fallback copy is also distributed among the AMPs in the system. However, the fallback copy of each row is stored on a different AMP from the one on which the primary copy is stored. This distribution ensures that a fallback copy of the stored data remains available on other AMPs if an AMP should fail.

As an example, in the eight-AMP system shown in Figure 11, the primary copy of data on AMP 4 is distributed for fallback on AMPs 5, 6, and 7. If AMP 4 were to fail, these data would be available through these other AMPs.

If AMPs 4 and 7 were to fail simultaneously, however, there would be a loss of data availability. For this reason, the system administrator may provide additional data protection by "clustering" the AMPs in groups. A cluster may consist of 2 to 16 AMPs.

Clustering increases the probability that all system data are available even if two or more AMPs fail simultaneously. In the example shown in Figure 12, if the AMPs are grouped into two clusters of four, the data on AMP 4 would be backed up by AMPs 1, 2, and 3 and the data on AMP 7 would be backed up by AMPs 5, 6, and 8. If both AMPs 4 and 7 were now to fail, all data would still be available.

The COP enables users to access the DBC/1012 from hosts or personal computers connected to an Ethernet-based LAN. The COP functions very much

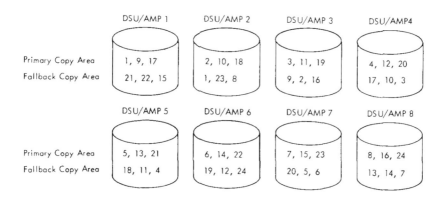

FIGURE 11. Distribution of primary and fallback data.

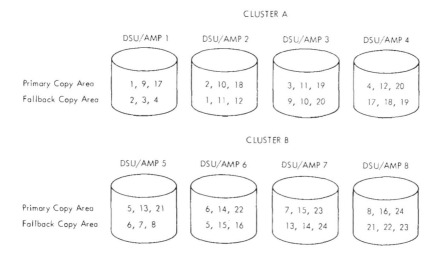

FIGURE 12. Distribution of data with clustered AMPs.

like an IFP, performing session management, parsing, and dispatching functions. The same LAN may have more than one DBC/1012 connected, and a single DBC/1012 may be attached to more than one LAN. See Figure 13 for an illustration of the COP/LAN relationship and Figure 14 for the COP interface. Instead of receiving requests from the host via a block multiplexer channel as does the IFP, the COP receives requests via messages over the LAN. The network protocols define how the messages are transported between the LAN-attached host(s) and the DBC/1012. The current release of the COP is available with two types of network protocol:

- ISO/OSI
- TCP/IP

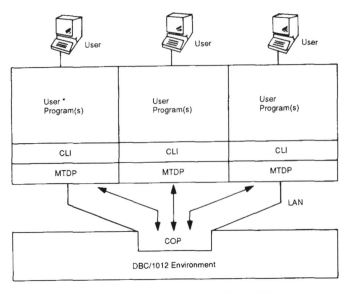

* User programs may include such programs as Batch, PC-SQL-Link, BTEQ, etc.

FIGURE 13. Workstation operating environment.

with additional protocols anticipated in the near future. In addition to the network software, Teradata provides software for each host or workstation that accesses the DBC/1012 via the COP.

The host/workstation software includes BTEQ, CLI Version 2, the Micro Teradata Director Program (MTDP), and the Micro Operating System Interface (MOSI).

CLI Version 2 is used in managing the blocking and deblocking of messages. MTDP is called by CLI to manage communications with the DBC/1012. MOSI is a library in which operating system- and communication protocol-dependent services are collected for ease in porting the interface to a new operating system. Calls to MOSI procedures are mapped onto calls to the relevant operating system or protocol. Teradata currently provides software versions for

- IBM PCs or PC-compatibles running MS-DOS or PC-DOS (Version 3.1 or higher within the 3.n series).
- AT&T 3B2/300 and 400 computers running under UNIX 5.2.
- DEC VAX and Micro VAX II running under VMS (Version 4.4 or higher within the 4.n series).

Using CLI, programmers may also write special-purpose applications, if they are required at the user's site.

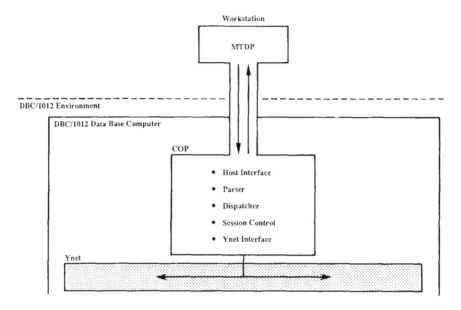

FIGURE 14. COP interface components.

Research Machines

12 THE J-MACHINE: A FINE-GRAIN CONCURRENT COMPUTER

OVERVIEW

The J-Machine is a distributed-memory, MIMD, concurrent computer.[11] It provides primitive mechanisms for communication, synchronization, and translation. Communication mechanisms are provided that permit a node to send a message to any other node in the machine in $<2\ \mu s$. No processing resources on intermediate nodes are consumed and buffer memory is automatically allocated on the receiving node. The synchronization mechanisms schedule and dispatch a task in $<1\ \mu s$ on message arrival and suspend tasks that attempt to reference data that are not yet available. The translation mechanism maintains bindings between arbitrary names and values. It is used to perform address translation to support a global virtual address space.

These mechanisms have been selected to be both general and amenable to efficient hardware implementation. They efficiently support many parallel models of computation including actors,[1] dataflow,[16] and communicating processes.[19] To support fine-grain concurrent programming systems, the mechanisms are designed to efficiently handle small objects (8 words) and small tasks (20 instructions).

The hardware is an ensemble of up to 65,536 message-driven processors (MDPs).[9] This limit is set by the addressability of the router and the bandwidth of the network. Each node contains a 36-bit processor, 4K 36-bit words of memory, 256K words of external memory, and a communications controller (router). The nodes are connected by a high-speed, three-dimensional mesh network. Each network channel has a bandwidth of 450 Mbits/s. A 128-node prototype has been operating since October 1991. The first medium-scale prototype will be a 1024-node system.

This design was chosen to make the most efficient use of available chip and board area. Packaging a small amount of memory on each node gives us an extremely high memory bandwidth (3 Gbits/s per chip or 200 Tbits/s in a fully populated system). Memory consumes most of the chip area: from one point of view, the system is a memory with processors added to each node to improve bandwidth for local operations. The fast communication and global address space prevent the small local memories from limiting programmability or performance. The 3-D network gives the highest throughput and lowest latency for a given wire density.[7,13] It allows the processing nodes to be packed densely and results in uniformly short wires.

The J-Machine project is driven by the following goals:

- To identify and implement simple hardware mechanisms for communication, synchronization, and naming suitable for supporting a broad range of concurrent programming models.
- To reduce the overhead associated with these mechanisms to a few instruction times so that fine-grain programs may be efficiently executed.
- To design an area-efficient machine: one that maximizes performance for a given amount of chip and wiring area.

The J-Machine is a fine-grain concurrent computer in that (1) it is designed to support fine-grain programs and (2) it is composed of fine-grain processing nodes.[14]

The *grain size* of a program refers to the size of the tasks and messages that make up the program. Coarse-grain programs have a few long ($\approx 10^5$ instruction) tasks, while fine-grain programs have many short (≈ 20 instruction) tasks. With more tasks that can execute at a given time — *viz.* more concurrency — fine-grain programs (in the absence of overhead) result in faster solutions than coarse-grain programs.

The *grain size* of a machine refers to the physical size and the amount of memory in one processing node. A coarse-grain processing node requires hundreds of chips (several boards) and has $\approx 10^7$ bytes of memory, while a fine-grain node fits on a single chip and has $\approx 10^4$ bytes of memory. Fine-grain nodes cost less and have less memory than coarse-grain nodes; however, because so little silicon area is required to build a fast processor, they need not have slower processors than coarse-grain nodes.

The J-Machine builds on previous work in the design of MP and shared memory machines. Like the Caltech Cosmic Cube,[25] the Intel iPSC,[21] the NCUBE,[23] and the Ametek,[2] each node of the J-Machine has a local memory and communicates with other nodes by passing messages. Because of its low overhead, the J-Machine can exploit concurrency at a much finer grain than these early MP computers. Delivering a message and dispatching a task in response to the message arrival takes <3μs on the J-Machine, as opposed to 5 ms on an iPSC. Like the BBN Butterfly[4] and the IBM RP3,[24] the J-Machine provides a global virtual address space. The same IDs (virtual addresses) are used to reference on and off node objects. Like the InMOS transputer[20] and the Caltech MOSAIC,[22] a J-Machine node is a single chip processing element integrating a processor, memory, and a communication unit.

The J-Machine is unique in that it extends these previous efforts with efficient primitive mechanisms for communication, synchronization and naming.

The J-Machine consists of an ensemble of (up to 64K) MDP-based processing nodes connected by a three-dimensional mesh network. The network delivers messages between nodes. All routing and flow control are handled in the network. The network consists of a communication controller or router on each node,[6,10,17] and wires connecting these routers to their neighbors in each of the three physical dimensions.

Each node contains a memory, a processor, and a communication controller. The communication controller is logically part of the network, but physically part of the node. The 260K-word by 36-bit memory is used to store objects and system tables. Each word of the memory contains a 32-bit data item and a 4-bit tag. In addition to the usual uses of tags to support dynamic typing and garbage collection, special tags are provided to synchronize on data presence and to indicate if an address is local or remote. Memory accesses to write messages or read code are made a row (144 bits) at a time to improve memory bandwidth. A part of the memory can be mapped as a set-associative cache. This cache is used to implement the processor's translation mechanism.

The processor is message driven. It executes user and system code in response to messages arriving over the network. A conventional processor is instruction driven in that it fetches an instruction from memory and dispatches a control sequence to execute the instruction. A message-driven processor receives a message from the network and dispatches a control sequence to execute the task specified by the message. The MDP uses an instruction sequence to *execute* a message. Hardware mechanisms for communications, synchronization, and translation are provided to accelerate the dispatch operation and the subsequent task execution.

To support communication over the network, the MDP provides a *SEND* instruction and performs automatic buffering of arriving messages. To synchronize execution with arriving messages, a primitive dispatch operation is provided that eliminates scheduling overhead. To synchronize on data, tags are provided to support futures. A general translation mechanism uses a set associative cache in the node memory to maintain arbitrary bindings.

To see how the system functions together, consider the example shown in Figure 1. In Figure 1A, a task executing in Context 37 on Node 124 sends a *Sum* message to an object, *point1* This message requests that the object sum its two fields x and y. The sending node translates the object name for *point1* (unique 32-bit pattern) into a node address, Node 262 (a 16-bit integer), using the MDP translation mechanism. A sequence of MDP *SEND* instructions is then used to inject the message into the network. The message includes (1) the node address of *point1* (Node 262), (2) the object name of *point1*, (3) message name or selector, *Sum*, and (4) a continuation (the node and ID of the sender's context and the slot into which the reply should be stored). The sending task continues to execute until it needs the result of the *Sum* image.

The network delivers the injected message to Node 262. At this node, the MDP buffering mechanism allocates storage for the message and sequences the

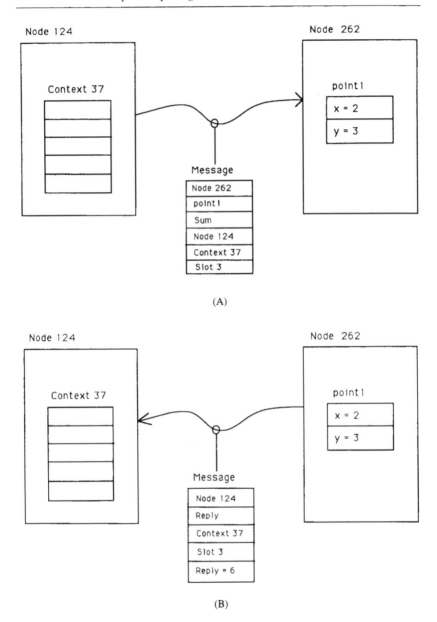

FIGURE 1. (A) A task executing in Context 37 on Node 124 sends a message to object *point1*. (B) A reply message is sent to Context 37.

message off the network into the memory of the node. When the node completes its current task (and any other tasks ahead in the queue), the MDP dispatch mechanism creates a new task in response to the message. This task translates the ID of *point1* into a segment descriptor for the object, adds the *x* and *y* fields of the object together, and uses a sequence of *SEND* instructions to inject a message containing the sum into the network. As shown in Figure 1B, this message contains (1) the node address of the sender's context, (2) the ID of the sender's context, (3) the context slot awaiting the reply, and (4) the result. This task then terminates.

The network delivers the reply message to Node 124, where it is buffered and eventually dispatched to create a task. This task translates the ID for Context 37 into a segment descriptor. The reply value is stored into the specified slot of this context. The sending task is then resumed by loading its context from this segment.

The round-trip time for this example message send and reply is $\approx 5\mu s$. The difficulty in building a concurrent system the scale of the J-Machine is not developing the mechanisms conceptually. It is implementing them efficiently so the overhead of accessing remote nodes is made small enough to permit the execution of fine-grain programs.

The J-Machine network has a three-dimensional mesh topology, as shown in Figure 2. Each node is located by a three coordinate address (x, y, and z). A node is connected to its six neighbors (if they exist) that have addresses differing in only one coordinate by ±1. All connections are bidirectional channels. Each channel requires 15 wires to carry 9 data bits, 1 tail bit, and 5 control lines.[10] Addressing is provided to support up to a $32 \times 32 \times 64$ cube of 65536 nodes. For a machine such as the J-Machine, where wire density is a limiting factor, this topology has been shown to give the lowest latency and highest throughput for a given wire density.[7,13]

The network topology is not visible to the programmer. The latency of sending a message from any node, i to any other node, j, is sufficiently low that the programmer sees the network as a complete connection. Zero-load[*] network latency is given by

$$T = T_d D + T_c \frac{L}{W}$$

where D is the distance (number of hops) the message must travel, L is the length of the message in bits, and W is the width of the channel in bits. The network has a propagation delay per stage, T_d, of 30 ns and a channel cycle time, T_c, of 30 ns. With these times, a 6-word ($L = 216$ bit) message traversing half the network diameter ($D = 24$) has a latency of 1.4 μs equally divided between the two components of latency.[13]. An average message travels one third of the network diameter for a latency of 1.2 μs.

The network provides all end-to-end message delivery services. The sending node injects a message containing the absolute address of the destination node. The network determines the route of the message, and sequences each flow-control digit (flit) of the message over the route. Flow control is performed as required to resolve contention and match channel rates. This flow control is performed in a manner that is provably deadlock free.[8]

---

[*]With a typical load of 30% capacity, latency is $\approx 10\%$ greater than zero-load latency.

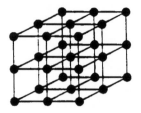

FIGURE 2. The J-Machine as a 3D mesh.

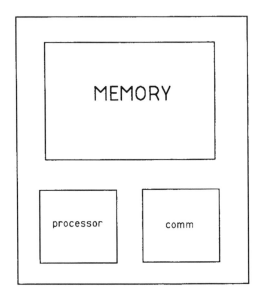

FIGURE 3. The message-driven processor chip.

There is no acknowledgement, error detection, or error correction on the network channels. The network wires are all short, contained within a single physical cabinet, and operated at low impedance. The error rate of a network channel is no higher than that of a properly terminated signal in a conventional CPU.

The MDP is a 36-bit, single-chip microcomputer specialized to operate efficiently in a multicomputer.[5,9,12] The MDP chip includes the processor, a 4K-word by 36-bit memory, and a router (Figure 3). An on-chip memory controller with ECC permits local memory to be expanded up to 1M-words by adding external DRAM chips.

Other machines have combined processor, memory, and communications on a single chip.[20,22,23] The MDP extends this work by providing fast, primitive mechanisms for synchronization, communication, and translation (naming) that allow the processor to efficiently support many parallel models of computation. A fast network is of little use if very large overheads are required to initiate and receive messages at the processing nodes. The mechanisms of the MDP reduce the overhead of interacting with other processors over the network to levels that make fine-grain parallelism efficient.

The following mechanisms are provided:

Communication mechanisms
- A *SEND* instruction injects messages into the network.
- Messages arriving from the network are automatically buffered in a circular queue.

Synchronization mechanisms
- A dispatch mechanism creates and schedules a task (thread of control and addressing environment) to handle each arriving message.
- Tags for *futures*[3] synchronize tasks based on data dependencies.

Translation
- *ENTER* and *XLATE* (translate) instructions make bindings between arbitrary 36-bit key and data values (*ENTER*) and retrieve a value given the corresponding key (*XLATE*).
- Segmented memory management provides relocation and protection for data objects stored in the memory of a node.

The processor is *message driven* in the sense that processing is performed in response to messages (via the dispatch mechanism). There is no receive instruction. A task is created for each arriving message to handle that message. A computation is advanced (driven) by the messages carrying tasks about the network.

The MDP injects messages into the network using a send instruction that transmits one or two words (at most one from memory) and optionally terminates the message. The first word of the message is interpreted by the network as an absolute node address (in x,y,z format) and is stripped off before delivery. The remainder of the message is transmitted without modification. A typical message send is shown in Figure 4. The first instruction sends the absolute address of the destination node (contained in R0). The second instruction sends two words of data (from R1 and R2). The final instruction sends two additional words of data (one from R3 and one from memory). The use of the *SENDE* instruction marks the end of the message and causes it to be transmitted into the network. In a Concurrent Smalltalk message,[15] the first word is a message header, the second specifies the receiver, the third word is the selector, subsequent words contain arguments, and the final word is a continuation. This sequence executes in four clock cycles (240 ns).

A FIFO buffer is used to match the speed of message transmission to the network. In some cases, the MDP cannot send message words as fast as the network can transmit them. Without a buffer, *bubbles* (absence of words) would be injected into the network pipeline, degrading performance. The *SEND* instruction loads one or two words into the buffer. When the message is complete or the eight-word buffer is full, the contents of the buffer are launched into the network.

SEND	R0	; send net address
SEND2	R1,R2	; header and receiver
SEND2E	R3,[3,A3]	; selector and continuation - end msg.

FIGURE 4. MDP assembly code to send a 4 word message uses three variants of the SEND instruction.

Previous concurrent computers have used DMA or I/O channels to inject messages into the network. First an instruction sequence composed a message in memory. DMA registers or channel command words were then set up to initiate sending. Finally, the DMA controller transferred the words from the memory into the network. This approach to message sending is too slow for two reasons. First, the entire message must be transferred across the memory interface twice, once to compose it in memory and a second time to transfer it into the network. Second, for very short messages, the time required to set up the DMA control registers or I/O channel command words often exceeds the time to simply send the message into the network.

The MDP maintains two message/scheduling queues (one for each priority level) in its on-chip memory. The queues are implemented as circular buffers. As messages arrive over the network, they are buffered in the appropriate queue. To improve memory bandwidth, messages are enqueued by rows. Incoming message words are accumulated in a row buffer until the row buffer is filled or the message is complete. The row buffer is then written to memory.

It is important that the queue have sufficient performance to accept words from the network at the same rate at which they arrive. Otherwise, messages would back up into the network, causing congestion. The queue row buffers in combination with hardware update of queue pointers allow enqueuing to proceed using one memory cycle for each four words received. Thus, a program can execute in parallel with message reception with little loss of memory bandwidth.

Providing hardware support for allocation of memory in a circular buffer on a multicomputer is analogous to the support provided for allocation of memory in push-down stacks on a uniprocessor. Each message stored in the MDP message queue represents a method activation much as each stack frame allocated on a push-down stack represents a procedure activation.

Each message in the queues of an MDP represents a task that is ready to run. When the message reaches the head of the queue, a task is created to handle the message. At any time, the MDP is executing the task associated with the first message in the highest priority, nonempty queue. If both queues are empty, the MDP is idle — *viz.* executing a background task. Sending a message implicitly schedules a task on the destination node. This simple two-priority scheduling mechanism removes the overhead associated with a software scheduler.

Messages become active either by arriving while the node is idle or executing at a lower priority, or by being at the head of a queue when the preceding message *suspends* execution. When a message becomes active, a task is created to handle it. Task creation, changing the thread of control, and creating a new addressing

```
MOVE      [1,A3],R0      ; get method id
ILATE     R0,A0          ; translate to segment descriptor
LDIP      INITIAL_IP     ; transfer control to method
```

FIGURE 5. MDP assembly code for the CALL message.

environment are performed in one clock cycle. Every message header contains a message *opcode* and the message *length*. The message opcode is loaded into the *IP* to start a new thread of control. The length field is used along with the queue head to create a message segment descriptor that represents the initial addressing environment for the task. The message handler code may open additional segments by translating object IDs in the message into segment descriptors.

No state is saved when a task is created. If a task is preempting lower priority execution, it executes in a separate set of registers. If a task, *A*, becomes active when an earlier task, *B*, at the same priority suspends, *B* is responsible for saving its live state before suspending.

The dispatch mechanism is used directly to process messages requiring low latency (e.g., combining and forwarding). Other messages (e.g., remote procedure call) specify a handler that locates the required method (using the translation mechanism described below) and then transfers control to it.

For example, a remote procedure call message is handled by the call handler code as shown in Figure 5. The first instruction gets the method ID (offset 1 into the message segment reference by *A3*). The next instruction translates this method ID into a segment descriptor for the method and places this descriptor in *A0*. The final instruction transfers control to the method. The method code may then read in arguments from the message queue. The argument object identifiers are translated to physical memory base/length pairs using the translate instruction. If the method needs space to store local state, it may create a context object. When the method has finished execution, or when it needs to wait for a reply, it executes a *SUSPEND* instruction passing control to the next message.

Every register and memory location in the MDP includes a 4-bit tag that indicates the type of data occupying the location. The MDP uses tags for synchronization on data availability in addition to their conventional uses for dynamic typing and run-time type checking. Two tags are provided for synchronization: *future* and *c-future*. A *future* tag is used to identify a named placeholder for data that is not yet available.[3] Applying a strict operator to a *future* causes a fault. A *future* can, however, be copied without faulting. A *c-future* tag identifies a cell awaiting data. Applying any operator to a *c-future* causes a fault. As they are unnamed placeholders, they cannot be copied.

The *c-future* tag is used to suspend a task if it attempts to access data that has not yet arrived from a remote node. When a task sends a message requesting a reply, it marks the cell that will hold the reply as a *c-future*. Any attempt to reference the reply before it is available will fault and suspend the task. When the reply arrives, it overwrites the *c-future* and resumes the task, if it was suspended.

For example, when the task executing in Context 37 in Figure 1 sends the *Sum* message, it marks Slot 3 of its context as a *c-future*. The reply message overwrites Slot 3 to indicate data presence.

The future tag is used to implement named futures as in Multilisp.[18] *futures* are more general than *c-futures* in that they can be copied. However, they are much more expensive than *c-futures*. A memory area and a name must be allocated for each future generated.

The MDP is an experiment in unifying shared-memory and MP parallel computers. Shared-memory machines provide a uniform global name space (address space) that allows PEs to access data regardless of their location. MP machines perform communication and synchronization via node-to-node messages. These two concepts are not mutually exclusive. The MDP provides a virtual addressing mechanism intended to support a global name space while using an execution mechanism based on MP.

The MDP implements a global virtual address space using a general translation mechanism. The MDP memory allows both indexed and set-associative access. By building comparators into the column multiplexer of the on-chip RAM, we are able to provide set-associative access with only a small increase in the size of the peripheral circuitry of the RAM.

The translation mechanism is exposed to the programmer with the *ENTER* and *XLATE* instructions. *ENTER Ra, Rb* associates the contents of *Ra* (the key) with the contents of *Rb* (the data). The association is made on the full 36 bits of the key so that tags may be used to distinguish different keys. *XLATE Ra, Rb* looks up the data associated with the contents of *Ra* and stores this data in *Rb*. The instruction faults if the look-up *misses*. This mechanism is used by our system code to cache ID to segment descriptor (virtual to physical) translations, to cache ID to node number (virtual to physical) translations, and to cache class/selector to segment descriptor (method look-up) translations.

Tags are an integral part of our addressing mechanism. An ID may translate into a segment descriptor for a local object, or a node address for a global object. The tag allows us to distinguish these two cases and a fault provides an efficient mechanism for the test. Tags also allow us to distinguish an ID key from a class/selector key with the same bit pattern.

Most computers provide a set-associative cache to accelerate translations. We have taken this mechanism and exposed it in a pair of instructions that a systems programmer can use for any translation. Providing this general mechanism gives us the freedom to experiment with different address translation mechanisms and different uses of translation. We pay very little for this flexibility, since performance is limited by the number of memory accesses that must be performed.

The J-Machine is a general-purpose parallel computer. It provides general mechanisms for communication, synchronization, and translation rather than hardwiring mechanisms for a specific model of computation. These mechanisms efficiently support many proposed models of computation. Using these mechanisms, the overhead of creating a task on a remote node is reduced to a few microseconds. This low overhead permits concurrency to be exploited at a fine-grain size.

The J-Machine is designed to make efficient use of silicon and wiring area. Each MDP node is a *jellybean* part. It can be fabricated in the same technology used to manufacture existing commodity semiconductor parts such as DRAMs. The network is designed to make efficient use of wires so the machine can be packaged densely, with processing nodes consuming most of the volume.

REFERENCES

1. **Agha, G. A.,** *Actors: A Model of Concurrent Computation in Distributed Systems,* MIT Press, Cambridge, MA, 1986.
2. Ametek Computer Research Division, Series 2010 Product Description, 1987.
3. **Baker, H. and Hewitt, C.,** The incremental garbage collection of processes, in *ACM Conference on AI and Programming Languages,* Rochester, New York, August, 1977, 55.
4. BBN Advanced Computers, Inc., Butterfly Parallel Processor Overview, BBN Report No. 6148, BBN, Cambridge, MA, March 1986.
5. **Dally, W. J.,** *A VLSI Architecture for Concurrent Data Structures,* Kluwer Academic, Hingham, MA, 1987.
6. **Dally, W. J. and Seitz, C. L.,** The Torus routing chip, *J. Distrib. Syst.,* 1(3), 187, 1986.
7. **Dally, W. J.,** Wire efficient VLSI multiprocessor communication networks, in *Proc. Stamford Conf. on Advanced Research in VLSI,* Losleben, P., Ed., MIT Press, Cambridge, MA, 1987, 391.
8. **Dally, W. J. and Seitz, C. L.,** Deadlock-free message routing in multiprocessor interconnection networks, *IEEE Trans. Comput.,* C-36(5), 547, 1987.
9. **Dally, W. J. et al.,** Architecture of a message-driven processor, in *Proc. of the 14th ACM/ IEEE Symp. on Computer Architecture,* June 1987, 189.
10. **Dally, W. J. and Song, P.,** Designs of a self-timed VLSI multicomputer communication controller in *Proc. International Conf. on Computer Design, ICCD-87,* 1987, 230.
11. **Dally, W. J.,** The J-Machine: system support for actors, in *Actors: Knowledge-Based Concurrent Computing,* Hewitt and Agha, Eds., MIT Press, Cambridge, MA, 1989.
12. **Dally, W. J., et al.,** *Message-Driven Processor Architecture, Version 11,* MIT Artificial Intelligence Laboratory Memo No. 1069, MIT, Cambridge, MA, August, 1988.
13. **Dally, W. J.,** Performance Analysis of k-ary n-cube interconnection networks, *IEEE Trans. Comput.,* in press.
14. **Dally, W. J.,** Fine-grain concurrent computers, in *Proc. 3rd Symp. on Hypercube Concurrent Computers and Applications,* ACM, 1988.
15. **Dally, W. J. and Chien, A. A.,** Object oriented concurrent programming in CST, in *Proc. 3rd Symp. on Hypercube Concurrent Computers and Applications,* ACM, 1988.
16. **Dennis, J. B.,** Data flow supercomputers," *IEEE Trans. Comput.,* 13(11), 48, 1980.
17. **Flaig, C. M.,** VLSI Mesh Routing Systems, Technical Report 5241:TR:87, Department of Computer Science, California Institute of Technology, Pasadena, 1987.
18. **Halstead, R. H.,** Parallel symbolic computation, *IEEE Trans. Comput.,* 19(8), 35, 1986.
19. **Hoare, C. A. R.,** Communicating sequential processes, *Commun. ACM,* 21(8), 666, 1978.
20. Inmos Limited, IMS T424 Reference Manual, Order No. 72 TRN 006 00, Bristol, U.K., November 1984.
21. Intel Scientific Computers, iPSC User's Guide, Order No. 175455-001, Santa Clara, CA, August, 1985.
22. **Lutz, C. et al.,** Design of the mosaic element, in *Proc. MIT Conf. on Advanced Research in VLSI,* Artech Books, 1984, 1.
23. **Palmer, J. F.,** The NCUBE family of parallel supercomputers, in Proc. IEEE Int. Conf. on Computer Design, ICCD-86, 1986, 107.

24. **Pfister, G. F. et al.,** The IBM research parallel processor prototype (RP3): introduction and architecture, in Proc. Int. Conf. on Parallel Processing, ICPP, 1985, 764.
25. **Seitz, Charles L.,** The cosmic cube, *Commun. ACM,* 28(1), 22, 1985.

13 PAX

Modern computing for the solution of scientific and engineering problems began in the 1940s. It was recognized by Von Neumann and others that simulation or modeling of real-world phenomena could be used to reduce the amount of physical experimentation in certain situations. When no experimentation was possible it was hoped that models would allow extrapolation and prediction of new results. Weather modeling provides an excellent example. Solving the differential equations describing the temporal evolution of pressures, temperature, and associated variables was one of the earliest motivations for computer modeling. Even today the problem is largely unsolved, although tremendous progress has been made. This growing desire to compute has fueled research in computers and computational science. There are many papers detailing the need for ever faster hardware and software. While some of the original problems can now be solved, science's appetite for computing has grown at least as rapidly as the technology has been able to provide solutions. There is no reason to think that this trend will abate.

Computers were originally sequential, each operation occurring in its own turn. This is essentially the "Von Neuman model", although early scientists appreciated that there is often simultaneity in nature that can be exploited. A substantial amount of simultaneous processing takes place in all computers today. For example, input and output are performed asynchronously with arithmetic processing; but it is only within the last 10 to 15 years that parallel or nonsequential numerical processing has been available.

Here, some aspects of a long-term parallel processing research project (PACS, PAX, and QCDPAX), begun in 1977 at Kyoto University and Hitachi Corporation's Nuclear Power Division, are summarized. The initial name, Processor Array for Continuum Simulation (PACS) was soon changed to Processor Array eXperiment (PAX). QCDPAX is the current running computer (see below). Unless it is necessary to distinguish between them, we refer to the family as PAX computers. Research is now centered at the Institutes of Engineering Mechanics and Information Sciences, University of Tsukuba, Ibaraki 305 Japan. The principal investigator is Professor Tsutomu Hoshino, although he has many long term collaborators, including Professors Shirakawa, Iwasaki, Oyanagi, and others.

The PAX idea was to capitalize on the obvious parallelism in many continuum problems, and the rapidly decreasing costs of computer hardware, along with its associated miniaturization. These ideas have been pursued by other researchers, and PAX was in many ways inspired by research on the ILLIAC IV. The latter was a SIMD computer in which a control unit forced a large number of identical processors to perform identical calculations on different data. ILLIAC IV was seminal for the subsequent development of parallel computers, but it was never commercialized. Even today it is extremely difficult to design a SIMD computer which can cope with data-dependent branching, and this has restricted the applicability of SIMD to special kinds of problems such as image processing. Current SIMD machines, for example, the AMT DAP, are often criticized for this restriction.

PAX also uses several hundred identical processors, but each of these can operate independently. Supercomputers such as Cray Y/MP or NEC SC-3 utilize very powerful asynchronous processors, but typically fewer than 16. Careful programming can sometimes lead to speed-ups nearly equal to the number of processors. Synchronous or asynchronous multiprocessing, sometimes called spatial replication,[2] is one approach for substantial performance improvements and is a major part of current research in high-performance computing. However, very few scientific computations can be performed completely asynchronously. That is generally limited to searches of tables for state variables, calculation of material properties, portions of ray tracing algorithms, etc., that are data driven and hence will be different for each process. Thus, a program may need to determine information about a function at a large number of points on the boundary of a region before further processing. Each of these computations may be done asynchronously, but it will also be necessary to globally synchronize before proceeding from these results. Similarly, in any computation involving iteration or time integration, a global synchronization of all parallel tasks must be performed. However, between these points the calculations are independent and may be performed asynchronously. Thus, a good model for real problems is "quasi MIMD".[3]

Supercomputers also incorporate "pipelining" within their arithmetic units. This means that an operation such as multiply is broken down into a number of steps where each can proceed concurrently on distinct operands. This provides a different type of parallelism and has successfully generated performance improvements of perhaps one order of magnitude, roughly equal to the number of sections or steps in arithmetic operations. Pipeline computers are often referred to as "vector" computers. Sometimes two or more operations, such as + and *, can be "chained" together to provide further speed-up. If there are independent arithmetic units, several operations can be done simultaneously. Breaking arithmetic operations into sections requires some additional hardware and results in its own increased overhead; it is necessary to "latch" between sections to synchronize. New hardware devices such as Josephson junctions might provide a means of increasing the number of steps without introducing too much additional latching overhead. There is also the possibility of pipelining from memory so that operands can be provided to the arithmetic units in greater number per unit time. Currently this is not being done, although the primary restriction seems to be economic. Pipelining is an essential ingredient in improving computer performance, but additional gains to be obtained from it appear to be limited. Most computer scientists think of spatial replication as true parallel processing and feel that it has the most potential for high performance.

The physical questions that originally motivated Hoshino were related to nuclear reactor models, and from these he became interested in "action through a medium" problems. For example, fluid flow equations are normally derived by describing various conservation laws on a small volume of material. Thus, the state at point P is mostly determined by local data around P. Similarly, in statics problems, such as solving the Laplace or Poisson equation, the usual approach is to discretize the region and compute approximations to the spatial derivatives at P by suitable averages of data adjacent to P. Finite differences in two dimensions take five- or nine-point averages; finite-element approximations can be described in the same way. On the other hand, many particle problems do not satisfy this proximity condition, nor do many important numerical methods, such as Gaussian elimination with partial pivoting.

All PAX computers have shared the same basic design philosophy:

1. A high-performance parallel computer to solve specific real engineering problems, rather than general problems.
2. Multiple processing units (PUs), arranged in a torus or two-dimensional rectangular lattice (nearest-neighbor mesh, NNM).
3. Fast connections for data between neighboring PUs.
4. Synchronization and broadcasting between many, or all, PUs whenever necessary, but otherwise independent MIMD processing.

Various PAX computers have been constructed; they are named according to the number of PUs: PAX-9, PAX-32, PAX-128, QCDPAX-240, and QCDPAX-488. The major differences among them are the number of processors, the specific hardware technology used in each PU, the connection network, the memory, and in the case of the last two, the application software that is being developed. QCDPAX-488 is still under construction as this is written. It will have peak performance of about 14 GFLOPS.

The PAX computer system consists of an array of PUs, a host computer, and an interface unit (HPI) to communicate between the host and the PUs. In QCDPAX, each PU contains a Motorola MC68020 (CPU) with a clock speed of 25 MHz. The host is a Sun 3-260 workstation with a color graphic display. In QCDPAX-488 the PUs are arranged in a 24 by 16 array; QCDPAX-288 has them arranged in a 24 by 12 array. The PUs are arranged toroidally. By that we mean that the eastern end of each row is also adjacent to the western end, similarly for north and south.

One particularly interesting design feature of all PAX computers is the provision for fast floating-point computation. This is an integral part of most scientific computing, but is not always included in multiprocessing computers of size comparable to PAX. (All supercomputers provide floating-point hardware.) In QCDPAX each PU contains a vector processing unit (VPU) with floating-point hardware, memory, and control circuitry.

For fast floating-point computation, Hoshino and associates decided that the usual floating-point processor, Motorola 68881, that is often provided with 68020 chips was not fast enough. Instead, they selected as a floating-point processor (FPU) the LSI L64133, which has a peak arithmetic performance of 33 MFLOPS. This chip can perform both a 32-bit, floating-point add and multiply in a single clock period. They have incorporated the L64133 into the VPU, which also has 2 Mbytes of high speed (25 ns) SRAM for data memory (DM), another 8 Kbytes of similar memory called writable control storage (wcs) for the microcode (software) to support the floating-point operations, and 1-Kbyte ROM for a look-up table. There is a three-part pipeline associated with instruction fetch–decode–execute. This pipeline, along with the two simultaneous arithmetic operations, accounts for the name "vector processor". Floating-point operations occur after the CPU sends instructions and data information to the VPU, which then operates independently. Computations such as evaluation or the elementary functions, random numbers, and some matrix computations have been programmed via microcode into the wcs. There are no wait states when the CPU addresses the wcs. The VPU also supports vector operations which are set up by the CPU, but run by the wcs and the FPU. Compared to the 68881, the FPU is difficult to program, and programming problems have been a significant time sink for the small PAX research group. Nevertheless, it accounts for the high peak performance of QCDPAX.

In addition to the 68020 CPU and VPU, each PU contains communication hardware and several different kinds of memory. The PU communicates rapidly to four PUs on the east, west, north, and south by communication memories. Each of these is an 8-kbyte, two-port RAM and thus can be accessed by the PUs that share it. Adjacent PUs thus communicate by reading and writing into the same memory location. Synchronization is by a flag register. The PUs also communicate rapidly with the host by another 8-kbyte, two-port RAM that is associated with the HPI and is mapped onto the address space of the host. As the host writes to its memory it also sets a switch, indicating which PU is to get the data, and that data item then becomes available at the appropriate PU. Communication between a PU and the host works in the same way. Also, by overlapping the addresses, i.e., having all the PUs associated with the same memory locations on the host, the latter can transmit identical data to all the PUs. This mechanism also permits transmission of data from one PU to all the others. In PAX descriptions this type of communication is known as a "ferry". Each PU also has

4 Mbytes of DRAM local memory; this is used to store the PU program and whatever local data it requires.

The host computer is used to compile and assemble source programs, load them into the array of PUs, initiate parallel tasks, send and receive data from the PUs, and output the results to a user via files and the color display. Each local PU, or node, runs whatever program is loaded into it. All PUs are normally loaded with the same program, but this is not a requirement. (This is similar to other asynchronous multiprocessors having many PUs.) Different data at each node means that the program execution path, or thread, is likely to be different for each PU.

To summarize, PAX has four levels of parallelism. First, each PU can compute independently. Second, the CPU can compute independently of the VPU. Third, in the VPU, instruction fetch, decode, and execute can occur simultaneously. Fourth, within the FPU, addition and multiplication can occur simultaneously. A user should only have to think about the first level. This determines the algorithm that is selected and how it is programmed. System software should be responsible for parallelizing on the lower levels. However, Hoshino and colleagues have resorted to programming at all four levels in order to generate good performance from QCDPAX.

The host computer on QCDPAX runs a standard version of UNIX. A user's job is to write a main program for the host and programs for the PUs. Originally, PAX programs were written in Fortran, with some extensions. Now C is used for QCDPAX. This has been extended to allow programmers to communicate between nodes, to the host, etc. Complex arithmetic is supported. Assembly language is also available for taking advantage of the additional levels of parallelism mentioned above, and using it seems to be an essential part of getting optimal performance.

The structure of programs written to run on a PU will be almost entirely understandable to any experienced C programmer. There are some differences due to the communication. Thus variables can be declared as "fast", meaning that they are to be stored in the VPU, "slow", meaning that they are to be stored in the local memory of the PU, "north", "east", "south", or "west", meaning that they are to be stored in the communication memory, and hence available to the adjacent processor, or "ferry", meaning that they are to be stored in the memory space that is mapped onto the memory of the host and hence are available to other PUs. The "vfor" loop forces computations to occur in the FPU; a more traditional "for" loop generates code that executes on the CPU.

```
    vfor (k=0; k<NMAXAL; k+=1) {
u[k] = a[k]*la[k];
    v[k] = 0.0;
}
    for (j=0; j<10; j+=1)
    l_ferry = l[j];
```

The "vfor" loop microcode instructions and data are sent to the VPU, where they begin executing. In this example, both loops execute simultaneously. Usually, a "vfor" loop is inside a "for" loop. In that case the same microcode is sent to the VPU during each iteration of the "for," resulting in substantial overhead. Hoshino is working to have this transmission only occur once.

Hoshino has always felt that it was a mistake to try and develop very general parallel computers. At every opportunity he emphasizes "simple is better." A good example is the two-dimensional lattice arrangement (NNM) of the PUs in PAX. In going from one to two to three dimensions, the mean distance between PUs goes from P/2 to sqrt(P) to 1.5*cbrt(P) for P total processors. If P = 4096 these values are 2048, 64, and 24. Thus, the incremental improvement in three dimensions is modest. Another well-known parallel computer architecture is a "hypercube", in which 2**n PUs are arranged at the corners of an n-dimensional lattice. Rapid communication is possible with PUs sharing an edge. Communication with PUs that are further away takes longer because on average, n/2 edges must be traversed. This is also an example of a nearest-neighbor architecture. PAX is a two-dimensional specialization, but Hoshino believes that the more complicated hardware design of a hypercube reduces its efficiency. For example, the ratio of maximum communication overhead for a 4096-node hypercube (2**12) to a 4096-node PAX computer (2**6 by 2**6) is 128 to 12, about a factor of 10; but if overall communication overhead is 15%, reducing this, even to zero, does not contribute much to program speed-up. It can also be easier to design, write, and debug programs for a "two-dimensional" computer compared to a higher dimensional one, but this depends upon the problem and the tools that are available for the user. On the other hand, communication time has been a serious problem for many multiprocessing computers, and in those cases reducing communication costs by a factor of ten would be extremely important. Further, some problems map better onto a higher dimensional arrangement; but Hoshino and other PAX researchers have shown in a number of papers that often the improvement is asymptotic, and it may be difficult to achieve the best rate in practice.

Hoshino has repeatedly stated that any parallel processing research must be verified by actual implementation. Although some of his papers estimate the efficiency of PAX by mathematical analysis, he has run and measured performance on calculations, including

- Poisson equation in two and three dimensions via successive overrelaxation (SOR), alternating direction implicit (ADI), and Fourier analysis cyclic reduction (FACR) methods.
- Navier-Stokes equations via Beam-Warming, MacCormack, and SOLA methods.
- Linear equations via Gauss-Jordan and conjugate gradient methods.
- Molecular dynamics.
- Plasma simulation via particle in cell (PIC) method.
- Two-dimensional Heisenberg spin model.
- Unsteady flow in flooding river.

TABLE 1

Update time per iteration (μs)	Communication time per iteration (μs)	Efficiency %	GFLOPS	Time to converge (s)
199294	265214	42.9	0.98	26.7

- Logic circuit analysis.
- Quantum chromodynamics (QCD).

Peak measured performance is 8.25 GFLOPS on the 288-node QCDPAX. Expected peak performance will be 13.75 GFLOPS on the 488-node QCDPAX. Table 1 indicates the measured performance of QCDPAX-240 on the solution of a three-dimensional Poisson equation, using a seven-point discretization stencil and red/black SOR iteration.[6] The mesh is $360 \times 370 \times 360$.

The figure of 42.9% efficiency given by Hoshino is computed as computation time divided by total time. The latter includes all the overhead associated with noncomputational tasks, primarily communication. The usual measure is the ratio of measured speed-up relative to maximum theoretical speed-up. Speed-up is the ratio of time for serial processing to solve the problem to time for parallel processing.

Computers with parallel processing capabilities that also require the user to formulate an algorithm in order to take advantage of parallelism are inherently special purpose. Some, such as the NASA Finite Element Machine (FEM) or the Massively Parallel Processor (MPP), have been designed from the ground up for a particular class of problems; in the case of MPP this is satellite image processing. Others, such as Cray YM/P-4, can be used for general computation without making use of parallel constructs. PAX, like a hypercube, is general purpose in the sense that the hardware and software are not restrictive. Both are special purpose in that some problems are much better suited to their architecture than others. They are also special purpose in that their power is from the large number of modestly powerful processors, rather than a modest number of very powerful ones. Within the context of this special-purpose environment, the PAX researchers have always tried to develop a general-purpose computer. The current QCDPAX is unusual because it alters this trend and was developed in response to funding pressures from the Japan Ministry of Education, Science and Culture.

QCDPAX joins together most of the original PAX researchers with physicists interested in a special problem, QCD, and its discrete model, lattice guage theory. QCD calculations involve massive amounts of random number generation, multiplication by 3 by 3 matrices, and computation of exponentials — all floating-point operations. Thousands of hours of supercomputer time are required to generate interesting results. The computations are highly parallel and satisfy the proximity property and thus are extremely well suited to PAX architecture. Special efforts have been made to speed up the software that is necessary for QCD. For example, 3 by 3 matrix multiplication has been written in microcode for the FPU, and complex floating point is supported in C. Several other research projects are attempting to develop special-purpose computers for QCD calculations, notably at Columbia, CalTech, IBM, Rome, Edinburgh, and at FermiLab. A summary discussion of the progress of these groups through 1987 is given by Christ.[1]

Versions of PAX have been under development since 1977. They do not represent any new architectural breakthroughs. Instead, PAX is a determined attempt to design a parallel computer that is simply built to work reliably, simply organized to reduce the chore of programming, and not too special purpose. The application areas are those continuum problems satisfying the proximity property and for which substantial amounts of rapid floating point computation are required. QCDPAX is the most special purpose version of PAX, but only in the application programs and microcode that are specialized for QCD calculations, not in the hardware. The lessons learned building and running QCDPAX can be used directly in other versions, if these are built.

Although many computations have been performed on PAX computers over the years, some of them do not use optimal algorithms. Further, the best PAX results are for two-dimensional problems, despite some extremely clever algorithmic implementations by Hoshino and others.[5] Nevertheless, their examples indicate the scope of potential applications. A number of specific types of computations seem very well suited to the PAX architecture, for example, solving two-dimensional Laplace or Poisson equations using five-point discretization and periodic boundary conditions, because the numerical solution at point P is determined by data at immediately adjacent points. Similarly, spin systems, because they are infinitely expandable homogeneous systems, are appropriate to the periodic boundaries of PAX computers. Further, these systems are controlled by the proximity property; hence most of the communication is between nearest neighbors.

The two-dimensional, nearest-neighbor approach of PAX has somewhat limited its applicability at the same time that it has made the design easier to build and to program. Its fast FPU is a great boost to performance, but has caused many systems programming headaches. The PAX research group is very small. Writing microcode is time consuming and does not produce new physics. The C language extensions that enable users access to PAX's parallelism are easy to understand, but their implementation still has bugs. Some of these have to be worked around via assembly language. Documentation at all levels is weak. What exists is entirely in Japanese. This is not as much of a disadvantage as it seems, because program segments are the best models for writing new programs, and all programs are in English. However, without one-on-one assistance, it would be difficult for a user outside of the PAX environment to use the machine. Hoshino states that he is interested in adding a global addressing capability. He notes, correctly, that in many problems it is convenient to have a single, large data set available to all the PUs.

REFERENCES

1. **Christ, N.,** QCD machines, *Nucl. Phys. B.*, 9 (Proc. Suppl.), 549, 1989.
2. **Hockney, R. and Jesshope, C.,** *Parallel Computers 2.* Adam Hilger, Bristol, 1988.
3. **Hoshino, T.,** An invitation to the world of PAX, *Computer*, May, 69, 1986.
4. **Hoshino, T., Hiromoto, R., Sekiguchi, S., and Majima, S.,** Mapping schemes of the particle-in-cell method implemented on the PAX computer, *Parallel Comput.*, 9(1), 53, 1988.
5. **Hoshino, T., Sato, Y., and Asamoto, Y.,** Parallel Poisson solver FAGECR — implementation and performance evaluation on PAX computer, *J. Inf. Process.*, 12(1), 20, 1988.
6. **Hoshino, T.,** Development of super parallel computer PAX and its applications, in Int. Conf. on Supercomputing in Nuclear Applications, SNA '90.
7. **Iwasaki, Y., Kanaya, K., Yoshie, T., Hoshino, T., Shirakawa, T., Oyanagi, Y., Ichii, S., and Kawai, T.,** Status of QCDPAX, presented by Iwasaki at Int. Symp. on Lattice Field Theory, LATTICE 89, Capri, Italy, September 1989.
8. **Iwasaki, Y., Hoshino, T., Shirakawa, T., and Oyanago, Y.,** QCDPAX: a parallel computer for lattice QCD simulation, *Comput. Phys. Commun.*, 49, 449, 1988.
9. **Shirakawa, T., Hoshino, T., Oyanagi, Y., Iwasaki, Y., Yoshie, T., Kanaya, K., Ichii, S., and Kawai, T.,** QCDPAX — an MIMD array of vector processors for the numerical simulation of quantum chromodynamics, in Proc. Supercomputing '89, Reno NV, November 13 to 17, 1989, ACM Order Number 415892.
10. **Hoshino, T.,** *PAX Computer, High-Speed Parallel Processing and Scientific Computing,* Addison-Wesley, Reading, MA, 1985.

14 CONCERT

OVERVIEW

Concert is a tightly coupled, shared-memory multiprocessor consisting of up to 64 processing modules being developed by MIT and Harris Corporation. A module is shown in Figure 1. These processing modules reside in a multibus chassis, as shown in Figure 2. A single chassis is called a "cluster" or a "slice". As can be seen, each processor is connected via a high-speed bus to at least 512 kbytes of local memory. These memories are dual ported — meaning that the memory module listens to both the multibus and the high-speed bus and that the memory of a given processor can also be accessed by other processors in the cluster. Each cluster has in it up to four processors (with 1 Mbyte each) in the MIT Concert and up to eight processors (with $^1/_2$ Mbyte each) in Harris's Concert.

At MIT, the individual multibusses are connected together in a ring topology, while Harris chose the crossbar interconnection for its increased performance. These interconnections are shown in Figure 3. However, both the machines provide an identical programming model and any software that runs on one machine can run on the other. It can be seen that Concert is a hierarchically organized machine in that it has three levels of memories. The first level is the local memory accessed over the high-speed bus; the second level is the multibus memory in the same cluster; and the third level is the Global memory.

As a program executes on the 68000, it generates addresses. All the addresses in the system are physical addresses. No virtual memory hardware is present. There is no on-chip or external "cache". All the memory in the system is shown in Figures 1, 2, and 3. Thus, the memory requests are satisfied by either the on-board memory, the local memory via the high-speed bus, the multibus memory, or by the globally shared memory. Both machines support 8 Mbytes of global memory. In the MIT machine, this memory is distributed around the ringbus as shown in Figure 3 (left) and is accessed via a ringbus interface board (RIB). At Harris this memory consists of a single block of 16-way interleaved memory called the global communications media (GCM). The Concert address map is shown in Figure 4. If a memory reference cannot be satisfied by any of these sources, a *bus error* will result. Varying access time to different parts of the address space classifies Concert as a NUMA machine. There are several levels

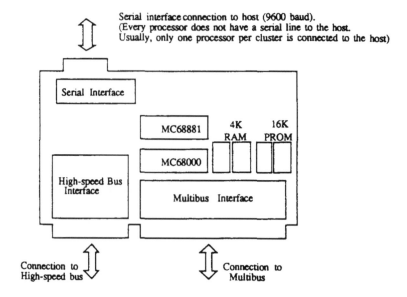

Serial interface connection to host (9600 baud).
(Every processor does not have a serial line to the host.
Usually, only one processor per cluster is connected to the host)

FIGURE 1. A processor module.

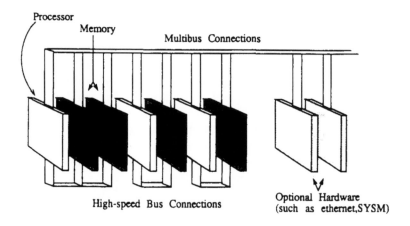

FIGURE 2. A Concert cluster.

of interconnections (busses, ring, or crossbar) that provide varying levels of access speed and are subject to different sources of contention. This is summarized in Figure 5. Naive program organization can easily result in too many processors trying to access some small set of addresses on a single memory module, and the resulting contention can slow the program execution by an order of magnitude or more.

The Concert execution environment may be viewed at several levels, and it is explicitly intended that implementors of programs on Concert should be able to choose among them. At the lower levels, the programmer receives less help

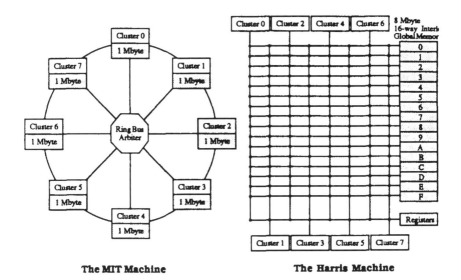

The MIT Machine   **The Harris Machine**

FIGURE 3. Concert interconnect topologies

from library software, but is rewarded with a reduction in overhead and system-imposed conventions.

At the lowest level, Concert can be thought of as consisting of *processors* connected to a memory system. A processor interacts with its environment by means of *messages* sent to other processors (one of which may be running various system servers), by means of areas of memory shared with other processors, and by means of a *serial port* that may be connected to a host machine.

The code executing on a processor may be viewed as having two components: the *bootstrap*, which exists in ROM on each processor, and *loadable code*, which exists in RAM. The bootstrap normally provides a first level of interface for messages and the serial port and handles several kinds of interactions directly, without recourse to the loadable code; loadable code is encouraged to use the subroutines provided in the bootstrap for MP and serial port I/O, rather than bypassing the bootstrap and dealing directly with these facilities.

In the usual case, loadable code is further subdivided into *library code* and *user code*. Library code provides the UNIX standard I/O library interface and convenient access to various levels of MP facilities. Some of the standard I/O functions are accomplished by access to the bootstrap facilities for serial port I/O, and some are accomplished by exchanging messages with other processors presumed to support certain services, such as a TCP/IP server.

The programmer has some flexibility in deciding how much of the available library code to use. If the user needs only processor-to-processor MP service, he or she need not use any of the library code as this type of service is provided by the v3.7 bootstrap. The standard library also includes the level 1 software which provides MP service from process to process. Level 1 involves a significant

000000-000FFF	On-board RAM. Used for interrupt vectors, private variables, and monitor stack.
010000-01FFFF	Global Register Space. Various GCM & Ringbus status registers exist here. These include the protection and Inter Processor interrupt registers.
020000-02FFFF	Local Register Space, only used on Ringbus
080000-3FFFFF	Multibus/High-Speed Bus Memories
400000-B7FFFF	Global Memory (GCM or Ringbus). Global vars.
C00000-DFFFFF	More Multibus-High Speed Bus Memories.
E00000-E7FFFF	More Multibus Mems. or Instrumentation SySM starting @ E10000 DLA starting @ E20000
EE8000-EEBFFF	On Board PROM. Bootstrap code
EF0000-EFFEFF	Multibus I/O Space, e.g. ethernet/disk controllers Interlan Ethernet Controller @ EFxx10-f Disk Controller @ EF0010-3
EFFF00-EFFFFF	On Board I/O Space. Includes: 8274 Serial Controller starting @ EFFF20 8253 Interval Timer starting @ EFFF40 8255 Parallel Ports starting @ EFFF60 8259 Interrupt Controller „ @ EFFF80 68881 FPU starting @ EFFFA0

Note: Missing addresses are reserved.

FIGURE 4. Concert address space.

RAM	Latency	Sources of Contention			
			Other Processor		
		Refresh	In Same Cluster	Using Multibus	In Other Cluster
High Speed Bus	500 ns	■	■		
On-board	750 ns				
Multibus	1250 ns	■	■	■	
Global (GCM)	1500 ns	■		■	■

FIGURE 5. Memory access differences.

amount of library code and it adds a run-time scheduler to multiplex the processor among several executable "threads" or processes. All the Concert programs linked with the standard library will have the level 1 code automatically included in them. While this means that all programs must accept the overhead of a run time scheduler, this overhead is small ($\approx 11\%$) for programs that do not use level 1 code for anything other than interacting with the TCP/IP server.

If user code contains a definition of a particular function, the correspondingly named function will not be loaded from the library; thus, user code may supersede particular modules of the library while still obtaining other modules from the library. Such behavior is acceptable, but will generally require detailed knowledge of the structure of the library, and may open a program to becoming obsolete when the implementation of any of the library functions is changed.

15 COMPUTER VISION APPLICATIONS WITH THE ASSOCIATIVE STRING PROCESSOR*

Real-time computer vision applications are characterized by a rapid response to information derived from a high-speed continuous data stream from a sensor, as indicated in Figure 1. The requirements for rapid response to visual information derived from high-speed continuous data streams and the performance limitations of general-purpose uniprocessors led to parallel processing architectures for computer vision applications.

Sensors include visual, thermal, electromagnetic, and nuclear radiation energy types; and applications include remote sensing, surveillance, tracking, avionics, autonomous navigation, and biomedical. In particular, real-time computer vision systems appear very attractive to any form of robotic applications.

Typical detection tasks range from image reconstruction or restoration (to compensate for sensor nonuniformities, and aging), to filtering (to improve signal-to-noise ratios and enhance particular image features), to correlation (to discriminate relevant image information from background clutter).

Typical analysis tasks range from grouping data associated with meaningful objects, to specific transforms for feature extraction and edge or contour following (to establish feature connectivity), to object quantification (for length or area measurement or center-of-mass coordination) and listing of relevant object properties.

Typical decision tasks range from pattern matching (correlation of invariant object descriptors with model descriptions) and navigation of semantic networks, to hypothesis analysis, to interpretation of object-related information to facilitate decision making.

Applying digital computers to perform detection tasks started over 40 years ago with the use of general-purpose uniprocessors. To overcome the limitations of single processors when operating on image data, two-dimensional mesh-connected arrays of simple processors were proposed. Similarities between the structure of these arrays and the retina served to commend a

* Taken from the article by A. Krikelis in *J. Parallel Distrib. Comput.*, 13, 170-184, 1991. Copyright © 1991 by Academic Press, Inc.

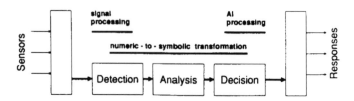

FIGURE 1. Real-time computer vision scenario schematic.

generation of mesh-connected machines. Indeed, the ideas originally proposed by Unger[36] for a two-dimensional array of processors with each processor interconnected to its neighboring processors formed the foundation for such machines as the ILLIAC IV,[34] the CLIP series,[8] the MPP,[3] the DAP,[29] and several others. Such machines are optimized for local transformations (usually associated with detection tasks), but are less efficient for analysis and decision tasks where data processing at the global level is required.

A number of architectures have been proposed with emphasis given to the interconnections among the computational units so as to increase the flexibility in supporting both static and dynamic hardware matching to the task. Such architectures use interprocessor communication paths and/or hierarchical organization which do not only map the image data structures, but also attempt to support computations which result from the particular data flow required in the computation, which in turn depends on the vision level of the task. Such architectures include partitionable SIMD/MIMD architectures,[33] hypercube architectures,[11,14] and homogeneous pyramidal architectures[4,32] and heterogeneous multiresolution architectures.[38] Although such machines can achieve reasonable performance for most of the computer vision tasks when they are performed independently, their performance is reduced considerably when used in integrated computer vision applications where interaction between the different data structures is required. In addition, a number of *logistic* limitations, such as I/O bottlenecking, extensive hardware resource requirements leading to high cost, and low reliability coupled with the general difficulty of developing software for parallel machines, render them inefficient for real-world computer vision applications.

The associative string processor (ASP) modules comprise highly versatile parallel processing building blocks for the simple construction of fault-tolerant, massively parallel processors[19,20] capable of TeraOPS performance (TeraOPS = 10^{12} operations per second) and support of continuous data input. Indeed, in response to application requirements, an appropriate combination of ASP modules would be plugged into the control bus of the data communication network of an appropriate modular ASP system, as indicated in Figure 2.

An ASP module comprises a multiple-instruction control of multiple SIMD nodes and a parallel processing structure of intercommunicating ASP substrings, each supported with an ASP data buffer (ADB) and an ASP control unit (ACU), as shown in Figures 3 and 4. The ASP module also incorporates a single CU and multiple data interfaces (DIs) to connect to the system rack.

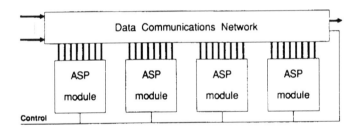

FIGURE 2. Modular ASP system.

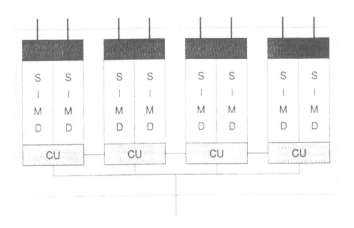

FIGURE 3. ASP module.

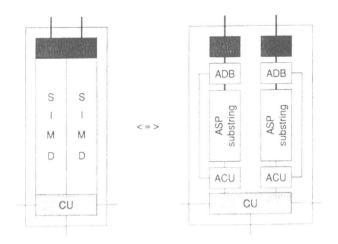

FIGURE 4. ASP module implementation.

Based on a fully programmable, reconfigurable, and fault-tolerant homogeneous parallel processing architecture, ASP modules offer considerable application flexibility, maintaining computational efficiency over a particularly wide range of applications, due to

1. Simple configuration of ASP modules to simplify the development of systems which are well matched to functional application requirements.
2. Pipelining and overlapping I/O data transfers (via the data communication-network) with parallel processing (within ASP substrings) by separating I/O from processing with the ADBs.
3. Overlapping of sequential processing with parallel processing (in ASP substrings).
4. Mapping different application data structures to a common string representation within ASP substrings (supporting content-addressing, massively parallel processing and a reconfigurable interprocessor communication network).

The numeric-to-symbolic data conversion process required in integrated computer vision applications is achieved by first distributing data elements over the identical processors of the ASP substrings and then executing successive tasks on the stored data. Such temporal partitioning of task parallelism ensures that task processing rates are independent of input data rate, and scaling of overall parallel processing power simply entails adjusting the number and length of ASP substrings. Indeed, the uniformity of ASP substrings provides a much cheaper medium for scaling than the expensive processors of a dedicated pipeline.

An ASP substring is a programmable, homogeneous, and fault-tolerant, fine-grain SIMD massively parallel processor incorporating a string of identical associative processing elements (APEs), a reconfigurable interprocessor communication network, and a vector data buffer for fully overlapped data I/O, as indicated in Figure 5.

As shown in Figure 6, each APE incorporates a 64-bit data register and a 6-bit activity register, a 70-bit parallel comparator, a single-bit full-adder, 4 status flags (*viz.* C to represent arithmetic Carry, M and D to tag Matching and Destination APEs, and A to activate selected APEs), and control logic for local processing and communication with other APEs.

In operation, the signal-to-data processing sequence indicated in Figure 1 is achieved by first distributing data elements over the APE data registers and then executing successive tasks on the stored data. Indeed, each ASP substring supports a form of set-associative processing, in which a subset of active APEs (i.e., those which associatively match scalar data and activity values broadcast from the ACU) support scalar-vector (i.e., between the ACU and the data registers) and vector-vector (i.e., within data registers) operations. Matching APEs are either directly activated or source inter-APE communications to indirectly activate other APEs. A match reply (MR) line in the control interface provides feedback on whether none or some APEs match.

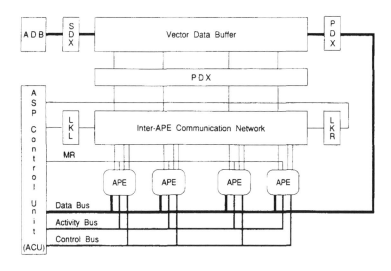

FIGURE 5. ASP substring.

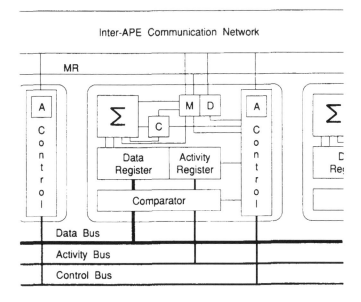

FIGURE 6. Associative processing element (APE).

The inter-APE communication network implements a simply scalable, fault-tolerant and dynamically reconfigurable, tightly coupled APE interconnection strategy, which supports cost-effective emulation of common network topologies, *viz.* arrays (e.g., vector, matrix, and binary *n*-cube), tree networks, graph (e.g., semantic) networks, address permutation (e.g., exchange, shuffle, and butterfly) networks, and shifting networks.[19]

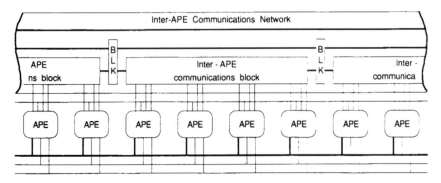

FIGURE 7. ASP block bypassing.

Most significantly, the APE interconnection strategy supports simple unlimited modular network extension, via the link left and link right (LKL and LKR) ports (shown in Figure 5), to enable tailoring of parallel processing power to match user requirements.

In fact, the APE interconnection strategy supports two modes of inter-APE communication:

1. **Circuit-switching** — Asynchronous bidirectional, single-bit communication to connect APE sources and corresponding APE destinations of high-speed activation signals, implementing a fully connected, dynamically configured (programmer transparent) permutation and broadcast network for APE selection and inter-APE routing functions.

2. **Packet-switching** — Synchronous bidirectional, multibit communication, via a high-speed bit-serial shift register for data-message transfer between APE groups.

Thus, the interconnection strategy adopted for ASP substrings supports a high degree of parallelism for local communication and progressively lower degrees of parallelism for longer distance communication.

At an abstract level, and assuming a programmable connection between the ends of a chain of ASP substrings (i.e., LKL and LKR ports shown in Figure 5), the inter-APE communication network can be considered as a hierarchical chordal-ring structure, with the chords bypassing APE blocks (and groups of APE blocks), as indicated in Figure 7, to

1. **Accelerate inter-APE communication signals** — APE blocks not including destination APEs are automatically bypassed for both circuit- and (if required) packet-switched modes of inter-APE communication and, if appropriate for the former mode, activated in a single step.

2. **Provide APE blocks and inter-APE communication network defect/fault-tolerance** — APE blocks and corresponding inter-APE communication networks failing a test routine, either in manufacture or service, are switched out of the string, such that defective or faulty blocks are simply bypassed.

The data communication network supports pipelining of continuous input data streams by routing input data through the (optional) hierarchy of double-buffered global and local ADBs. The network and global ADBs also enable data transfer between parts of selected global and local ADBs, respectively.

At the lowest level of the data-buffering hierarchy, the vector data buffer (shown in Figure 5) supports a dual-port exchange of vector data (i.e., output dumped and input loaded in a single step) with alternating primary data exchange (PDX) and secondary data exchange (SDX), as illustrated in the diagram below.

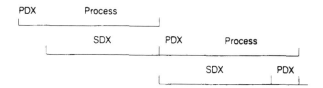

The bit-serial PDX, between the vector data buffer and APEs, performs a very high-bandwidth (i.e., APE-parallel) data exchange via the inter-APE communication network. With a lower bandwidth (i.e., APE-sequential), the bit-parallel SDX provides a data exchange between the vector data buffer and a local or global ADB or the data communication network, which is overlapped with parallel processing and therefore does not present a sequential processing overhead.

Thus, ADBs fulfill a dual purpose: to create a processing window (i.e., decoupling task processing rates from input data rates at the expense of a latency delay in the output response) and to enable data transfer to be fully overlapped with parallel processing in the ASP substrings (thereby eliminating sequential processing overheads which could otherwise be responsible for significant loss of computational efficiency).

Edge detection is an important first task in many image detection approaches. Its objective is to generate a concise, compact description of image features suitable for manipulation.

The most recent work in stereopsis[10,24] and motion detection[25] and the increasing availability of computing power have induced the use of edge operators based on zero crossings of band-passed images. There are many different approaches to edge detection, with each technique attempting to exploit the presence or not of salient image features. In this article the ASP implementation of the Marr-Hildreth edge detector[23] is analyzed. The tasks involved are two-dimensional convolution, zero-crossing detection, and border following.

The two-dimensional convolution of an image results in a weighted average of each pixel with its neighbors.

For example, given the value of each pixel, $p_{x, y}$, the two-dimensional convolution computes the following quantity for each pixel:

W_{11}	W_{12}	W_{13}
W_{21}	W_{22}	W_{23}
W_{31}	W_{32}	W_{33}

FIGURE 8. Definition of a 3×3 convolution window.

$$P_{x,y} = \sum_{i=y}^{n} \sum_{j=1}^{n} W_{i,j} * p_{x+i,y+j}$$

This is called a convolution of radius n, where the $W_{i,j}$ terms are constants representing the function with which the image is being convolved. It is essentially a blurring step that filters out noise with spatial frequency less than n.

For edge detection, in many instances, simple convolution functions, e.g., Gaussian and difference of Gaussian (DOG), or their approximations (e.g., binomial approximation) are used if *a priori* knowledge for the captured scene is available. Such assumptions ease the burden of implementing the convolution tasks since the computation involved is simplified (i.e., separable one-dimensional convolution and additions instead of multiplications).

For the ASP implementation, the general-case two-dimensional convolution is detailed. The special convolution cases can be effortlessly adapted from the following description.

The implementation of a discrete two-dimensional convolution is based on a mask of multipliers, e.g., Figure 8 depicts a 3×3 mask that each pixel applies to its local neighborhood, forming the sum of the pair-wise products of the pixel's neighbors with their corresponding mask values. Typically the sum is then scaled in some manner and the resulting value is used to update the original pixel value. On a parallel processor the application of the mask can be performed simultaneously by all the cells on the grid, as can the update operation. Thus, the algorithm for the convolution can be described as the actions of a single pixel with the understanding that each action is performed simultaneously by all of the pixels.

The majority of two-dimensional convolution algorithms are in the form of systolic processing. A parallel two-dimensional convolution algorithm, described in Reference 21, for mesh-connected array processors (MCAPs) performs the required computation in n^2 computational steps (where $n \times n$ is the convolution mask), emulating systolic-like processing along a convolution path ending at the center of the kernel. The algorithm transfers the partial convolution result along the convolution path, which is a Hamiltonian path, with each PE along the path adding its contribution to the final convolution sum. A shortcoming of the algorithm when implemented on bit-serial SIMD mesh computers (e.g., the Connection Machine, DAP, MPP, and MasPar) is the variable length of the partial results, which has to be transferred from PE to PE. The length of the partial result increases, by $\log_2 m$ (where m is the number of multiply accumulate operations up to that stage), along the convolution path and, therefore, the communication overhead of the algorithm increases, decreasing the overall system performance.

The ASP implementation of the convolution algorithm follows the systolic concept of the algorithm in Reference 21, but it minimizes the communication overhead and uses scalar-vector processing techniques to enhance performance.

On the ASP architecture, pixels are distributed one per APE by concatenating image lines/columns. The basic mechanism of the ASP parallel two-dimensional convolution algorithm is illustrated with the following example of the 3×3 convolution kernel shown in Figure 8. Initially it is assumed that the image is loaded, by concatenating image lines (i.e., pixels $p_{i,j-1}, p_{i,j}$, and $p_{i,j+1}$ are stored in neighboring APEs and, if N is the line length in pixels) $p_{1,N}$ and $p_{i+1,1}$ are stored in neighboring APEs), and a serial field r_{conv} (corresponding to partial result) is declared in every APE. The length of the field r_{conv} is equal to $a + b + \log_2 9$, where a and b are the bit length of image pixels and convolution weights, respectively, and are initially reset to zero. Initially, all APEs calculate the first partial result

$$r_1 = r_o + W_{22} * p_{i,j} = W_{22} * p_{i,j}$$

where $p_{i,j}$ indicates for all image pixels. The multiply-accumulate operation is performed as a scalar-vector type of operation (*viz.* all vector elements p_i are multiplied with the scalar W_{22}). The scalar is analyzed by the ASP controller, and depending on its bit pattern, the appropriate instructions are broadcasted to the ASP system (e.g., accumulation of partial result and pixel and/or arithmetic shifting). The accumulation of the partial results is bit serial, but the analysis of the scalar bit patterns can be done on the multibit basis. Indeed, fast multiplication algorithms have been developed, e.g., modified Booth's multiplication, which is twice as fast as the bit-serial multiplication, based on the analysis of multibit sequences of the multiplier. In addition, multiplication by weights equal to zero can be completely ignored (no computation time is necessary), since it will not alter the already calculated partial result. Each APE is transferring its pixel value to its left APE neighbor (one bit at a time). In the next step all APEs calculate the following partial result:

$$r_2 = r_1 + W_{32} * p_{i,j} = W_{22} * p_{i,1} + W_{23} * p_{i+1,j}$$

Each APE transfers the received pixel value to the APE, which is $p - q + 1$ (where p is the horizontal length in pixels of the processed image patch and q is the corresponding length of the convolution kernel) APEs to its left.

The above-described sequence of operation is repeated two more times for the neighbors of each APE, which is approximately half of the total convolution path. In five steps the following partial result has been calculated by each APE:

$$r_5 = W_{22} + p_{i,j} + W_{23} * p_{i+1,j} + W_{31} * p_{i-1,j-1}$$
$$+ W_{32} * p_{1,j-1} + W_{33} * p_{i+1,j-1}$$

If the same sequence of operations is repeated to the right of each APE, the remaining terms of the convolution result are computed:

$$r_{conv} = r_5 * W_{21} * p_{i-1,j} + W_{13} * p_{i+1,j+1}$$
$$+ W_{12} * p_{i,j+1} + W_{11} * p_{i-1,j+1}$$

Therefore, a convolution using a 3×3 kernel is completed in nine steps independent of the size of the image path processed by the ASP system.

It is evident that the implemented convolution algorithm is more efficient than the one described in Reference 21 for bit-serial MCAPs.

The ASP implementation minimizes the amount of information transferred between APEs (a-bit pixel information in contrast to $a + b + \log_2 m$-bit partial result information), reducing the communication overhead to that which is absolutely necessary for the algorithm to be performed. Although in cases where large images are processed the physical distance between APEs corresponding to pixels of successive image rows is long, there is no severe communication overhead (only graceful degradation) since electrically (for information transfer) they are close due to ASP segment bypassing mentioned in Lea.[19,20]

The ASP implementation enhances the performance of the convolution algorithm using scalar-vector computation to calculate the required sum. Instead of broadcasting the convolution weights to each processing element and then performing bit-serial accumulation as is suggested in Reference, the scalar-vector technique allows multibit analysis of the weights, which can considerably speed up the execution time (at least twice the improvement can be expected).

Zero crossings are detected by scanning the convolved image along horizontal lines for the pixel occurrences:

+	−		+	0		−	+		−	0		0	+		0	−

along vertical columns for the pixel occurrences:

+		+		−		−		0		0
−		0		+		0		+		−

and along the 45° direction for the pixel occurrences:

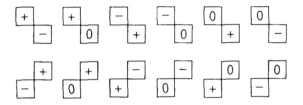

where + denotes a positive pixel value, – a negative pixel value, and 0 a zero pixel value. When one of the above occurrences is found, a zero crossing is designated to the position of the left/top pixel in cases where nonzero pixels are examined and to the zero pixel in all other cases.

The zero-crossing detection algorithm, if all pixel occurrences illustrated above are combined, will detect a zero crossing at the central pixel location for a 3 × 3 pixel neighborhood, if any of the neighborhood pixels are of different sign/nonzero values.

The ASP implementation of the zero-crossing detection is very much simplified using the associative property of the architecture. Initially, information from the 3 × 3 neighborhood of each pixel (for all pixels simultaneously) is collected using the synchronous mode of the inter-ASP communication network. The information comprises only the sign bit of each of the neighbors (i.e., 0 if the neighbor is positive and 1 if the neighbor is negative). Information from neighbors which are 0 is not transferred, since it will not have any effect, as was indicated above in the algorithm description.

The transferred information is used to set two of the bits in the activity register of each APE. One activity bit is set if any of the neighbors is positive and the other bit is set if any of the neighbors is negative. With the neighborhood information local to each pixel (i.e., APE), four simple associative search operations are employed to identify all zero-crossing locations in the image.

In some cases, in order to provide some measure of noise immunity, all zero crossings with contrast less than a given threshold are ignored. This can be implemented on the ASP if, in the four associative search operations used to identify the zero crossings, instead of the convolved pixel value the operations use the value incremented by the perceived threshold, i.e., a scalar-vector addition will increase all convolved pixel values by the threshold value.

Connected sets of edge points are useful for many analysis and decision-level computer vision tasks. Border following is the task of turning a set of edge points into connected curves. Subsequent operation may allocate a common label to all points of a curve, which can be used by later computer vision tasks that need a unique tag or pointer to refer to the edge of an object as a unit.

In the ASP, each edge point finds the neighboring edge points with which it forms a connected line by examining the eight neighbors in its 3 × 3 neighborhood. Following this operation, all edge points calculate pointers to the edge pixels they are connected to. This is achieved by adding and/or subtracting one from their own row and column addresses according to the position of the linked edge points in the 3×3 neighborhood.

The labeling of all edge points with a common tag can be achieved in two ways:

1. *Propagation*. With the propagation technique, a label is relayed to all connected edge points through the inter-APE communication network. Starting from an arbitrary single edge point, a single bit is propagated to the edge points in its neighborhood. The propagation must negotiate all edgepoints and therefore the procedure takes n steps on a curve n points long. If the points of the curve form a ring, the labeling is achieved in $n/2$ steps.
2. *Broadcasting and associative matching*. In this case, starting from an arbitrary single edge point, the addresses to the edge points in the neighborhood are broadcasted through the ASP global data bus and used to associatively search for the next edge point. The step is repeated until there are no pointers left directed to nonlabeled edge points. The procedure requires n steps to label a curve n points long, independent of the shape of the curve.

In the case of the ASP, the second method of labeling the edge points is the most cost effective, since it makes use of the associative nature of the architecture and its capability to search for byte- and word-wide information and avoids interprocessing element communication, a computationally expensive feature in all parallel architectures.

The Hough transform[2,7,12] is a powerful technique for accumulating evidence for parameterized objects from features in an image. In computer vision applications, and especially in the industrial field, the features are edge points and the parameterized objects are lines. The generalized Hough transform is also used to determine the parameters of the position and orientation of a known object from an image.

An image line (or, in general, a curve) is represented by a single point (d, θ) in the Hough parameter space, where d is the distance of the line from the origin of the coordinate system and θ is the angle of the normal of the line relative to the positive x-axis. If (x_i, y_i) are the coordinates of nonzero pixels of the line, then Hough mapping can be expressed as follows:

$$d = x_i * \cos(\theta) + y_i * \sin(\theta)$$

Therefore, the Hough transform is a two-dimensional histogram indexed by (d, θ).

The above definition of the Hough transform is suitable for sequential machines and data-driven pipeline architectures, but rather ineffective for parallel implementation.

A more efficient way for parallel implementation is based on the observation that the Hough transform can be regarded as based on projections of the image along many different directions.[31] The image projection in a direction θ is obtained by accumulating the nonzero pixels along the family of lines perpendicular to θ. The strength of the image lines is defined as the sum of the accumulated nonzero pixels.

According to the above observation, the Hough transform can be calculated by tracing the image lines on a parallel architecture along natural image *corridors* which can be adapted to the features of the traced line (e.g., line width).

In the ASP the APEs corresponding to the bottom, left, and right borders of the image (namely accumulating APEs) constitute a serial field C (equal to zero in the beginning of the operation), which will accumulate the counts for a single direction . All (d, θ) for a given θ will be accumulated simultaneously. The accumulation of the Hough histogram is reduced to a series of shift and add operations, as described below:

1. In all accumulating APEs, if pixel $< > 0$, then $C = C + 1$.
2. Shift all image pixels to the right until the next line pixel reaches its accumulating APE. The shift distance depends on the direction θ and is *a priori* known.

The above implementation achieves a first-order parallelism with still many sequential elements evident in the computations (i.e., the steps are repeated sequentially for all possible directions).

An improved algorithm[37] based in the above approach achieves a greater degree of parallelism by calculating the Hough transform of an image in only two steps. In the first step the Hough transform bins for all lines with direction between the angles $-45°$ and $+45°$ are calculated, and in the second step the bins for all lines between the remaining angles are calculated (i.e., $+45°$ and $+135°$). The improvement in the algorithm stems from the fact that all available APEs are used as accumulating APEs.

Geometrical computations in computer vision applications form the backbone of the processing required by the analysis stage. Detected scene objects must be examined to abstract and extract information effective for the decision stage.

Computational geometry provides the principal mathematical tools that aid in the design of efficient approaches. More specifically, algorithms of combinatorial geometry and the concepts of convexity, proximity, and inversive geometry enable global analysis of object properties and object classification into clusters.

In this discussion, the implementation of two geometrical computation algorithms on the ASP is detailed:

1. Planar convex hull.[26]
2. Planar Voronoi diagram.[26]

The convex hull of a set of points/features/objects in the plane is the smallest convex region enclosing all of the points/features/objects. Convex hull algorithms are useful for pattern normalization, topological feature extraction, shape decomposition, etc.

As indicated below, a number of planar convex hull algorithms have been implemented on the ASP.

Graham's algorithm — This is a parallel version of Graham's sequential algorithm,[9] which for p planar points requires $O(\log p)$ steps. The algorithm selects a random point q within the extremes of the set of the points of interest and sorts the remaining points with respect to the angle they make with q. Once sorted, neighboring pairs of points in the sorted order are merged into partial hulls (convex arcs). These merged pairs are then merged, and so forth, until all points of S are merged into one convex hull. Each merging step require $O(\log_2 p)$ steps, and in all $O(\log_2 p)$ merges will be required.

QUICKHULL algorithm — This is a parallel version of the QUICKHULL sequential algorithm,[26] which, depending on the distribution of the points in the plane, can be completed in the range of $O(\log_2 n)$ up to $O(n)$ steps for n hull points, depending on the distribution of the points. For the set of the points of interest, the algorithm first finds the two X extrema and partitions the points into two subsets, one above and one below the line. Each subset will contain a polygon chain. The polygon in each subset is approximated by a number of triangles, which, if they were merged, will construct the polygon for the corresponding subset. Linking the two polygon chains will give the convex hull polygon of the set of points.

Jarvis's algorithm — This is a modified parallel version of the Jarvis sequential algorithm,[15] which is completed in $O(n)$ steps for n hull points. The concept of the algorithm is to start from a point of the set of interest which is known to be a hull point (e.g., extremum points are always hull points) and find the point that makes a maximum angle with this point; this point will be the next point on the hull. The step will be repeated until the next hull point found is the original hull point. Although the algorithm requires $O(n)$ steps to be completed, the operations performed at each step are very simple and in many cases can be faster than the other convex hull algorithms mentioned.

Below, the implementation of the Jarvis convex hull algorithm is detailed. Each image point in the ASP can be considered as a record comprising a tag associated with the point representing the key of the record, and the X, Y planar coordinates, associated with each point, as its elements.

A new hull point is identified by evaluating the angles made between the known hull point and the rest of the points of interest (e.g., all marked with a given bit in their corresponding activity register).

Initially, the coordinates X_h and Y_h of the hull point are broadcasted to all other points. In the set of the points of interest, for each point i (except the hull points), calculate the values $(X_i - X_h)$ and $(Y_i - Y_h)$.

The angle (with the hull point) of each point is calculated in two phases:

1. In the first phase the quadrant of the angle for each point is calculated by sensing the signs of the differences $(X_i - X_h)$ and $(Y_i - Y_h)$.
2. The tangent of each angle inside the corresponding quadrant is computed by calculating the ratio $abs(X_i - X_h)/abs(Y_i - Y_h)$, where $abs(A)$ is the absolute value of A.

Finally the point with the maximum angle, in the expected quadrant, is identified using minimum or maximum search and marked as a convex hull point. A pointer is set to indicate the tag of the next hull point.

The above computation is repeated until the pointer for the next hull point refers to an image point which is already classified as a convex hull point. The associative nature of the ASP supports all the conditional operations involved.

The implementation of Jarvis's algorithm for the computation of a planar convex hull for computer vision applications does not involve any inter-APE communications. Instead, the global data bus of the ASP architecture is used to globally broadcast point coordinates and tag information with the APEs, performing actual computation almost continuously.

The Voronoi diagram of a set of planar points (e.g., image points) portrays a tessellation of the plane into Voronoi regions. Each region surrounds a point and delimits the portion of the plane closer to that point than to any other point in the set.[26] The Voronoi diagram is an efficient computational structure for tackling proximity problems (e.g., two objects with the smallest mutual distance in a scene and nearest neighbor of an object).

The algorithm selected to be implemented on the ASP makes use of the Voronoi diagram properties:

- The Voronoi regions partition the image plan into a convex net. The vertices of the net are called Voronoi vertices and its line segments are called Voronoi edges.
- Every vertex of the Voronoi diagram is the common intersection of exactly three edges of the diagram.

Equivalently, the above property means that the Voronoi vertices are the centers of circles defined by three points of the original set of points. Each of these circles has the property that it contains no other points of the set.

Therefore, if the area defined by the points is triangulated in such a way that the circumcircle of each triangle contains no other points, then the centers of the circles define the Voronoi vertices and the lines connecting these circles are the Voronoi edges. All the triangles are encompassed by the convex hull defined by the set of the image points of interest.

The ASP implementation of the algorithm has two stages.

Initially, the algorithm simultaneously builds all the triangles necessary to triangulate the plane by identifying the three vertices defining each triangle (using a brush fire-like technique). Each point on the image propagates its identity, through the inter-APE communication network, to its neighborhood (*viz.* eight-way connected). In this way pairs of pixels will be formed which will not propagate any further. If a single pixel is propagated to a pixel where a pair already exists, then a triple will be formed which will represent a triangle. Pixels with triplets will neither propagate nor accept any information. The algorithm terminates when no further propagation is possible. Usually the algorithm requires M steps, where M is the diameter of the largest propagation region.

In the second stage, the algorithm obtains a graph representation of the dual of the Voronoi diagram, the Delauney triangulation, from the triplets formed in the previous stage.

The algorithm calculates the center of the circumcircle of each triangle, which is the Voronoi vertex, for all triangles in parallel.

Each triangle is stored in the ASP as a record with the coordinates of its vertices as the record elements.

The main steps of this part of the algorithm are

1. The line equations corresponding to two of the edges of each triangle are computed. Using homogeneous coordinates if $Ax + By + C = 0$ is a line equation, then for the line defined by the points $(x_i, y_i, 1)$ and $(x_j, y_j, 1)$,

$$A = y_j - y_i$$
$$B = x_i - x_j$$
$$C = x_j * y_i - x_i * y_j$$

   The line equations for all required edges can be calculated with one vector-vector multiplication and two subtractions.

2. The bisector lines of the edges (for which the line equations have been determined) are computed next.

   The bisector line (e.g., $Dx + Ey + F = 0$) of a line segment, between the points $(x_i, y_i, 1)$ and $(x_j, y_j, 1)$, on the line defined by the equation $Ax + By + C = 0$ is computed as follows:

$$D = B$$
$$E = -A$$
$$F = A * (y_i + y_j)/2 - B * (x_i + x_j)/2$$

   The bisector lines for all edges simultaneously can be calculated with three additions and two vector-vector multiplications.

3. A way of calculating the center of the circumcircle of a triangle (the Voronoi vertices are defined as the centers of such circles) is by computing the intersection point between the bisectors of two of its edges. The intersection of two lines defined by the equations $Ax + By + C = 0$ and $Dx + Ey + F = 0$ is the point with coordinates

$$x = (F * B - E * C)/(E * A - D * B)$$
$$y = (C * D - A * F)/(E * A - D * B)$$

   The coordinates of all centers of circumcircles can be calculated with three scalar-vector multiplications, three subtractions, and one division.

When all Voronoi vertices have been generated, the Voronoi edges can be computed by connecting vertices corresponding to triangles which share a common edge. The Voronoi rays correspond to the Voronoi vertices which belong to a triangle, with one of its edges being a convex hull edge. The Voronoi ray has a line equation identical to the bisector of the corresponding convex hull edge. This last step of the algorithm is data dependent (*viz.* how the points corresponding to set *S* are distributed on the plane). The worst-case performance is when all $2N - 5$ possible vertices of the Voronoi diagram must be calculated.

Finding subgraphs of a given graph that are isomorphic to another given graph is very useful in the decision stage of a computer vision scenario for detecting relationally described objects embedded in a scene.

The problem of identifying a single isomorphic embedding is an NP-class problem (i.e., it can be solved in predictable time on a nondeterministic machine). A subgraph isomorphism algorithm can identify all the possible embeddings, but the estimated time can be very long. The specified variation of the algorithm, where the sum of absolute differences between the identified subgraphs and reference subgraph is minimum, is a non-NP-complete problem. However, it requires the identification of all possible subgraphs which is an NP-complete problem.

Assuming that the input is a graph *G* having *M* vertices, each joined by the edge to *n* other selected at random, and another graph *H* having *p* vertices, each joined by an edge to *q* other vertices selected at random, the output is a list of the occurrences of (an isomorphic image of) *H* as a subgraph of *G*.

The ASP algorithm for the graph isomorphism problem is based on Boolean matrix manipulation using Ullmann's refinement procedure.[35] This method reduces unnecessary graph searching by eliminating mappings that are impossible because of connectivity requirements.

Initially a permutation Boolean matrix *M* is defined, with dimensions $m \times p$, which contains all possible isomorphisms (*viz.* all matrix elements are set to 1). G_M and H_M are defined as the adjacent matrices corresponding to graphs *G* and *H*, respectively.

The adjacent matrix of a graph $G = <V, E>$ (where *V* is the set of vertices and *E* is the set of edges) is defined as a $\| V \| \times \| V \|$ Boolean matrix, *A*, of 0s and 1s, where the *ij*th element of matrix *a* is 1 if, and only if, there is an edge from vertex *i* to vertex *j*.

Ullmann's method is iterative, where in the *k*th iteration, the following Boolean matrix equations are calculated:

$$P = M_k \times H_M \tag{1}$$

$$Q = G_M \times P_{comp} \tag{2}$$

$$M_{k+1} = M_k \cdot Q_{comp} \tag{3}$$

where "$\times$" indicates the Boolean product, "$\cdot$" indicates logical AND, *k* indicates the iteration number, and P_{comp} and Q_{comp} indicate the complement matrices of *P* and *Q*, respectively.

The matrix M_{k+1} is tested for isomorphism (*viz*. each row contains exactly one 1 and no column contains more than one 1). If no isomorphism is found, the same matrix is tested for the stop condition (*viz*. any row of the matrix containing no 1s), which terminates the algorithm execution. If an isomorphism is found (the matrix M_{k+1} is dumped) or no stop condition is found, the algorithm carries on to the next iteration by reevaluating the matrix Equations 1 to 3. In the case of an isomorphism, each 1 in matrix M_{k+1} (corresponding to element m_{ij}), the coordinate i corresponds to the ith vertex of G, and the coordinate j corresponds to the jth vertex of H, between which isomorphism exists.

For the defined task, G_M is a $m \times m$ matrix and H_M is a $p \times p$ matrix.

Initially, in the ASP implementation of a graph-matching algorithm, it is assumed that the matrix M is loaded into the ASP. The assumption of the loading is that M is loaded in a row-major manner: a row of M (one element in each APE) as well as a complete copy of H is stored in each virtual APE. In other words, matrix H is replicated with every row of M. The following steps describe one iteration of the ASP implementation of the graph-matching algorithm.

1. Calculate the matrix $P = M_k \times H_M$. Each ASP section will, in parallel, generate one row of p by performing a series (*viz*. p) of operations, each comprising
    *   Logical AND between the row of M and a column of H using pattern searching.
    *   Logical OR of the resulting vector using the inter-APE communication network to obtain each element of matrix M.
    The matrix p calculated above is concatenated in the ASP chain row-by-row.
2. Calculate the matrix $Q = G_M \times P_{comp}$. P_{comp} is obtained by inverting 0 elements of P to 1s and 1 elements to 0s. Computation of row of matrix Q is achieved by distributing one row of G_M with each row of P_{comp} and performing logical ANDs between matrix elements using pattern searching. Each element of Q is then derived as the OR of corresponding subelements of the intermediate result vector. This is performed as a single global search operation. The matrix Q is concatenated row-by-row in the ASP chain.
3. Calculate the matrix $M_{k+1} = M_k Q_{comp}$. Q_{comp} is obtained by inverting 0 elements of Q to 1s and 1 elements to 0s. The elements of M_k and Q_{comp} are in place and a logical AND (implemented using pattern searching) will generate the matrix M_{k+1}.

Following the refinement procedure, the matrix M_{k+1} is tested for isomorphism (*viz*. each row contains exactly one 1 and no column contains more than one 1). The test for isomorphism is accomplished by counting the number of 1s along the matrix rows and using global searches along the matrix columns.

If a matrix M_{k+1} passes the isomorphism test, it is dumped to the output buffer. If it fails the isomorphism test, then it is tested for the stop condition: any row of the matrix containing no 1s. This is accomplished by inspecting the row counters used in the previous step with a single global search.

REFERENCES

1. **Annaratone, M., Arnould, E., Gross, T., Kung, H. T., Lam, M., Menzilcioglu, O., and Webb, J. A.,** The warp computer: Architecture, implementation and performance, *IEEE Trans. Comput.*, C-36(12), 1523, 1987.

2. **Ballard, D. H.,** Generalizing the Hough transform to detect arbitrary shapes, *Pattern Recognition*, 3(2), 11, 1981.

3. **Batcher, K.,** Design of a massively parallel processor, *IEEE Trans. Comput.*, C-29(9), 836, 1980.

4. **Cantoni, V., Ferreti, M., Levialdi, S., and Stefanelli, R.,** PAPIA: pyramidal architecture for parallel image analysis, in *Proc. 7th IEEE Symp. on Computer Arithmetic*, IEEE Computer Society, 1985, 237.

5. *DATACUBE: MaxVideo User Manual*

6. **Davis, R. and Thomas, D.,** Systolic array chip matches the pace of high-speed processing, *Elec. Des.*, 31, 207, 1984.

7. **Duda, R. O. and Hart, P. E.,** Use of the Hough transformation to detect lines and curves in pictures, *Commun. ACM*, 15(1), 11, 1972.

8. **Duff, M. J. B., Watson, D. M., Fountain, T. J., and Shaw, G. K.,** A cellular logic array for image processing, *Pattern Recognition*, 5, 229, 1973.

9. **Graham, R. L.,** An efficient algorithm for determining the convex hull of a finite planar set, *Inf. Process. Lett.*, 1, 132, 1972.

10. **Grimson, W. E. L.,** A computer implementation of a theory of human stereo vision, *Philos. Trans. R. Soc. London Ser. B*, 292, 1981.

11. **Hillis, D.,** *The Connection Machine*, MIT Press, Cambridge, MA, 1986.

12. **Hough, P. V.,** Methods and Means to Recognize Complex Patterns, U.S. Patent 3.069.654, 1962.

13. **Ibrahim, H. A. H., Kender, J. R., and Shaw, D. E.,** Low-level image analysis tasks on fine-grained tree-structured SIMD machines. *J. Parallel Distrib. Comput.*, 4(3), 546, 1987.

14. Intel Corporation, iPSC System Overview, January 1986.

15. **Jarvis, R. A.,** On the identification of the convex hull of a finite set of points in the plane, *Info. Process. Lett.*, 2, 18, 1973.

16. **Krikelis, A. and Lea, R. M.,** Performance of the ASP on the DARPA architecture benchmark, in *Proc. Frontiers 88. 2nd Symp. on the Frontiers of Massively Parallel Computation*, IEEE Computer Society, October 1988, 483.

17. **Krikelis, A.,** Benchmarking the ASP for computer vision, in Proc. SPIE VCIP '90, SPIE Conference, Lausanne, Switzerland, October 1990.

18. **Kushner, T., Wu, A. Y., and Rosenfeld, A.,** Image processing on MPP, *Pattern Recognition*, 15(3), 121, 1982.

19. **Lea, R. M.,** The ASP: a cost-effective parallel microcomputer, *IEEE Micro.*, October, 10, 1988.

20. **Lea, R. M.,** ASP modules: cost-effective building blocks for real-time DSP systems, *J. VLSI Signal Process.*, 1(1), 61, 1989.

21. **Lee, S.-Y. and Aggarwal, J. K.,** Parallel 2-D convolution on a mesh connected array processor, *IEEE Trans. Pattern Anal. Mach. Intell.*, PAMI-9, 590, 1987.

22. **Lone, S., Bock, R. K., Ermolin, Y., Krischer, W., Ljuslin, C., and Zografos, K.,** Fine-grain parallel computer architectures in future triggers CERN Report CERN-LAA RT/89-05, Sept. 1989; *Nucl. Instrum. Methods Phys. Res.*, submitted.

23. **Marr, D. and Hildreth, E.,** Theory of edge detection, *Proc. R. Soc. London Ser. B*, 207, 187, 1980.

24. **Marr, D. and Poggio, T. A.,** computational theory of human stereo vision, *Proc. R. Soc. London Ser. B*, 204, 301, 1979.

25. **Marr, D. and Ullman, S.,** Directional selectivity and its use in early visual processing, *Proc. R. Soc. London Ser. B*, 211, 151, 1981.

26. **Preparata, F. P. and Shamos, I. M.,** *Computational Geometry — An Introduction,* Springer-Verlag, Berlin, 1985.

27. **Preston, K., Jr.,** Comparison of parallel processing machines: a proposal, in *Languages and Architectures for Image Processing,* Duff, M. J. B. and Levialdi, S., Eds., Academic Press, New York, 1981, 305.

28. **Preston, K., Jr.,** The Abingdon cross-benchmark survey, *IEEE Comput.,* 22(7), 9, 1989.

29. **Reddaway, S. F.,** DAP-A distributed array processor, in Proc. First Annu. Symp. on Computer Architecture, Florida, 1973, 61.

30. **Rosenfeld, A.,** A report on the DARPA image understanding architectures workshop in *Proc. DARPA Image Understanding Workshop, Los Angeles, CA, Feb. 1987,* Kaufmann, San Mateo, CA, 1987, 298.

31. **Rosenfeld, A. and Kak, A.,** *Digital Picture Processing,* Academic Press, New York, 1982.

32. **Schaefer, D. H., Ho, P., Boyd, P., and Vallejos, C.,** The GAM pyramid, in *Parallel Computer Vision,* Uhr, L., Ed., Academic Press, New York, 1987, 15.

33. **Siegel, H. J.,** PASM: a reconfigurable multicomputer system for image processing, in *Languages and Architectures for Image Processing,* Duff, M. J. B. and Levialdi, S., Eds., Academic Press, New York, 1981, 257.

34. **Slotnick, D. L.,** The fastest computer, *Sci. Am.,* 224, 76, 1987.

35. **Ullmann, J. R.,** An algorithm for subgraph isomorphism, *J. Assoc. Comput. Mach.,* 23(1), 31, 1976.

36. **Unger, S. H.,** A computer oriented towards spatial problems, *Proc. IRE,* 46, 1744, 1958.

37. **Vesztergombi, G.,** "Iconic" Tracking Algorithms for High Energy Physics Using the Trax-1 Massively Parallel Processor, Max-Plank Institut fur Physik and Astrophysik, Germany, Document MPI-PAE/Exp.E1. 216, 1989.

38. **Weems, C. C., Levitan, S. P., Hanson, A., Riseman, E., Shu, D. B., and Nash, G. J.,** The image understanding architecture, *Int. J. Comput. Vision,* 2, 251, 1989.

39. **Weems, C. C., Riseman, E., Hanson, A., and Rosenfeld, A.,** An integrated image understanding benchmark: recognition of a 2-1/2 D mobile, in *Proc. DARPA Image Understanding Workshop, Cambridge, MA, Apr. 1988,* Kaufmann, New York, 1988, 111.

40. **Weems, C. C., Riseman, E., Hanson, A., and Rosenfeld, A.,** A report on the results of the DARPA integrated image understanding benchmark exercise, in *Proc. DARPA Image Understanding Workshop, Palo Alto, CA, May 1989,* Kaufmann, New York, 1989, 165.

41. **Weems, C. C., Riseman, E., Hanson, A., and Rosenfeld, A.,** The DARPA image understanding benchmark for parallel computers, *J. Parallel Distrib. Comput.,* 11(1), 1, 1991.

Part 2
MIMD Software

This part of *Parallel Supercomputing in MIMD Architectures* briefly examines MIMD software, a topic too large and complex to cover thoroughly here. In keeping with the survey character of the book, four major facets of the topic are addressed: operating systems, languages, translating sequential programs to parallel, and semiautomatic parallelizing. In each case the exposition is undertaken through the explanation of an example: Trollius, Apply, Linda, and Ptool.

One facet of MIMD software that would be expected were this a survey of vector-style supercomputers but that is not discussed here is commercial off-the-shelf (COTS) applications software from third-party vendors. Regretably, there isn't much available as yet.

16 OPERATING SYSTEMS: TROLLIUS

OVERVIEW

Trollius was conceived because the best multicomputer software in existence at the time of its birth, Autumn 1985, was not very appealing even to the adventurous computational scientist. In particular, the transputer, building block of a new generation of multicomputers, had essentially no operating system. Trollius started with both a general and a specific purpose.

An opportunity appeared to build a strong operating system foundation suitable for future multicomputers. Contemporary system software from the hypercube generation appeared to be hastily constructed solutions from companies under pressure to reach the market fast and to focus attention on the promise of the architecture. Since the first commercial introduction of multicomputers, hardware concerns have been progressively addressed, while software has lagged behind. Standards have not appeared. Trollius did not anticipate a multicomputer as a stand-alone workstation, but virtually every other possible usage is supported. New operating systems continue to be built, often for proprietary reasons, and the lowest common denominator remains stalled at the MP multicomputer architecture.

The Trollius Project is structured for maximum impact on the young field of concurrent processing, within the university environment at Ohio State University, Columbus. Trollius is the only operating system portable to any multicomputer incarnation and available in source form. In particular, getting the source into the hands of students and faculty at other universities has been its formula. People can use other systems and become expert users. People can perform surgery on Trollius and become developers.

Trollius was born, raised, and lives in U.S. supercomputer centers. From this perspective, the ultimate purpose of scalable parallel architectures is to extend the limits of supercomputing and permit attacks on the grand challenges of computational science. Thus, the first concern of Trollius has been to support

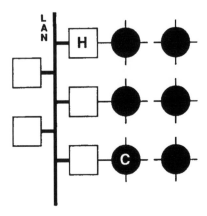

FIGURE 1. The general multicomputer.

data parallel applications. Many other systems and languages emphasize elegance and efficiency at solving function parallel applications. Trollius has concentrated higher level development on data parallelism, but the building blocks are in place for any strategy of multicomputer problem solving.

Is executing the designated application the only role an operating system must play? If the answer is yes, then a network loader and a communication library might be the final solution, because multicomputer applications are typically not very complex — today. Trollius anticipates that answer is no. Monitoring, debugging, machine maintenance, and machine portability must also be considered. If third-party vendors and university research groups are essential to software progress on new architectures, then an operating system must not set limits and must not attempt to predict the future, but instead must concentrate on the accepted basics. The alternative is more new operating systems starting from scratch.

Trollius is a uniform foundation for message passing on every multicomputer node. Host nodes (H) run Trollius with the local operating system while compute nodes (C) run it natively.

The basic activity of a multicomputer (see Figure 1) is passing messages between processes running on nodes. Other activity is an abstraction, a paradigm, and a package. No one abstraction is the answer to every problem solving strategy on the multicomputer architecture. A thorough package of the basics supporting any possible abstraction is the answer.

The previously discussed motivation and purpose spawned the principles of Trollius's design, listed below:

- Focus on supercomputer applications, scalable problems, and UNIX.
- Expand the multicomputer scope to include the local area network.
- Utilize an onion-like design, in which each layer can be peeled off for less functionality but better performance.
- Always allow the user to access the basics: nodes, processes, links and messages, and finally, the hardware.

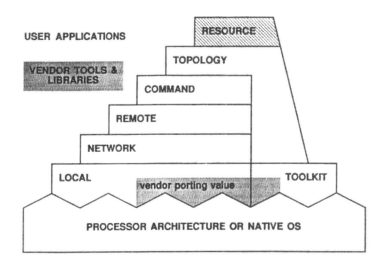

FIGURE 2.   Trollius layered structure.

- Maintain node uniformity; host and compute nodes run the same Trollius.
- Engineer modular software that facilitates other development. Structure with multiple processes above a spartan MP local node kernel.
- Don't bet on any one compiler, processor or topology.

To dispel the suspicion that Trollius is a primitive operating system, the functional description that follows begins at the current top of the evolutionary tree and works downward.

Crystalline (CrOS) is a communication library for domain decomposition developed at the California Institute of Technology, Pasadena. It bypasses the annoying problems of grid programming: simple deadlock and global singularities. CUBIX is a file access method, also developed at CalTech, that collects and distributes data among multiple nodes in an orderly manner. CUBIX often precludes the need to write a control program for the host or root node. Together, CrOS and CUBIX are the recommended tools for grid domain decomposition; Trollius offers both.

Ready to boot the multicomputer, the user confronts the highest layer of Trollius 2, the topology layer (see Figure 2). The topology layer tools interpret a simple grammar called node and interconnect language (NaIL), in which an exact description of the multicomputer to be booted has been recorded. Given this topology description, called the boot schema, Trollius endeavors to solve the general problem of link configuration, booting and routing with very few limits. In particular, topology is not limited. The boot schema describes the number, type (compute node, host, disk connected, CRT connected, wasted interface), and identification (any 32-bit quantity) of multicomputer nodes. The one restriction, removed in Version 3, is that only one host node may exist. Also described are the links between nodes, explicit routes to be taken between two points, priority route assignment between two points, and the correspondence

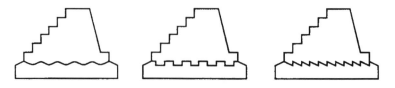

FIGURE 3. Portable node interface.

between Trollius node identifiers and vendor-specific naming conventions (so that vendor-specific link configuration tools may be invoked). Most common configurations for a particular installation will be created by the system administrator and saved.

Given multiple links between the host node and the compute node pool, Trollius permits multiple-user access to the pool. Trollius does not allow sharing of single links or single compute nodes. Multiple user access Version 2 is managed by personal communication. This illustrates an important policy of Trollius development — each release selects a cut-off point and provides the implementation of the basics to that point. Trollius 1 had no topology layer and the cut-off point for complete solution was the local node operating system. The route table for each node was hard wired. Booting was accomplished by invoking the individual node booting tools of the command layer (hboot [host node], sboot [device connected compute node], and vboot [remote compute node]) by hand. Today, the topology tools drive these command layer tools, and in Version 3, the resource layer tools will drive the topology tools. If a user ever needs to bypass the topology layer, the multicomputer can still be operated from the command layer. Each layer has a standard interface accessible to the user. Vendors can add proprietary value in porting to specific hardware and in bundling tools and libraries. The resource layer is part of the third generation of Trollius. The local and tool kit layers are portable to any compute node architecture. Other layers are unaffected by the underlying environment. (See Figure 3.)

Consider a two-dimensional problem in computational fluid dynamics solved with a finite difference method on a cartesian grid. Convinced of the merits of concurrent programming, the scientist parallelizes the code using CrOS and CUBIX. He has access to a 64-node transputer array connected to a UNIX workstation. First, he must compile the program, as demonstrated below:

```
% tf77 -o cfd cfd.f -lcros
%
```

CUBIX is an integral part of the Trollius remote file system and is accessed through the standard library. CrOS is an optional library linked explicitly. This compilation can take place on any UNIX machine with an installed and licensed tf77 cross-compiler supported by Trollius.

The user selects or prepares a boot schema for a multicomputer configuration suitable for the CFD application. Trollius CrOS demands consecutive node identifiers beginning with 0 and continuing through the grid in row major order. Perhaps the boot schema for an 8 × 8 grid is called "bnail64.cros", a slight misnomer because there are 65 Trollius nodes, including the host node. The following three topology tools establish a running multicomputer:

```
% map -c bnail64.cros
% solder -c bnail64.cros
% spread -c bnail64.cros
%
```

The map step generates a file containing a route table for each node using a greedy algorithm based on accumulating link weights. The weights and more may be fine tuned or overridden in the boot schema. Once a route file exists for a given boot schema, the map step can be skipped. Map makes the general problem of routing completely detached from the Trollius operating system. It is an off-line tool that can be run anywhere. With well-defined input and output files, the inventions of individual researchers can replace map.

The solder step invokes the vendor-specific software necessary to configure the communication links to match the topology described in the boot schema. Solder is not used with fixed links.

The spread step actually boots the multicomputer, invoking the right command layer booting tool for the right type of node and loading route and general network information into each node. Like all Trollius host programs, the user invokes spread from the UNIX interface. Control returns to the same interface; no special interface or shell exists. After booting, the user's UNIX workstation is also the origin host node in a Trollius multicomputer.

A host node program written with the Trollius communication library can cooperate with processes running anywhere in the multicomputer. Trollius supplies control and monitoring programs that make up the command layer. These programs supplement the suite of familiar UNIX development tools.

The user is ready to run the program, cfd, on the compute nodes using the loadgo command:

```
% loadgo -v C cfd
[1] 348013 cfd running on all compute nodes
%
```

Hands-on information gathering gives Trollius its flavor. For example, is the program running?

```
% state n0,63
NODE    INDEX   PID     KPRI    KSTATE      PROGRAM
n0      [1]     348     0       A(6,0)      cfd
n63     [1]     348     0       A(6,0)      cfd
```

Are the buffers getting clogged?

```
% bfstate -l n32
NODE    DEST    EVENT   TYPE    LENGTH          STATE
n32     n32     6       0       2048            blocked
n32     n32     6       0       2048            waiting
n32     max workers = 4, used workers = 2, full workers = 1
n32     max space = 262144, used space = 4096
%
```

How much memory is free? Perhaps the granularity of the data decomposition can be adjusted.

```
% mstate n7 -f
NODE        PID         ADDRESS         SIZE
n7          free        80000480        B80
n7          free        800535A8        1AD258
%
```

The application either terminates or the user may decide to deliver a termination signal with the doom command:

```
% doom C %1
% sweep C
%
```

Killing an MP parallel application in mid-flight will often result in a number of unreceived messages in lingering transit. These messages will settle into buffers and can be inspected with the bfstate command shown above. Lingering messages can affect the next program run. The debris can be cleaned up with the sweep command. Then, the program can be run again without rebooting. Afterwards, the host node Trollius is terminated and the compute nodes released by the tkill command.

```
% tkill
%
```

The topology and command layers form the operational level of Trollius. The programming interface begins with the remote layer.

Commands process options and present data surrounding one or more invocations to remote layer functions found in the remote library, trillium (Trollius's original name). Thus, any user program can accomplish the function of any command program. A remote function exchanges messages with a remote daemon. For example, the command bfstate calls the function rbfstate, which talks with the daemon, bufferd, on the indicated node.

The daemon called upon most often by user programs is filed, which provides a remote interface to the UNIX filesystem. Any compute or host node can run filed if it operates a disk. The service includes a large number of standard UNIX functions supporting the C stdio package, the Fortran Unit I/O system, or direct access of portable UNIX programs. (Note: this illustrates another Trollius policy concerning UNIX. Trollius is real UNIX where it must be [the host nodes], UNIX compliant where it makes sense [the filesystem], and not UNIX where it would serve badly [communications].)

Filed only runs on nodes with attached disks. Other daemons only run on compute nodes, where Trollius is the native operating system and has more responsibility for basic services. The following table lists the other remote daemons, their responsibilities, and where they run.

Daemon	Node	Function
mallocd	Compute	Memory allocation, deallocation, and monitoring
kenyad	Compute	Process creation, signaling, monitoring and spawning
memd	Compute	Fetch/load specific memory addresses
flatd	Compute	Load memory and assign tag to address
bootd	Compute	Boot neighboring compute nodes
loadd	Host	Load data and program files on compute nodes
echod	All	Echo messages

This excursion to the depths of Trollius has reached the layer at which CrOS is written, the network layer. If not the heart, the network layer is certainly the hub of the operating system. A group of processes that run on every node provides a network MP service. At the network layer, Trollius functionality is completely uniform among all node types.

The smartest process is the router, which keeps the identifier, type, and route to every other node in the system. One will be able to change this information at runtime in Version 3. A small cache enables local clients to avoid continuous, expensive communication with the router for each network message sent.

When a process wants to send a message to an arbitrary node, it invokes nsend, which first consults its own route cache or the router (see Figure 4). The obtained "route" indicates the local forwarding process, a process running on the same node that will receive the message and forward it towards its destination. Typically, the forwarding process is a datalink output process, the proprietor of an output link. Whenever a datalink output gets a message, it simply pumps the message out its link, which could be a transputer link, a UNIX device driver, or a TCP/IP connection. Datalink input processes wait for a Trollius network message to arrive on the same links. It calls nsend on every message it receives and the procedure just described repeats itself. When a message has reached the correct node, the forwarding process is the receiving destination process specified by the original caller of nsend.

Once the forward process is known at each stage, nsend moves the message to the forwarding process by calling dsend. If a process knows the forwarding process in advance, it need not consult the router and can call dsend directly.

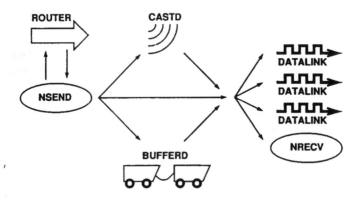

FIGURE 4. Network forwarding processes.

At no point does any process consider dropping a network message. If dsend detects that a forwarding process is busy, it must either wait or pick another forwarding process. If a datalink input must wait, no other message can be read from its link. A clogged link can be a bottleneck and can cut off control over a node by the user seated at the host node. A need exists for message buffers.

Bufferd is network layer forwarding process that manages a pool of message buffers and a small group of worker processes. Unless forbidden by its caller, dsend can forward its message to bufferd. The buffer daemon forwards the message to an idle buffer worker, which then blocks until the original forwarding process is free. If no worker is available, the message waits in a large pool inside bufferd. When this is exhausted, callers to dsend wait on bufferd. Users can tune the number of workers and the size of the message pool from the command line. After consulting the router, a process calling nsend transfers the message to a local forwarding process. Bufferd can hold multiple messages, up to a space limit tuned by the user.

Another forwarding process is the cast daemon. Within the domain of physical node identifiers are cast identifiers which refer to groups of nodes. A few useful casts, like broadcast, are predefined, but all others are explicitly chosen and registered by using processes, which then use it as a normal destination node with nsend. When nsend consults the router about a cast identifier, it is directed to forward its message to castd. Castd then performs the necessary nsend operations to the correct node neighbors, which reflect an efficient spanning tree to the cast members. Castd is completely topology independent. Castd demonstrates the modularity of the Trollius network layer. Services can be unplugged, modified, and reinserted with little concern for the rest of the system. Forwarding processes can be implemented for new network services, and only the programmable router is the wiser.

The real messages in Trollius are the local or kernel layer messages. A network message comprises two kernel messages: the first contains the network message descriptor and the second contains the actual data. Forwarding a message with dsend really means sending kernel message to a local process with ksend. Processes synchronize on two arbitrary numbers: a message event and a message type. Synchronizing events must equal in value and the types must match in at least one bit (bitwise AND is true) for a message to be transferred. The "indication of forwarding process" alluded to earlier is the correct event and type. Local MP calls block a process until it synchronizes unless the associated ktry_send polling calls are used.

Synchronization is managed by the kernel, which in Trollius is a server not an executive. The kernel is considered a server because it is not in complete control of the processor. Final scheduling and ultimate process creation/destruction is handled by the underlying environment: the host node's native operating system or the transputer's microcode. On other compute node processors, these responsibilities are handled by a software subkernel.

Although the rendezvous between two processes occurs through the kernel, actual data are transferred directly, without the aid of the kernel. The rendezvous tells each process what it needs to know in the underlying environment to communicate directly with its peer (on the transputer, this would be a channel address). Directed by a flag, the ksend run time function can remember the results by synchronizing on multiple events and then bypass the kernel the next time a known event is indicated. The capability is called a virtual circuit which effectively welds a dynamic group of processes into a static group. When used between callers of nsend and datalink processes, an efficient virtual circuit can exist between any two points in the multicomputer. The lower overhead of virtual circuits makes them the preferred communication mechanism for nearest-neighbor, domain-decomposed applications seeking finer granularity. (Note: CrOS has an option for virtual circuits.) A circuit that crosses a link reserves that link for the caller of dsend. This facilitates the lowest level call in the network layer, psend, which drives the hardware directly.

The predominant responsibility of the spartan local kernel is MP among dynamic local node processes. It does not know of multiple processors. Since many underlying environments do not supply a fully developed, predictable priority scheduling service, the Trollius kernel has an option to schedule processes (by controlling replies to kernel requests) to the underlying environment. When a client process attaches to the kernel it can decide to be uncontrolled, in which case scheduling is completely handled below. Alternatively, the client process can supply a priority, in which case Trollius guarantees that only the highest priority unblocked process runs.

Another kernel feature is asynchronous signal delivery modeled on UNIX. Signals may only by caught during kernel requests because the kernel is a server. Clients can detach from the kernel and go about their business independently. The kernel's server design is the key to Trollius uniformity and portability between host and compute nodes.

The tool kit layer is not formally part of the operating system because it does not deal in Trollius messages. Instead, the tools concentrate on booting the host nodes and device connected compute nodes. It would be possible to burn Trollius into PROMs on compute nodes, but again, a modular approach is followed. As with all other layers, the philosophy guiding the tool kit layer is to support other solutions that could fit into the next ascending layer, not just the one chosen by

Trollius. Since the tool kit layer boots Trollius onto certain nodes, the same tools could presumably be used to boot anything.

A configuration file called a process schema indicates the programs that constitute the operating system on each type of node. The hboot tool uses one process schema to start processes on a host node. The sboot tool uses another processs chema to start processes on a device connected compute node. Sboot invokes the patch tool to relocate addresses (given a compute node with no dynamic address translation) and then the cboot tool to actually load a program. Cboot communicates with the Trollius bootstrap loader, called moses, running on the device connected compute node. Moses is loaded by the export tool, which talks to whatever boot mechanism exists after reset. Moses can also be burned into PROM. The reset is achieved with the fish tool, the simplest, lowest order of functionality in Trollius and a very long way from the spread tool.

The following performance numbers were obtained on a T800 (20 MHz) using the LSTT C compiler under the latest version of Trollius (2.0 He). Both code and data resided in external memory. The measurements made between two nodes involved directly connected neighbors (and no C004 or other link reconfiguration switch).

The time per message per hop T is

$$T = Ts + (Tc * L) + (Tp * Ep)$$

where   T   is   time per message per hop (μs);
         Ts   is   set up charge (μs);
         Tc   is   communication time (μs/byte);
         Tp   is   packetization cost (μs/extra-packet);
         L   is   message length (bytes); and
         Ep   is   number of extra packets (after the first one); and

$$Ep = ceil(L/Lp) - 1 \quad \text{for } L > 0$$
$$= 0 \quad \text{for } L = 0$$

where ceil() is the integer ceiling function, and p is the packet size (bytes) (4096).

Network messages are limited in size. Larger messages are packetized. The default packet size is 4096 bytes, although this can be changed before compiling Trollius.

A high overhead is incurred when a message passes through a buffer. Buffers can prevent deadlock and reduce blocking. In practice, they tend to enhance the robustness of the system. Trollius does not choke if a process accidently sends 1000 messages to a node with no receiving process, but at a substantial cost. Message buffering can be disabled by setting the NOBUF flag in the nh_flags field of the message header. The cost per message buffered is Tb=2375+(0.20 * L) (in μs).

At each communication layer, regular and virtual circuit MP have been measured between two directed connected nodes (n0-n1) and within one node (n0). To help in comparing and deciding which of the communication layers is most suited for a particular application, a quick summary of the functionality provided by each layer is given below.

- Network layer (nsend)
    - Routed: from/to any node
    - Multitask: from/to any node process
    - Packetized: any message size
- Network datalink layer (dsend)
    - Not-routed: nearest neighbor only
    - Multitask: from/to any node process
    - Not-packetized: message size limited to Lp (default: 4096)
- Local layer (ksend)
    - Not-routed: local communication only
    - Multitask: from/to any process on the same node
    - Not-packetized: no limits on message size, but with protection against mismatched lengths
- Network physical layer (psend)
    - Not-routed: nearest neighbor only
    - Not-multitask: single-process communication
    - Not-packetized: no limits on message size, but no protection against mismatched lengths
    - Restricted use: can only be used after a virtual circuit has been set up

Regular			Virtual Circuit			
Ts	Tc	Tp	Ts	Tc	Tp	
480	.56	552	168	.56	140	nsend n0-n1
388	.10	374	211	.10	195	nsend n0
453	.56	—	128	.56	—	dsend n0-n1
355	.10	—	177	.10	—	dsend n0
230	.10	—	69	.10	—	ksend n0
—	—	—	36	.56	—	psend n0-n1

During the development and debugging of an application program, use of nsend is recommended. When the program is working, nsend can be substituted for dsend, ksend, or psend where possible. Although this provides enough flexibility for most users, it is possible that none of the above suits a particular application. For such requirements, the project has developed the custom network gearbox approach to tailor the communication functionality/performance to applications that fall into the cracks of the regular Trollius communication layers.

THE PERFORMANCE/FUNCTIONALITY DILEMMA OF MULTICOMPUTER MP

Since their commercial introduction in 1985, multicomputer hardware and operating systems have been judged by their communication performance.[1,2] Low bandwidth or high latency restrict the number of algorithms that can make efficient use of concurrency among a large number of processors. Operating systems that provide one mechanism for message transfer are forced, by the research community as well as the commercial market, to adopt a high-performance, low-functionality solution. Simplistic MP can limit the scope of debugging and monitoring tools and force some applications to implement their own higher level paradigms.

Multicomputer operating systems can accommodate a wider variety of applications by offering several different combinations of performance and functionality. This discussion begins with the trade-offs at different layers of a production operating system. Even with four or five choices, however, some applications fall into the cracks. An example in the field of circuit fault simulation is given along with a new approach to MP performance — the custom network gearbox.

The specific example of MP is the Trollius operating system,[3,4] now in its second generation as a production operating system for distributed memory architectures. All performance numbers were obtained on an array of 20MHz T800 transputers.[5]

How are the common MP functions implemented? The big decision that affects all else is whether to gather all functionality in an omnipotent kernel, or to employ extensive multiprocess structuring above a spartan kernel. Trollius uses the latter approach, but with several methods to bypass processes for direct access to devices and system data structures. The philosophy is to start with functionality and tune for performance.

- **Multitask access** — If a processor can be multiprogrammed, then the communication links become resources which must be guarded with a proprietor process. The base price of a message passed between neighboring nodes is just over 500 μs, mostly due to kernel synchronization with link processes on both nodes. Trollius processes have the capability to establish semipermanent communication channels to link processes which bypass the kernel. These "virtual circuits" effectively lock the link device and reduce the base price to 172 μs. Once a device is locked, a few error checking features can be traded away in return for a base price of 36 μs. These lower cost access points preclude the use of any of the following functions.
- **Routing** — A route is either calculated (fixed topology) or read from a table (random topology) at a fixed memory location or within a local proprietor process. Quick calculations are possible with hypercubes, but the future of fault tolerant dynamic routing demands a more closely guarded data structure. Trollius uses a separate route proprietor process and a run-time cache within each client, but has a difficult problem with cache invalidation. The initial route acquisition is expensive, but subsequent cache hits take less than 10 μs.
- **Buffering** — Buffers can either be managed by a separate process or built into the link proprietor processes. Trollius link processes have one buffer each for store and forward operations. The main buffer pool is in a separate process for a very high degree of buffer control and monitoring, but at a very great

expense. It was difficult to measure the exact buffer cost because forcing messages into buffers changes what is wanted to be measured in the first place. Calculations show an added cost of 2 ms. Buffers are turned off with a flag bit on each message.

- **Flow control** — The simplest form of flow control for multicomputers is an "invitation to send" message from the receiver to the sender, thus ensuring that all messages are consumable. Add this to deadlock-free routing for a robust debugging environment. This protocol effectively doubles the message transfer overhead and is only recommended for debugging.
- **Broadcasting/multicasting** — Again, a distinct process makes multicasting a modular addition to Trollius, but at an added expense of just under 400 μs. Combining and distributing messages in a higher level library such as CrOS is hard coded for grids and costs less.

Functionality	Performance penalty	Applications
Transport layer End to end flow controlled, routed, buffered, multitask accessed	1000 ms and up, depending on buffer use, route cache hits, hop count	Mostly debugging any applications
Network layer Routed, buffered, multitask accessed	550 to 3000 ms depending on buffer use, route cache hits, 170 μs with virtual circuit	Imperfectly embedded structures, farmer/worker, Linda support
Datalink layer Buffered, multi-task accessed, nearest neighbor	500 to 2500 ms depending on buffer use, 150 μs with virtual circuit.	Domain decomposition
Physical layer Nearest neighbor, single task accessed, no system messages	36 μs	Synchronous domain decomposition

- **The trade-offs** — Most computational models can find a suitable mix of features and speed, but some cannot find the right combination. Development and debugging flexibility is also to be weighed against performance when only common communication links are present.

Another broad generalization is that more powerful asynchronous behavior requires more decoupled software processes and more internal coordination overhead. Performance is paid for by introducing restrictions. Restrict the topology and routing is faster. Restrict the number of user processes and resource management is easier. Trollius delivers performance by caching information and temporarily locking resources.

What performance levels are required by typical models of multicomputer computation? The Trollius network sublevels were chosen to optimize domain decomposition algorithms.

- **Domain decomposition** — This model is best suited for multicomputers because communication is primarily between nearest neighbors, though occasional global synchronization may be required. Buffering is generally not necessary during execution. Programmers should be able to utilize the 36-μs functionality level.
- **Task allocation** — Linda[6] presents a useful paradigm for programming the task allocation model. Since tuples may be stored anywhere, routing is necessary. Buffering may relieve congestion at hot nodes.

Code development, debugging, and performance monitoring further complicate the balance of performance due to the extra functions they require beyond the application. If an application process directly accesses the communication hardware, the user must give up software monitoring capability or build the added complexity into the application to handle monitoring messages.

A few multicomputer implementations have separate communication links for system-specific message traffic from the user interface, or perhaps from secondary storage. This allows the operating system to place more restrictions on the communication links reserved for the application and, hence, buy more speed.

The front-line solution to the varying requirements of typical algorithms is simply to present as many choices of performance and functionality combinations as the implementation will allow. This can be accomplished by building features which can be bypassed and incur no performance penalty when they are turned off.

The various combinations of MP protocols, as discussed in the previous sections, offer ample flexibility for the majority of parallel processing applications. Domain decomposition algorithms exhibiting nearest-neighbor communication may take full advantage of the hardware message transfer performance by using the low-functionality protocols. Multiple-process farmer/worker algorithms and applications based on shared memory emulation require the added functionality of multihop routing, buffering, and multiprocess support. Nevertheless, some applications would not be adequately served by any of the above communication layers, since they may require different trade-offs between performance and functionality. These requirements, particular to each application, would best be handled by a customized solution (*gearbox*) that would coexist with the underlying operating system as well as with other customer gearboxes.

The *custom network gearbox* approach solves the problem of inadequate trade-off choices in MP protocols while providing a standardized interface, enabling access to the full functionality of the operating system it serves. This method enables users to develop and debug their applications relying on the regular communication protocol offered by the operating system, and later on, with no code modification, switch gears and run the developed code using the protocol customized to fit the trade-offs that are advantageous to the particular application. Hence, topics such as routing, buffering, message packetization, deadlock and multiple-user processes can be addressed or ignored depending on the requirements of the application both from the functionality and performance standpoints.

Following is the description of the communication structure of a fault simulation algorithm that would benefit from the custom gearbox approach, and the gearbox implementation suited for it.

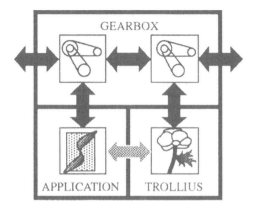

Customer Network Gearbox — A standardized custom message passing solution enables full access to Trollius features and the right mix of functionality and performance.

The application in question is a gate-level zero-delay circuit fault simulator.[7] Fault simulation is used in automatic test pattern generation (ATPG) tools in order to speed up the task of testing a VLSI circuit and improve its quality. The simulation consists of generating the lists of modeled defects (*fault lists*) observable at the circuit outputs for a large number of input test patterns. This requires the evaluation of the fault list at the output of each gate in the circuit, for each input test pattern, computed using set operations on the fault lists of the gate inputs. Faults detected at the outputs of the circuit are removed from the circuit specification (*fault dropping*) and the percentage of faults detected (*fault coverage*) is computed. The simulation is terminated when a specified fault coverage is reached, typically in the 97 to 100% range.

The parallelization approach followed decomposes the circuit and assigns each subcircuit onto a different processor, giving each the location of the gates it needs to communicate the fault list results to. One processor, the host workstation, reads the test patterns from a file and starts the simulation by sending the input signals to the processors handling the primary inputs of the circuit. These processors evaluate the output fault lists of their subcircuits, sending the results to other processors requesting them. The simulation proceeds in a medium-grain dataflow fashion until the fault lists of the primary outputs of the circuit are available, at which time fault dropping is performed by a host. Following this procedure, test patterns are evaluated until an adequate fault coverage is achieved.

The partitioning of the circuit being NP-complete, a linear time heuristic is used. Linear complexity of this preprocessing step is desirable due to the large number of gates in a typical VLSI chip. Thus, the resulting assignment of gates to processors leads to poor load balancing and imperfectly embedded communication patterns. In addition, the volume of data transferred (size of the fault lists) fluctuates widely from one test pattern to the next and is nondeterministic. For these reasons, decentralized dynamic load balancing is used to enhance the performance of the application.

The amount of message exchanges in this application, their varying sizes, and their irregular communication patterns require performance/functionality trade-offs not offered by the mentioned MP protocols.

The gearbox was designed to have the lowest communication setup cost permissible by the underlying hardware, in this case a T800 transputer,[6] while still allowing user-transparent multi-hop communication. Separate datalink processes guard each hardware communication link and run at high priority in order to take full advantage of the inherent parallelism in the transputer hardware. Fast routing is done through table look-up and crossbar of internal communication channels. The kernel and the router of the operating system are therefore bypassed, reducing contention and amount of context switching.

Achieving this performance/functionality mix required some compromises to be made. Bypassing the kernel required a static process structure with a predetermined number (currently two) of user processes connected to the gearbox. Reducing the message setup cost precluded packetization of the messages, limiting the size of the message to that of the buffers. This is acceptable in this application since, although message sizes fluctuate widely, their maximum size is limited by the number of faults in the circuit. The irregularity of the communication patterns meant that the possibility of routing deadlock had to be considered. The *virtual networks* approach outlined by Yautchev and Jesshope[8] was used, thus incurring the extra cost of updating buffer flags.

The design decisions taken in the implementation of this gearbox clearly reflect the requirements of the application at hand. By tailoring the MP protocol to the user algorithm, significant performance gains are achieved while maintaining all the flexibility of a full operating system.

Any custom gearbox can recognize a message bound for an operating system process rather than an application process. Once identified, the system message is forwarded to a Trollius/gearbox "plug" process. Cut-through messages are not seen by Trollius because the gearbox owns the links. Within Trollius, there are no link proprietor processes and all messages are routed to the Trollius/gearbox plug, which looks like a link proprietor on the Trollius side and looks like something the custom gearbox understands on the other side. It is a gateway in every sense.

The custom gearbox reads standard Trollius network frames from the links of the node so that its neighbors can run the standard gearbox or any other custom gearbox. Trollius booting configuration makes it very easy to insert different gearboxes into different types of nodes.

REFERENCES

1. **Dunigan, T. H.,** Hypercube performance, in Proc. Second Conf. on Hypercube Multiprocessors, SIAM, 1987, 178.
2. **Bergmark, D. et al,** On the performance of the FPS T-Series hypercube, in Proc. Second Conf. on Hypercube Multiprocessors, SIAM, 1987, 193.
3. **Burns, G. et al,** Trillium operating system, in Proc. Third Conf. on Hypercube Concurrent Computers and Applications, ACM, 1988, 374.
4. **Burns, G.,** A local area multicomputer, in Proc. Fourth Conf. on Hypercubes, Concurrent Computers and Applications, 1989.
5. **Whitby-Strevens, C.,** The transputer, in Proc. 12th Int. Symp. on Computer Architecture, ACM, 1985, 292.
6. **Carriero, N. and Gelernter, D.,** Linda on Hypercube Multicomputers, in Proc. First Conf. on Hypercube Multiprocessors, SIAM, 1986, 45.
7. **Huisman, L., Nair, I., and Daoud, R.,** Fault simulation on message passing parallel processors, in Proc. Fifth Distributed Memory Computing Conference.
8. **Yantchev, J. and Jesshope, C. R.,** *Adaptive Low Latency Deadlock-Free Packet Routing for Networks of Processors.*

17 APPLY: A PROGRAMMING LANGUAGE FOR LOW-LEVEL VISION ON DIVERSE PARALLEL ARCHITECTURES

In computer vision, the first, and often most time-consuming, step in image processing is *image to image* operations. In this step, an input image is mapped into an output image through some local operation that applies to a window around each pixel of the input image. Algorithms that fall into this class include edge detection, smoothing, convolutions in general, contrast enhancement, color transformations, and thresholding. Collectively, we call these operations low-level vision. Low-level vision is often time consuming simply because images are quite large — a typical size is 512×512 pixels, so the operation must be applied 262,144 times.

Fortunately, this step in image processing is easy to speed up through the use of parallelism. The operation applied at every point in the image is often independent from point to point and also does not vary much in execution time at different points in the image. This is because at this stage of image processing, nothing has been done to differentiate one area of the image from another, so that all areas are processed in the same way. Because of these two characteristics, many parallel computers achieve good efficiency in these algorithms, through the use of *input partitioning*.[12]

We define a language, called *Apply*, which is designed for implementing these algorithms. Apply runs on the Warp machine, which has been developed for image and signal processing. We discuss Warp, and describe its use at this level of vision. The same Apply program can be compiled either to run on the Warp machine, or under UNIX, and it runs with good efficiency in both cases. Therefore, the programmer is not limited to developing programs just on Warp, although they run much faster (typically 100 times faster) there; the programmer can do development under the more generally available UNIX system.

We consider Apply and its implementation on Warp to be a significant development for image processing on parallel computers in general. The most critical problem in developing new parallel computer architecture is a lack of software which efficiently uses parallelism. While building powerful new computer architectures is becoming easier because of the availability of custom VLSI and powerful off-the-shelf components, programming these architectures is difficult.

Parallel architectures are difficult to program because it is not yet understood how to "cover" parallelism (hide it from the programmer) and get good performance. Therefore, the programmer either programs the computer in a specialized language which exploits features of the particular computer, and which can run on no other computer (except in simulation), or he uses a general-purpose language, such as Fortran, which runs on many computers, but which has additions that make it possible to program the computer efficiently. In either case, using these special features is necessary to get good performance from the computer. However, exploiting these features requires training, limits the programs to run on one or at most a limited class of computers, and limits the lifetime of a program, since eventually it must be modified to take advantage of new features provided in a new architecture. Therefore, the programmer faces a dilemma: he must either ignore (if possible) the special features of his computer, limiting performance, or he must reduce the understandability, generality, and lifetime of his program.

It is this thesis of Apply that *application dependence,* in particular, *programming model dependence,* can be exploited to cover this parallelism while getting good performance from a parallel machine. Moreover, because of the application dependence of the language, it is possible to provide facilities that make it easier for the programmer to write his program, even as compared with a general-purpose language. Apply was originally developed as a tool for writing image processing programs on UNIX systems; it now runs on UNIX systems, Warp, and the Hughes HBA. Since we include a definition of Apply as it runs on Warp, and because most parallel computers support input partitioning, it should be possible to implement it on other supercomputers and parallel computers as well.

Apply also has implications for benchmarking of new image processing computers. Currently, it is hard to compare these computers, because they all run different, incompatible languages and operating systems, so the same program cannot be tested on different computers. Once Apply is implemented on a computer, it is possible to fairly test its performance in an important class of image operations, namely low-level vision.

Apply is not a panacea for these problems; it is an application-specific, machine-independent language. Since it is based on input partitioning, it cannot generate programs which use pipelining, and it cannot be used for global vision algorithms[11] such as connected components, Hough transform, FFT, and histogram.

We begin by reviewing the structure of the Warp machine and then discuss our early work on low-level vision, where we developed the input partitioning

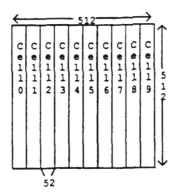

FIGURE 1. Input partitioning method on Warp.

method on Warp. Then we define and discuss Apply. Following this, we describe how Apply might be implemented on other computers.

We describe Warp only in order to discuss the different Warp-like architectures on which Apply has been implemented. A complete description of Warp is available elsewhere.[2]

Warp is a short linear array, typically consisting of ten cells, each of which is a powerful 10 MFLOPS processor. The array has high internal bandwidth, consistent with its use as a systolic processor. Each cell has a local program and data memory and can be programmed in a Pascal-level language called W2. The systolic array is attached to an external host, which sends and receives data from the array from a separate memory. The external host in turn is attached to a Sun computer, which provides the user interface.

Fault-tolerant (FT) Warp is a two-dimensional array, typically a five-by-five array, designed by Carnegie Mellon. Each cell is a Warp cell. Each row and column can be fed data independently, providing for a very high bandwidth. As the name suggests, this array has as a primary goal, fault tolerance, which is supported by a virtual channel mechanism mediated by a separate hardware component called a switch.

iWarp is an integrated version of Warp designed by Carnegie Mellon and Intel. In iWarp each Warp cell is implemented by a single chip, plus memory chips. iWarp includes support for distant cells to communicate as if they were adjacent, while passing their data through intermediate cells.

We map low-level vision algorithms onto Warp by the *input partitioning* method. On a Warp array of ten cells, the image is divided into ten regions, by column, as shown in Figure 1. This gives each cell a tall, narrow region to process: for 512×512 image processing, the region size is 52 columns by 512 rows. To use technical terms from weaving, the Warp cells are the "warp" of the processing; the "weft" is the rows of the image as it passes through the Warp array.

The image is divided in this way using a series of macros called *GETROW*, *PUTROW*, and *COMPUTEROW*. *GETROW* generates code that takes a row of an image from the external host and distributes one-tenth of it to each of ten cells. The programmer includes a *GETROW* macro at the point in his program where he wants to obtain a row of the image; after the execution of the macro, a buffer in the internal cell memory has the data from the image row.

The *GETROW* macro works as follows. The external host sends in the image rows as a packed array of bytes — for a 512-byte wide image, this array consists of 128 32-bit words. These words are unpacked and converted to floating point numbers in the interface unit. The 512 32-bit floating point numbers resulting from this operation are fed in sequence to the first cell of the Warp array. This cell takes one tenth of the numbers, removing them from the stream, and passes through the rest to the next cell. The first cell then adds a number of zeros to replace the data it has removed, so that the number of data received and sent are equal.

This process is repeated in each cell. In this way, each cell obtains one tenth of the data from a row of the image. As the program is executed, and the process is repeated for all rows of the image, each cell sees an adjacent set of columns of the image, as shown in Figure 1.

We have omitted certain details of *GETROW* — for example, usually the image row size is not an exact multiple of ten. In this case, the *GETROW* macro pads the row equally on both sides by having the interface unit generate an appropriate number of zeros on either side of the image row. Also, usually the area of the image each cell must see to generate its outputs overlaps with the area of the next cell. In this case, the cell copies some of the data it receives to the next cell. All this code is automatically generated by *GETROW*.

PUTROW, the corresponding macro for output, takes a buffer of one tenth of the row length from each cell and combines them by concatenation. The output row starts as a buffer of 512 zeros generated by the interface unit. The first cell discards the first one tenth of these and adds its own data to the end. The second cell does the same, adding its data after the first. When the buffer leaves the last cell, all the zeroes have been discarded and the data of the first cell have reached the beginning of the buffer. The interface unit then converts the floating point numbers in the buffer to zeros and outputs it to the external host, which receives an array of 512 bytes packed into 128 32-bit words. As with *GETROW*, *PUTROW* handles image buffers that are not multiples of ten, this time by discarding data on both sides of the buffer before the buffer is sent to the interface unit by the last cell.

During *GETROW*, no computation is performed; the same applies to *PUTROW*. Warp's horizontal microword, however, allows input, computation, and output at the same time. *COMPUTEROW* implements this. Ignoring the complications mentioned above, *COMPUTEROW* consists of three loops. In the first loop, the data for the cell are read into a memory buffer from the previous cell, as in *GETROW*, and at the same time the first one tenth of the output buffer is discarded, as in *PUTROW*. In the second loop, nine tenths of the input row is passed through to the next cell, as in *GETROW*; at the same time, nine tenths of the output buffer is passed through, as in *PUTROW*. This loop is unwound by

COMPUTEROW so that for every nine inputs and outputs passed through, one output of this cell is computed. In the third loop, the outputs computed in the second loop are passed on to the next cell, as in *PUTROW*.

There are several advantages to this approach to input partitioning:

- Work on the external host is kept to a minimum. In the Warp machine, the external host tends to be a bottleneck in many algorithms; in the prototype machines, the external host's actual data rate to the array is only about one fourth of the maximum rate the Warp machine can handle, even if the interface unit unpacks data as they arrive. Using this input partitioning model, the external host need not unpack and repack bytes, which it would have to if the data were requested in another order. On the production Warp machine, the same concern applies; these machines have DMA, which also requires a regular addressing pattern.
- Each cell sees a connected set of columns of the image which are one tenth of the total columns in a row. Processing adjacent columns is an advantage, since many vision algorithms (e.g., median filter[8]) can use the result from a previous set of columns to speed up the computation at the next set of columns to the right.
- Memory requirements at a cell are minimized, since each cell must store only one tenth of a row. This is important in the prototype Warp machines, since they have only 4K words memory on each cell.
- The image is processed in raster order, which has for a long time been a popular order for accessing data in an image. This means that many efficient algorithms, which have been developed for raster-order image processing, can be used.
- An unexpected side effect of this programming model was that it made it easier to debug the hardware in the Warp machine. If some portion of a Warp cell is not working, but the communication and microsequencing portions are, then the output from a given cell will be wrong, but it will keep its proper position in the image. This means that the error will be extremely evident — typically a black stripe is generated in the corresponding position in the image.

The Apply programming model is a special-purpose programming approach which simplifies the programming task by making explicit the parallelism of low-level vision algorithms. We have developed a special-purpose programming language called the Apply language which embodies this parallel programming approach. When using the Apply language, the programmer writes a procedure which defines the operation to be applied at a particular pixel location. The procedure conforms to the following programming model:

- It accepts a window or a pixel from each input image.
- It performs arbitrary computation, usually without side effects.
- It returns a pixel value for each output image.

The Apply compiler converts the simple procedure into an implementation which can be run efficiently on Warp, or on a uni-processor machine in C under UNIX.

The idea of the Apply programming model grew out of a desire for efficiency combined with ease of programming for a useful class of low-level vision operations. In our environment, image data are usually stored in disk files and accessed through a library interface. This introduces considerable overhead in accessing individual pixels, so algorithms are often written to process an entire row at a time. While buffering rows improves the speed of algorithms, it also increases their complexity. A C language subroutine implementation of Apply was developed as a way to hide the complexities of data buffering from the programmer while still providing the efficiency benefits. In fact, the buffering methods which we developed were more efficient than those which would otherwise be used, with the result that Apply implementations of algorithms were faster than previous implementations.

After implementing Apply, the following additional advantages became evident.

- The Apply programming model concentrates programming effort on the actual computation to be performed instead of the looping in which it is embedded. This encourages programmers to use more efficient implementations of their algorithms. For example, a Sobel program gained a factor of four in speed when it was reimplemented with Apply. This speed-up primarily resulted from explicitly coding the convolutions. The resulting code is more comprehensible than the earlier implementation.
- Apply programs are easier to write, easier to debug, more comprehensible, and more likely to work correctly the first time. A major benefit of Apply is that it greatly reduces programming time and effort for a very useful class of vision algorithms. The resulting programs are also faster than the programmer would probably otherwise achieve.

The Apply language is designed for programming image to image computations where the pixels of the output images can be computed from corresponding rectangular windows of the input images. The essential feature of the language is that each operation is written as a procedure for a single pixel position. The Apply compiler generates a program which executes the procedure over an entire image. No ordering constraints are provided for in the language, allowing the compiler complete freedom in dividing the computation among processors.

Each procedure has a parameter list containing parameters of any of the following types: *in, out,* or *constant*. Input parameters are either scalar variables or two-dimensional arrays. A scalar input variable represents the pixel value of an input image at the current processing coordinates. A two-dimensional array input variable represents a window of an input image. Element (0,0) of the array corresponds to the current processing coordinates.

Output parameters are scalar variables. Each output variable represents the pixel value of an output image. The final value of an output variable is stored in the output image at the current processing coordinates.

Constant parameters may be scalars, vectors, or two-dimensional arrays. They represent precomputed constants which are made available for use by the procedure. For example, a convolution program would use a constant array for the convolution mask.

The reserved variables *ROW* and *COL* are defined to contain the image coordinates of the current processing location. This is useful for algorithms which are dependent in a limited way on the image coordinates.

Figure 2 is a grammar of the Apply language. The syntax of Apply is based on Ada;[1] we chose this syntax because it is familiar and adequate, and because we do not wish to create yet another new language syntax, nor do we consider language syntax to be an interesting research issue. However, as should be clear, the application dependence of Apply means that it is not an Ada subset, nor is it intended to evolve into such a subset.

Apply does not allow assignment of fixed expressions to floating variables or floating expressions to fixed variables. Expressions mixing fixed and floating values are also disallowed. A fixed expression may be explicitly converted to floating point by means of the pseudofunction *REAL* and a floating expression can be converted to fixed by using the pseudofunction *INTEGER*.

Variable names are alpha-numeric strings of arbitrary length, commencing with an alphabetic character. Case is not significant, except in the preprocessing stage, which is implemented by the *m4* macro processor.[10]

BYTE, INTEGER, and REAL refer to (at least) 8-bit integers, 16-bit integers, and 32-bit floating point numbers. *BYTE* values are converted implicitly to *INTEGER* within computations. The actual size of the type may be larger, at the discretion of the implementor.

As a simple example of the use of Apply, let us consider the implementation of Sobel edge detection. Sobel edge detection is performed by convolving the input image with two 3×3 masks. The horizontal mask measures the gradient of horizontal edges, and the vertical mask measures the gradient of vertical edges. Diagonal edges produce some response from each mask, allowing the edge orientation and strength to be measured for all edges. Both masks are shown in Figure 3.

An Apply implementation of Sobel edge detection is shown in Figure 4. The lines have been numbered for the purposes of explanation, using the comment convention. Line numbers are not a part of the language.

Line 1 defines the input, output, and constant parameters to the function. The input parameter *inimg* is a window of the input image. The constant parameter *thresh* is a threshold. Edges which are weaker than this threshold are suppressed in the output magnitude image, *mag*. Line 3 defines *horiz* and *vert*, which are internal variables used to hold the results of the horizontal and vertical Sobel edge operator.

Line 1 also defines the input image window. It is a 3×3 window centered about the current pixel processing position, which is filled with the value 0 when

procedure	::=	**PROCEDURE** *function-name* (*function-args*)
		IS
		variable-declarations
		BEGIN
		statements
		END *function-name;*
function-args	::=	*function-argument* [, *function-argument*]*
function-argument	::=	*var-list* : **IN** *type*
		[**BORDER** *const-expr*]
		[**SAMPLE** (*integer-list*)]
	|	*var-list* : **OUT** *type*
	|	*var-list* : **CONST** *type*
var-list	::=	*variable* [, *variable*]*
integer-list	::=	*integer* [, *integer*]*
variable-declarations	::=	[*var-list* : *type* ;]*
type	::=	**ARRAY** (*range* [, *range*]+) **OF** *elementary-type*
	|	*elementary-type*
range	::=	*int-expr* .. *int-expr*
elementary-type	::=	*sign object*
	|	*object*
sign	::=	**SIGNED**
	|	**UNSIGNED**
	|	*Empty*
object	::=	**BYTE**
	|	**INTEGER**
	|	**REAL**
statements	::=	[*statement* ;]*
statement	::=	*assignment-stmt*
	|	*if-stmt*
	|	*for-stmt*
	|	*while-stmt*
assignment-stmt	::=	*scalar-var* := *expr*
scalar-var	::=	*variable*
	|	*variable* (*subscript-list*)
subscript-list	::=	*int-expr* [, *int-expr*]*
expr	::=	*expr* + *expr*
	|	*expr* - *expr*
	|	*expr* * *expr*
	|	*expr* / *expr*
	|	(*expr*)
	|	*pseudo-function* (*expr*)
	|	*variable* (*subscript-list*)

FIGURE 2. Grammar of the Apply language.

if-stmt   : :=   **IF** *bool-expr* **THEN**
                               statements
                           END IF
                   | **IF** *bool-expr* **THEN**
                               statements
                           ELSE
                               statements
                           END IF

bool-expr   : :=   *bool-expr* **AND** *bool-expr*
                   | *bool-expr* **OR** *bool-expr*
                   | **NOT** *bool-expr*
                   | (*bool-expr*)
                   | *expr* < *expr*
                   | *expr* <= *expr*
                   | *expr* = *expr*
                   | *expr* >= *expr*
                   | *expr* > *expr*
                   | *expr* /= *expr*

for-stmt   : :=   **FOR** *int-var* **IN** *range* **LOOP**
                               statements
                           END LOOP

while-stmt   : :=   **WHILE** *bool-expr* **LOOP**
                               statements
                           END LOOP

FIGURE 2. (continued)

```
|  1  2  1 |          |  1  0 -1 |
|  0  0  0 |          |  2  0 -2 |
| -1 -2 -1 |          |  1  0 -1 |

  Horizontal            Vertical
```

FIGURE 3. The Sobel convolution masks.

the window lies outside the image. This same line declares the constant and output parameters to be floating-point scalar variables.

The computation of the Sobel convolutions is implemented by the straight-forward expressions on lines 5 through 7. These expressions are readily seen to be a direct implementation of the convolutions in Figure 3.

Border handling is always a difficult and messy process in programming kernel operations such as Sobel edge detection. In practice, this is usually left up to the programmer, with varying results — sometimes borders are handled in one way, sometimes another. Apply provides a uniform way of resolving the difficulty. It supports border handling by extending the input images with a constant value. The constant value is specified as an assignment. Line 1 of Figure 4 indicates that the input image *inimg* is to be extended by filling with the constant value 0.

```
procedure sobel (inimg  : in array (-1..1, -1..1) of byte    -- 1
                          border 0,
                 thresh : const real,
                 mag    : out real)
is                                                            -- 2
    horiz, vert : integer;                                    -- 3
begin                                                         -- 4
    horiz := inimg(-1,-1) + 2 * inimg(-1,0) + inimg(-1,1) - -- 5
             inimg(1,-1) - 2 * inimg(1,0) - inimg(1,1);
    vert := inimg(-1,-1) + 2 * inimg(0,-1) + inimg(1,-1) -   -- 6
            inimg(-1,1) - 2 * inimg(0,1) - inimg(1,1);
    mag := sqrt(REAL(horiz)*REAL(horiz)                       -- 7
               + REAL(vert)*REAL(vert));
    if mag < thresh then                                      -- 8
        mag := 0.0;                                           -- 9
    end if;                                                   -- 10
end sobel;                                                    -- 11
```

FIGURE 4.   An Apply Implementation of thresholded Sobel edge detection.

If the programmer does not specify how an input variable is to be extended as the window crosses the edge of the input image, Apply handles this case by not calculating the corresponding output pixel.

We plan to extend the Apply language with two other methods of border handling: extending the input image by replicating border pixels, and allowing the programmer to write a special-purpose routine for handling border pixels.

Apply allows the programmer to process images of different sizes, for example, to reduce a 512×512 image to a 256×256 image, or to magnify images. This is implemented via the *SAMPLE* parameter, which can be applied to input images, and by using output image variables which are arrays instead of scalars. The *SAMPLE* parameter specifies that the Apply operation is to be applied not at every pixel, but regularly across the image, skipping pixels as specified in the integer list after *SAMPLE*. The window around each pixel still refers to the underlying input image. For example, the following program performs image reduction, using overlapping 4×4 windows, to reduce an $n \times n$ image to an $n/2 \times n/2$ image:

```
procedure reduce(inimg  : in array (0..3, 0..3) of byte sample (2, 2),
                 outimg : out byte)
is
    sum : integer;
    i,j : integer;
begin
    sum := 0;
    for i in 0..3 loop
        for j in 0..3 loop
            sum := sum + inimg(i,j);
        end loop;
    end loop;
    outimg := sum / 16;
end reduce;
```

Magnification can be done by using an output image variable which is an array. The result is that, instead of a single pixel being output for each input pixel, several pixels are output, making the output image larger than the input. The following program uses this to perform a simple image magnification, using linear interpolation:

```
procedure magnify(inimg : in array(-1..1, -1..1) of byte border 0,
                  outimg: out array(0..1, 0..1) of byte)
is
begin
    outimage(0,0) := (inimg(-1,-1) + inimg(-1,0)
                      + inimg(0,-1)+ inimg(0,0)) / 4;
    outimage(0,1) := (inimg(-1,0) . + inimg(-1,1)
                      + inimg(0,0) + inimg(0,1)) / 4;
    outimage(1,0) := (inimg(0,-1)  + inimg(0,0)
                      + inimg(1,-1)+ inimg(1,0)) / 4;
    outimage(1,1) := (inimg(0,0)   + inimg(0,1)
                      + inimg(1,0) + inimg(1,1)) / 4;
end magnify;
```

The semantics of *SAMPLE* ($s1, s2$) are as follows: the input window is placed so that pixel ($0, 0$) falls on image pixel (0,0), (0,$s2$), . . ., (0,$n \times s2$), . . ., ($m \times s1$, $n \times s2$). Thus, *SAMPLE* (1, 1) is equivalent to omitting the *SAMPLE* option entirely. If only one *SAMPLE* parameter exists, it applies to the last image dimension.

Output image arrays work by expanding the output image in either the horizontal or vertical direction, or both, and placing the resulting output windows so that they tile the output image without overlapping. If only one dimension is specified, it applies to the last image dimension, as with *SAMPLE*.

It is a topic of current research to allow Apply to efficiently implement multiple functions. The current version of Apply requires a separate pass, producing intermediate output images, for each Apply function. If multiple Apply functions can be compiled together in a single pass, it will be possible to perform some operations much more efficiently. For example, many median filter algorithms use results from an adjacent calculation of the median filter to compute a new median filter when processing the image in raster order. This cannot be done with a single Apply function, since it requires the algorithms to make no restrictions on the order pixels are processed. However, we can define an efficient median filter using multiple Apply functions and allow the compiler to figure out how to efficiently execute this program on a particular machine by taking advantage of adjacent results.[5] The following 3×3 median filter has been carefully optimized for speed.

The algorithm works in two steps. The first step produces, for each pixel, a sort of the pixel and the pixels above and below that pixel. The result from this step is an image three times higher than the original, with the same width. The second step sorts, based on the middle element in the column, the three elements produced by the first step, producing the following relationships among the nine pixels at and surrounding a pixel:

From this diagram, it is easy to see that none of pixels g, h, b, or c can be the median, because they are all greater or less than at least five other pixels in the neighborhood. The only candidates for median are a, d, e, f, and i. Now we observe that $f < \{e, h, d, g\}$, so that if $f < a$, f cannot be the median since it will be less than five pixels in the neighborhood. Similarly, if $a < f$, a cannot be the median. We therefore compare a and f and keep the larger. By a similar argument, we compare i and d and keep the smaller. This leaves three pixels: e and the two pixels we chose from $\{a, f\}$ and $\{d, i\}$. All of these are median candidates. We therefore sort them and choose the middle element; this is the median.

This algorithm computes a 3×3 median filter with only 11 comparisons, comparable to many techniques for optimizing median filter in raster-order processing algorithms.

```
procedure median1(image in array(-1..1, 0..0) of byte,
                  si out array(-1..1, 0..0) of byte)
is
        byte a, b, c;
begin
    if image(-1,0) > image(0,0)
        then if image(0,0) > image(1,0)
            then si(1,0)  := image(-1,0);
                 si(0,0)  := image(0,0);
                 si(-1,0) := image(1,0); end if;
        else if image(-1,0) > image(1,0)
            then si(1,0)  := image(-1,0);
                 si(0,0)  := image(1,0);
                 si(-1,0) := image(0,0);
            else si(1,0)  := image(1,0);
                 si(0,0)  := image(-1,0);
                 si(-1,0) := image(0,0);
            end if;
        end if;
    else if image(0,0) > image(1,0)
        then if image(-1,0) > image(1,0)
            then si(1,0)  := image(0,0);
                 si(0,0)  := image(-1,0);
                 si(-1,0) := image(1,0);
            else si(1,0)  := image(0,0);
                 si(0,0)  := image(1,0);
                 si(-1,0) := image(-1,0);
            end if;
        else si(1,0)  := image(1,0);
             si(0,0)  := image(0,0);
             si(-1,0) := image(-1,0);
        end if;
    end if;
end median1;
```

```
procedure median2(si in array(-1..1, -1..1) of byte sample (3, 1),
                  median out byte)
-- Combine the sorted columns from the first step to give the median.

is
    int l, m, h;
    byte A, B;
begin
    if si(-1, 0) > si(0, 0)
        then if si(0, 0) > si(1, 0)
                then h := -1; m := 0; l := 1; end if;
                else if si(-1,0) > si(1,0)
                        then h := -1; m := 1; l := 0;
                        else h := 1; m := -1; l := 0; end if; end if;
            else if si(0, 0) > si(1, 0)
                then if si(-1,0) > si(1,0)
                        then h := 0; m := -1; l := 1;
                        else h := 0; m := 1; l := -1; end if;
                    else h := 1; m := 0; l := -1; end if; end if;

    if si(l, -1) > si(m, 1)
                then A := si(l, -1);
                else A := si(m, 1); end if;
    if si(m, -1) < si(h, 1)
                then B := si(m, -1);
                else B := si(h, 1); end if;

    if A > si(m, 0)
        then if si(m, 0) > B
                then median := si(m, 0); end if;
                else if A > B
                        then median := B;
                        else median := A; end if; end if;
            else if si(m, 0) > B
                then if A > B
                        then median := A;
                        else median := B; end if;
                    else median := si(m, 0); end if; end if;

end median2;
```

The implementation of Apply on Warp employs straightforward raster processing of the images, with the processing divided among the cells as described above. The Sobel implementation in Figure 4 processes a 512×512 image on a 10 cell Warp in 330 ms, including the I/O time for the Warp machine.

The iWarp implementation of Apply uses the logical pathway mechanism to allow each cell to process only data intended for that cell. This eliminates much of the complication of Apply on Warp; there is no need for a cell to explicitly pass data on to other cells; instead it can simply direct the rest of the data to pass on to later cells without further intervention.

Our description of Apply on iWarp will be clear if we describe the action of *GETROW* and *PUTROW* on this machine. In *GETROW*, each cell accepts data intended for that cell and then releases control of the data to be passed on to the next cell automatically, until the arrival of the start of the next row. After releasing control, it goes on to process the data it has just received. In the meantime, it is allowing data to pass by on the output channel until the end of the output row arrives. It then tacks on its computed output to the end of this output row, completing *PUTROW*.

We expect this method of implementing Apply to be at least as efficient as the *COMPUTEROW* model on Warp. We are currently investigating techniques to make the implementation fully systolic, so that it does not require buffering data in memory before processing.

The same Apply compiler that generates Warp code can also generate C code to be run under UNIX. We have found that an Apply implementation is usually at least as efficient as any alternative implementation on the same machine. This efficiency results from the expert knowledge which is built into the Apply implementation, but which is too verbose for the programmer to work with explicitly. In addition, Apply focuses the programmer's attention on the details of his computation, which often results in improved design of the basic computation.

The Apply implementation for uniprocessor machines relies upon a subroutine library which was previously developed for this purpose. The routines are designed to efficiently pass a processing kernel over an image. They employ data buffering which allows the kernel to be shifted and scrolled over the buffer with a low constant cost, independent of the size of the kernel. The Sobel implementation in Figure 4 processes a 512×512 image on a Vax 11/785 in 30 s.

The buffering technique which we developed for Apply on uni-processor machines operates as follows. Initially, a buffer is allocated and indexed by an Illiffe vector of pointers. For an $N \times N$ input image which will be processed with an $M \times M$ kernel, $N \times M + (N + M - 1)/M - 1$ pointers are required. The cost of computing these pointers is negligible compared to the N^2 cost of the actual computation being performed at all pixel locations.

After establishing the pointers, which remain unchanged during the remainder of the algorithm, the first M rows of the image are copied into the buffer in preparation for processing.

Figure 5 displays the specific pointer arrangement for processing a 3×3 kernel. When the pointer into the Illiffe vector is as shown in the figure, C language subscripting can be used to directly access the elements of the kernel surrounding the first pixel location. Two pointer dereference operations, possibly with small offsets, are needed for each access.

After the first pixel location has been processed, the base pointer is incremented by M. The 3×3 kernel surrounding the second pixel location can then be directly accessed as before. It is thus possible to shift the kernel across the entire buffer of data with a cost of only one addition per pixel. The cost of relocating the kernel is independent of the size of the kernel, so large kernels can be processed very efficiently by Apply.

When processing of an entire row is completed, the base pointer is set back to its original position and then incremented by one. This has the effect of rolling the individual rows of the buffer upwards. The row which was previously in the center of the 3×3 kernel is now at the top and the row which was previously at the bottom is now in the center. The row which was previously at the top is now the bottom row, but because it is being indexed by a new pointer, its origin has been shifted right one word. This shifting is not a problem, because it affects the row into which new data must be read. The only constraints imposed are that there must be additional buffer space available and that the rows must be organized in memory so that there are no overlaps when the buffer is rolled in this manner. Figure 5 shows an arrangement which satisfies these constraints: $(N + M - 1)/M$ additional words of buffer space are provided, and the initial top row, which is shifted first, is placed last in memory.

Notice that, once again, the cost of relocating the kernel is a single addition and does not depend on the size of the kernel being processed.

Apply has been implemented on the Hughes HBA computer[15] by Richard Wallace of Carnegie Mellon and Hughes. In this computer, several MC68000 processors are connected on a high-speed video bus, with an interface between each processor and the bus that allows it to select a subwindow of the image to be stored into its memory. The input image is sent over the bus and windows are stored in each processor automatically using DMA. A similar interface exists for outputting the image from each processor. This allows flexible real-time image processing.

The Hughes HBA Apply implementation is straightforward and similar to the Warp implementation. The image is divided in "swaths", which are adjacent sets of rows, and each processor takes one swath. (In the Warp implementation, the swaths are adjacent sets of columns, instead of rows.) Swaths overlap to allow each processor to compute on a window around each pixel. The processors independently compute the result for each swath, which is fed back onto the video bus for display.

The HBA implementation of Apply includes a facility for image reduction, which was not included in earlier versions of Apply. The HBA implementation subsamples the input images, so that the input image window refers to the subsampled image, not the original image as in our definition. We prefer the approach here because it has more general semantics. For example, using image reduction as we have defined it, it is possible to define image reduction using overlapping windows.

Here we briefly outline how Apply could be implemented on other parallel machine types, specifically bit-serial processor arrays and distributed-memory, general-purpose processor machines. These two types of parallel machines are very common; many parallel architectures include them as a subset or can simulate them efficiently.

Bit-serial processor arrays[3] include a great many parallel machines. They are arrays of large numbers of very simple processors which are able to perform a single bit operation in every machine cycle. We assume only that it is possible to load images into the array such that each processor can be assigned to a single pixel of the input image, and that different processors can exchange information locally, that is, processors for adjacent pixels can exchange information efficiently. Specific machines may also have other features that may make Apply more efficient than the implementation outlined here.

In this implementation of Apply, each processor computes the result of one pixel window. Because there may be more pixels than processors, we allow a single processor to implement the action of several different processors over a period of time, that is, we adopt Connection Machine's idea of *virtual processors*.[7]

The Apply program works as follows:

- Initialize: For $n \times n$ image processing, use a virtual processor network of $n \times n$ virtual processors.
- Input: For each variable of type *IN*, send a pixel to the corresponding virtual processor.
- Constant: *Broadcast* all variables of type *CONST* to all virtual processors.
- Window: For each IN variable, with a window size of $m \times m$, shift it in a spiral, first one step to the right, then one step up, then two steps to the left, then two steps down, and so on, storing the pixel value in each virtual processor the pixel encounters, until a $m \times m$ square around each virtual processor is filled. This will take m^2 steps.
- Compute: Each virtual processor now has all the inputs it needs to calculate the output pixels. Perform this computation in parallel on all processors.

Because memory on these machines is often limited, it may be best to combine the "window" and "compute" steps above to avoid the memory cost of prestoring all window elements on each virtual processor.

Machines in this class consist of a moderate number of general-purpose processors, each with its own memory. Many general-purpose parallel architectures implement this model, such as the Intel iPSC[9] or the Cosmic Cube.[14] Other parallel architectures, such as the shared-memory BBN Butterfly,[4,13] can efficiently implement Apply in this way; treating them as distributed memory machines avoids problems with contention for memory.

This implementation of Apply works as follows:

- Input: If there are n processors in use, divide the image into n regions and store one region in each of the memories of the n processors. The actual shape of the regions can vary with the particular machine in use. Note that compact regions have smaller borders than long, thin regions, so that the next step will be more efficient if the regions are compact.
- Window: For each *IN* variable, processors exchange rows and columns of their image with processors holding an adjacent region from the image, so that each processor has enough of the image to compute the corresponding output region.
- Compute: Each processor now has enough data to compute the output region. It does so, iterating over all pixels in its output region.

We have described our programming techniques for low-level vision on Warp. These techniques began with simple row-by-row image-processing macros, which are still in use for certain kinds of algorithms, and led to the development of Apply, which is a specialized programming language for low-level vision on Warp.

We have defined the Apply language as it is currently implemented and described its use in low-level vision programming. Apply is in daily use at Carnegie Mellon for Warp and vision programming in general; it has proved to be a useful tool for programming under UNIX, as well as an introductory tool for Warp programming.

The Apply language crystallizes our ideas on low-level vision programming on Warp. It allows the programmer to treat certain messy conditions, such as border conditions, uniformly. It also allows the programmer to get consistently good efficiency in low-level vision programming by incorporating expert knowledge about how to implement such operators.

One of the most exciting characteristics of Apply is that it may be possible to implement it on diverse parallel machines. We have outlined such implementations on bit-serial processor arrays and distributed-memory machines. Implementation of Apply on other machines will make porting of low-level vision programs easier, should extend the lifetime of programs for such supercomputers, and will make benchmarking easier.

We have shown that the Apply programming model provides a powerful simplified programming method which is applicable to a variety of parallel machines. Whereas programming such machines directly is often difficult, the Apply language provides a level of abstraction in which programs are easier to write, more comprehensible, and more likely to work correctly the first time. Algorithm debugging is supported by a version of the Apply compiler which generates C code for uniprocessor machines.

REFERENCES

1. *Reference Manual for the Ada Programming Language*, MIL-STD 1815 edition. U.S. Department of Defense, AdaTEC, SIGPLAN Technical Committee on Ada, New York, AdaTEC, 1982, Draft revised MIL-STD 1815, Draft proposed ANSI Standard document.
2. **Annaratone, M., Arnould, E., Gross, T., Kung, H. T., Lam, M., Menzilcioglu, O., Sarocky, K., and Webb, J. A.,** Warp architecture and implementation, in Conf. Proc. of the 13th Annu. Int. Symp. on Computer Architecture, June, 1986, 346.
3. **Batcher, K. E.,** Bit-serial parallel processing systems, *IEEE Trans. Comput.,* C-31 (5), 377, 1982.
4. BBN Laboratories, *The Uniform System Approach to Programming the Butterfly Parallel Processor,* 1st ed., BBN, Cambridge, MA, 1985.
5. **Fisher, A. J. and Highnam, P. T.,** Communications, scheduling and optimization in SIMD image processing, in *Computer Architectures for Pattern Analysis and Machine Intelligence,* IEEE, 1987, submitted.
6. **Hillis, W. D.,** *The Connection Machine,* MIT Press, Cambridge, MA, 1985.
7. **Huang, T. S., Yang, G. J., and Tangm G. Y.,** A fast two dimensional median filtering algorithm, in Int. Conf. on Pattern Recognition and Image Processing, IEEE, 1978, 128.
8. *iPSC System Overview,* Intel Corporation, 1985.
9. **Kernighan, B. W. and Ritchie, D. M.,** The M4 Macro Processor, *UNIX Programmer's Manual,* Bell Laboratories, Murray Hill, NJ, 1979.

10. **Kung, H. T. and Webb, J. A.,** Global Operations on the CMU Warp Machine, in Proc. of 1985 AIAA Computers in Aerospace V Conf., American Institute of Aeronautics and Astronautics, October 1985, 209.

11. **Kung, H. T. and Webb, J. A.,** Mapping image processing operations onto a linear systolic machine, *Distrib. Comput.*, 1(4), 246.

12. **Olson, T. J.,** An Image Processing Package for the BBN Butterfly Parallel Processor. Butterfly Project Report 9, Department of Computer Science, University of Rochester, NY, August 1985.

13. **Seitz, C.,** The cosmic cube, *Commun. ACM*, 28(1), 22, 1985.

14. **Wallace, R. S. and Howard, M. D.,** HBA vision architecture: built and benchmarked, in *Computer Architectures for Pattern Analysis and Machine Intelligence*, IEEE Computer Society, Seattle, WA, 1987.

18 TRANSLATING SEQUENTIAL PROGRAMS TO PARALLEL: LINDA

In the areas of science, engineering, and even business, computationally intensive problems and applications are requiring more processing power than ever before. Despite increasing processor speed, traditional sequential computers can't supply the power needed to solve these problems. Parallel computers, an area where many processors take on a problem simultaneously, hold the potential answers to many of these problems.

Previously, only companies with vast resources could afford parallel-processing systems. However, with recent advances in hardware and software, parallel processing is becoming more economical. Multiprocessor systems can now house more power than traditional supercomputers at a fraction of the cost. Yet, without the right software tools, programs can't easily take advantage of the parallel hardware. If there was a simple way to translate a sequential program into a parallel program, the parallel hardware could be maximized. Linda, a parallel-programming methodology, can do just that.

Even with the latest parallel hardware, parallel processing is left unexploited in areas where it could be helpful. One barrier to parallel applications is the hardware-specific nature of most parallel programming languages. Each parallel computer vendor has its own computers and language extensions to support the particular architecture of its products. Consequently, a parallel program created for a specific-hardware configuration couldn't be ported to another parallel architecture.

For this reason, it was difficult to try and justify translating a sequential program to a parallel machine. In addition, the software investment is lost if the program won't run on future-generation parallel systems.

Parallel computers can usually be broken into two architectural groups: shared memory and message passing (MP). Each has its strengths and weaknesses. Shared-memory systems are generally considered easier to program because access to shared data is made with simple memory references. However, because all processors can access the shared data simultaneously, extra care must be taken to protect shared resources. Alternatively, in an MP system, all data are shared between processes using explicit messages.

While this reduces the programming errors common to shared-memory systems, it requires significantly more programming effort to compose, send, and receive the messages.

With traditional parallel-programming methods, a program written for a shared-memory system would differ considerably from the same program written for an MP system. Because it has been implemented on both MP and shared-memory architectures, Linda has the potential to become a truly portable standard for parallel programming.

Linda, a parallel-processing modeling language, offers an elegant yet powerful way to create parallel programs. Linda is sometimes referred to as the "portable parallel" because it's been implemented on a wide variety of parallel-computer architectures.

With Linda, parallel programs can be written in conventional programming languages, such as C, Fortran, and Lisp by supplying — in a language-independent manner — primitive operations for interprocess communication. The advantage of adding these operations to an existing language is that it creates a parallel dialect of the language.

At the heart of this mechanism is a tuple space — an abstract object through which programs place and remove objects called tuples. A tuple is a collection of related data, containing a key that retrieves the tuple.

There are four fundamental Linda operations on a tuple space: "out", "in", "eval", and "rd". For example, the tuple

```
("hello", 5, true);
```

contains three items; a string, an integer, and a Boolean value. The operation

```
out("hello", 5, true);
```

places the tuple into the tuple space. The "out" operation never blocks, nor does it return from the "in" operation, and there can be any number of copies of the same tuple in the tuple space.

Tuples are removed from the tuple space by matching them against a template. For example, the operation

```
int i; boolean b;
in("hello",?i, ?b);
```

will match any tuple whose first element is the string "hello", second element is an integer, and third element is a Boolean value. If a matching tuple is found, the variables "i" and "b" are assigned the corresponding values from the matching tuple and the tuple is removed from the tuple space. If no match is found, the process that makes the "in" call blocks until a matching tuple becomes available. If the template matches more than one tuple, an arbitrary one is picked. The "rd" operation is similar to the "in" operation except that the matching tuple isn't removed from the tuple space. "Inp" and "rdp" are variations of "in" and "rd" that don't block, even if a matching tuple isn't available.

The "eval" operation is similar to "out" except that it creates an active tuple. For example, if foo is a function that returns an integer, then

```
eval("hello", foo(z), true);
```

creates a new process to evaluate foo. It then proceeds concurrently with the process that made the "eval" call. When foo returns, it leaves a passive data tuple in the tuple space, identical to an "out".

Because it can be implemented on many different parallel computers and used with numerous standard computer languages, Linda can increase portability, ease development, and increase software reliability.

The language supplies a high level of abstraction to accommodate various architectures. The same Linda program will run on either of the two architectures. As a result, the user's software investment isn't wasted if the hardware is upgraded at some point in the future.

Most sequential computer programs contain some inherent parallelism that can be exploited for faster execution on parallel computers. When analyzing a sequential program, one must first look at the structure of the results and ask if the result can be broken into smaller pieces that can be computed separately, such as the elements of a matrix or array. Second, examine the dependencies between the elements of the result. In the best case, the elements can be computed independently, yielding a high degree of parallelism. Even if dependencies exist, computation may still take on a parallel form, relying on synchronization code to coordinate the dependencies. Finally, one must examine the sequential code for nested loops, which often indicate computational chunks that could be processed in parallel rather than sequentially.

Much research has gone into parallelizing compilers, which are compilers that automatically transform sequential code into parallel code. However, a growing consensus agrees that to achieve the highest performance level, it's necessary to recode programs to use explicitly parallel algorithms, rather than extract the partial parallelisms of sequential programs.

Often, coding a program with a parallel algorithm results in a speed-up, even when the program is run on one processor. In this case, a parallel algorithm is better than a sequential one. With Linda, the extra effort for hand coding parallel algorithms can be used repeatedly.

Parallel solutions always involve a trade-off between communication and computation, which are characterized by their granularity. A fine-grain solution creates many small tasks, while a coarse-grain solution creates a few large ones. If the task is too small, the processes spend too much time communicating results and getting new tasks. If it's too big, then some processes might sit idle waiting for others to finish. Therefore, the programs must be carefully balanced. Linda makes it easy to adjust these factors by allowing users to alter the size of the task.

Parallel programs must be designed to minimize the synchronization delays created when one process is required to wait for the result of another before proceeding. Most problems in the real world involve data dependencies requiring synchronization processes.

Creating a parallel program, either from scratch or from an existing sequential program, includes decomposing the tasks into pieces that can be computed concurrently. There are two main types of parallel problem decomposition: data and functional. Both can easily be implemented in Linda. The type chosen depends on the nature of the computational problem. The two categories actually overlap somewhat; a parallel solution may incorporate aspects of both methods.

In data decomposition, data are divided into parts and multiple copies of an operation are applied to different parts of the data concurrently. This model is appropriate when the solution to a problem is seen as applying a process repeatedly to a collection of data objects. For example, when performing a smoothing algorithm on an image, it would be possible to divide the image into rectangular regions and have multiple smoothing processes computing in parallel.

The parts could then be recombined into the resulting image. Data decomposition is typically implemented as one master process that divides the data and multiple worker processes, which perform the computation Figure 1. The master-worker model is also referred to as a "processor farm".

In Linda, all communication between the master and workers is done through the tuple space. A tuple space is an abstract object through which programs place and remove objects called tuples. As indicated earlier, a tuple is a collection of related data containing a key that retrieves the tuple. The master divides the data and packages them as tuples which are "outed" (placed in a tuple in the tuple space) as tasks.

The workers retrieve the tasks from the tuple space with "in" or "inp" operations (operations that remove a tuple from the tuple space), compute the results, and then "out" the results as tuples. Meanwhile, the master is "ining" the results as they become available from the workers. Linda also synchronizes because a process will block if it tries to "in" a tuple that isn't yet in the tuple space.

A collection of similar tasks is often called a bag, because workers withdraw tasks relatively indiscriminately. This model is quite flexible because each worker continues to retrieve tasks until all are consumed — it works with any number of worker processes. The program could be written and tested on a system with one processor, using one or two worker processes.

Once debugged, it could be moved to a multiple-processor system, making it possible for more workers to take advantage of the extra processors. This model will also automatically balance the processing load among the processes. For example, one process could perform many smaller tasks while another is executing one larger task.

Functional decomposition applies different processes to a block of data. It's often the best method when a solution can be modeled as a network of nodes, such as a system simulation. Each node can be represented as a routine that processes

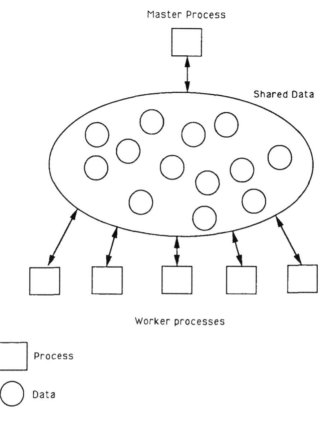

FIGURE 1. One large process can be broken up into many smaller processes to help divide the work load to perform a computation.

some data and sends the results to another node for further computation. Functional decomposition usually involves some sort of explicit message passing between the processes for synchronization purposes Figure 2.

Functionally decomposed solutions often take the form of a sequential pipeline. In the case of a pipeline, a series of processes is applied to incoming data, with the results of each process passed as input to the next process in the pipe. To achieve more concurrency, small chunks of data are continually fed into the pipe so that as each process finishes its computation, more data is waiting. The danger in this structure is that the process with the heaviest computational load becomes a bottleneck, slowing all of the processes in the pipe. To avoid this, more processes must be allocated to the heaviest computation task Figure 3.

Linda supports functional decomposition and message passing by allowing messages to form as tuples with the destination node name as one of the keys. A node then waits for messages by "ining" on its node name. In an example with individual processes named crunch1 and crunch2, and three processes named bigCrunch, crunch2 finishes processing a block of data and "outs" the result with the Linda operation:

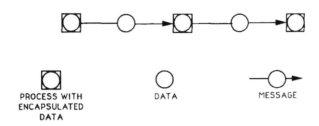

FIGURE 2. A block of data is sent along a string of nodes, whereby each node does a portion of the processing.

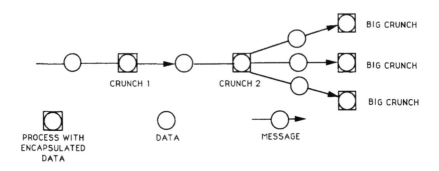

FIGURE 3. If one computational load requires an abundance of processing, that load is allocated to more processes.

```
out ("bigCrunch", theResults)
```

Meanwhile, all three bigCrunch processes would be waiting for messages from crunch2 by "ining":

```
in ("bigCrunch", ?theData)
```

Each bigCrunch process withdraws a message tuple as it becomes available, then computes the result, and "outs" it to the next stage in the pipe. Because of the time-intensive nature of bigCrunch, one crunch2 process could keep three bigCrunches supplied with data.

The MP model becomes inefficient when an abundance of traffic is needed to support the computation. Supplementing the MP model with some shared-data access can ease the communication load and increase the efficiency of the program.

Consider the problem of mapping a function onto a two-dimensional matrix of values. The results of the function at each location in the matrix produce a topology of the function (Figure 4). The third dimension of the topology is derived from the resulting arguments of the matrix. The sequential solution

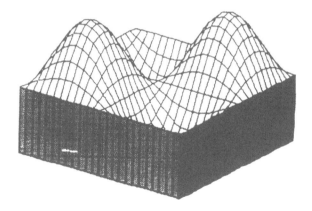

FIGURE 4. A functional topology is the outcome of a two-dimensional matrix and the resulting arguments of that matrix.

consists of two nested loops, one each for the rows and columns of the matrix (Table 1). Each point in the matrix is evaluated in sequence, and the results accumulate in a result matrix.

TABLE 1
Function Mapping (Sequential Solution)

```
#define N 300

double results [N] [N];
extern double f(int, int, int);

void main () {
    int row, column;
    // loop through the rows
    for (row = 0; row < N; row + + ) {
        // and do each item in each row
        for (column = 0; column < N; column + + ) {
                results [row] [column] = f(row, column, N);
            {
        {
    {
```

Using the techniques described earlier, it can be seen that the program contains a nested loop where each element in the result is computed independently of all other elements. This problem easily converts to a parallel solution.

The inner loop in the sequential solution computes the values for one row of the matrix. In the parallel version, many rows can be computed simultaneously by different processes (Table 2). Using the data-decomposition model, a master program launches multiple worker processes with Linda's "eval" primitive (creating an active tuple) and distributes tasks to the workers with the "out" primitive.

Table 2
Function Mapping (Parallel Solution)

```
#include <linda.h>

#define N 300

double results [N] [N];
extern double f(int, int, int);

int worker () {
    int row, column;
    double line [N]

    while (inp("task", ?row)) {
            for (column + ); columns < N; column + + ){
                    line[column] = f(row, column, N);

            {
            out ("line", row, line:N);
    {
    return 0;
{
void Imain (int argc, char ** argv)
{
    int row, workers;

    /*put tasks into tuple space*/
    for (row = 0; row < N; row + + ) {
            out("task", row);
    {

    /*create workers*/
    for (workers = 0; workers < atoi(argv [1]; workers + + ) {
            eval("worker", worker ());
    {

    /*remove results from tuple space*/
    for (row = 0; row < N; row + + ) {
            in("line", row, ?result[row] :N);
    {
    /*clean up workers*/
    for (row = 0; row < atoi(argv[1]); row + + ) {
            in("worker",0);
```

Each task represents one row of the matrix. Each worker repeatedly retrieves a task tuple from the shared tuple space, computes the values for the corresponding row of the result matrix, and "outs" the result row back to the tuple space. The workers continue to retrieve tasks with the "inp" primitive until no tasks remain. Meanwhile, the master process is retrieving the result rows from the tuple space to assemble the complete result matrix.

Once a parallel solution is coded, it is often necessary to evaluate the performance of the program and modify the design for greater efficiency. It is useful to track the performance of a parallel program as processes are added.

Ideally, a linear speed-up should be seen as more processes are added, but in practice this improvement won't go on indefinitely. The programs typically show a near-linear speed-up for the first several processors, and then fall off sharply. The reason the fall-off takes place is that communication overhead begins to dominate the computation time. Each program will have its own particular point where communication and computation are optimally balanced, which can typically be found by experimentation.

The function-mapping example divided the problem into tasks where each task represented the computation of one row of the result matrix. It could have been constructed so that each task represented either one element of the result matrix or a block of rows. Linda offers the flexibility to try different options without major program modification. As a result, it's easier to arrive at the optimal parallel solution. Lastly, the program can be moved to other parallel machines to ensure that the investment in software development won't become obsolete as hardware improves.

19 PTOOL: A SEMIAUTOMATIC PARALLEL PROGRAMMING ASSISTANT

Humans and machines bring very different strategies to bear on the problem-solving process. Humans make effective use of abstraction to develop broad strategies and are adept at simplification of complex situations and at generalization from simple principles. Unfortunately, they are not good at exhaustive search and seem unable to pay rigorous attention to details — areas where the computer excels. As a result, programmers often fail to precisely specify the details of their intended solutions to the machine, inevitably leading to "bugs" in their programs.

The advent of parallel machine architectures has added a new level of complexity to the debugging process. Parallelism implies nondeterminism; nondeterminism, in turn, implies nonrepeatability. In other words, program sections that are executed in parallel do not necessarily follow the same execution path every time the program is run — even when the program is run on the same data. This additional complexity requires additional precision from the programmer. Not only must he ensure that the computations performed by the sequential sections are those that he intended, but he must also ensure that the computations indicated in parallel regions perform the desired calculations regardless of the order in which they are executed. This increased complexity permits some very elusive bugs.

Given the difficulties that can arise in parallel programming, it seems clear that new methods and tools will be needed.[18] There are several possible lines of attack. At one extreme, a parallel program might be repeatedly run on the same test data to ensure that it computes the same answers on each execution. Although this approach is easy to implement, it is inelegant and probably would not be effective, since the conditions required to produce an error need not arise during the test. At the opposite extreme, formal correctness proofs could be applied to a parallel program. In spite of significant levels of activity on verification over the past decade, complete automatic proofs for realistic programs appear impractical for the time being.

A third approach would be to develop an automatic system to convert implicit parallelism within a sequential program into an explicit form. The advantage of this method results from the ability to debug the program using standard techniques on sequential machines prior to parallel conversion. Since the conversion involves probably correct transformations, the parallel version should execute correctly, thereby avoiding the need for a complex parallel debugging phase. Additionally, this approach provides a mechanism for converting "dusty deck" programs.

In spite of these advantages, fully automatic conversion to parallelism is not practical using current language processing technology. Although many promising techniques have been developed,[6,10,21,22] the programmer is still an essential ingredient in the exploitation of parallelism. There are several reasons for this. First, most programs contain a large number of potential parallel regions. Searching all regions for parallelism can be an extremely expensive task. In addition, synchronization overhead on many machines can easily overwhelm the advantages of parallel execution,[12] unless the granularity of the parallel regions is very large. Often, this determination depends upon values that are known only at run-time. Hence, the programmer's judgment is an important part of the parallel programming process.

This discussion describes a parallel programming assistant, called PTOOL, that combines the advantages of the manual and automatic approaches. When using PTOOL, a programmer develops and debugs a program on a sequential machine using traditional techniques. Once the program is debugged, the programmer identifies the regions, usually separate iterations of loops, that should be executed in parallel. PTOOL then reports whether the results of execution *depend in any way on the order of execution of the regions.* If not, the regions can be safely executed in parallel. Otherwise, the programmer can ask PTOOL for diagnostic information to help identify problems. This information is presented as reports on potential conflicts in the use of variables shared by different parallel regions.

This discussion initially introduces a model for the parallel programming process that PTOOL is intended to support. Fundamentally, PTOOL is a system designed to help programmers transform sequential loops in Fortran to parallel loops. Although this model is not universal, it covers many interesting cases.

Multiprocessors are the most efficient when each processor is able to compute at full speed, without requiring synchronization with other processors. To achieve this ideal, we not only must identify a collection of program fragments that can be executed in parallel, but we must also schedule the computation in a manner that balances the execution load across processors. In Fortran, the *lingus franca* of scientific computation, the construct most likely to give rise to a large number of parallel regions with comparable execution times is the DO loop. There are two reasons why. First, if they can be made to execute correctly in parallel, the separate iterations of a DO loop can provide enough computation to keep a significant number of processors busy. Second, since each iteration of a DO loop should take roughly the same time to execute, the load will be automatically balanced across the processors. Experience with actual multiprocessors strongly supports the importance of loops in parallel programming.[19]

In light of these observations, we have designed PTOOL to determine the conditions that inhibit parallel execution of loops. In doing so, we have been careful not to restrict the analysis to DO loops, because iterative regions in Fortran are often coded using backward GOTO statements.

In deciding whether or not the iterations of a given loop can be executed in parallel, it is important to decide which variables mentioned in the loop body must be shared by parallel tasks and which can be allocated to storage local to the processor. This decision is important because shared variables can be accessed in different orders by different processors. If we wish to compute the same results every time the program is run, all possible schedules for a set of parallel tasks must lead to a functionally equivalent access sequence for shared variables. For example, it should not be possible to have two schedules in which the order of a load and store of a shared variable are reversed. If we are to avoid the insertion of explicit synchronization between loop iterations, we must establish that, without synchronization, such a reversal can never happen. In performing the requisite analysis, we need not concern ourselves with local variables, since the relevant loads and stores happen on the same processor.

Since PTOOL was developed to support parallel programming research at Los Alamos National Laboratory, where Fortran code was being converted to a Cray, the methods for expressing parallelism on that machine strongly influenced its design. The standard method on that machine is to express a parallel body of code as a subroutine, because that expression provides a natural encapsulation of the code to be broadcast to the different processors. Furthermore, reentrant code is straightforward to generate.

In the subroutine model, the loop body is replaced by a parallel subroutine invocation that takes as parameters an indication of the loop iteration to execute and variables that have to be globally available to all processors. PTOOL assumes that subroutine parameters and variables in COMMON are shared, but all other variables used in the loop body are assigned to storage local to each processor. As we shall see, PTOOL offers a suggestion, based upon dependence analysis, of the variables which should be parameters. The programmer is free to ignore this suggestion and select his own parameters. PTOOL will accept this modification and tailor its advice to the set of shared variables specified by the user.

Under the model we have introduced, the principal requirement that must be preserved when converting a program for parallel execution is the access order to shared variables. The main tool for analyzing patterns of loads and stores in a program is a powerful theory of dependence, the subject next addressed.

In a sequential language such as Fortran, the execution order of statements is well defined. Therefore, the behavior of the program under sequential execution order can be used as a basis for evaluating other execution orders. Specifically, it is possible to examine alternative execution orders to see whether they will also produce the same results. Parallel execution implies that the statements in a collection of parallel regions (iterations of a loop, for our purposes) can be executed in any order so long as the statements within a particular region are executed in sequence. Hence, for a collection of parallel regions to be suitable for parallel execution, there must be no dependences between statements of different parallel regions. In our model, there must be no dependences that cross loop iterations.

To determine whether this condition holds, we must identify the pairs of statements whose relative execution order *must* be preserved under any program transformation if the results are to be preserved. This relationship is represented by a collection of *dependences* among the statements of the program. A statement S_2 is said to *depend upon* a statement S_1 if S_2 follows S_1 in dynamic execution of the sequential program and it must follow S_1 in any reordering that preserves the correct results.

There are two ways for a statement S_2 to depend on statement S_1. First, S_1 can cause a change in the control flow that determines whether S_2 is executed, creating a *control dependence* of S_2 on S_1. Second, the two statements may access the same variable in a way that requires that their order be preserved. A dependence created to prevent incorrect access order to a variable is called a *data dependence*. Although both types of dependence must be considered when rearranging statements, programs can always be transformed so that all control dependences become data dependences.[1] Therefore, we restrict the remainder of this discussion to data dependences.

Data dependence arises most naturally when one statement defines a variable that is later used by a second statement. However, Kuck has identified three types of data dependence:[17]

1. True dependence — S_1 stores into a variable that S_2 later uses.
2. Antidependence — S_1 fetches from a variable that S_2 later stores into.
3. Output dependence — two statements both store into the same variable.

All three types of data dependence must be considered to safely reorder a program.

It is also useful to distinguish dependences that cross loop iteration boundaries from those that do not. To see this, consider the following example:

```
               DO 100 I = 1, N
   S₁              A(I) = . . .
   S₂                 . . . = A(I)
       100     CONTINUE
```

S_2 quite obviously has a true dependence upon S_1, implying that S_1 must be executed before S_2 in order for the computation to be correct. However, this dependence in no way precludes the separate iterations of the loop from executing in parallel, because values created within the loop are used on the same iteration, and need not be saved for later iterations. Such a dependence is said to be *loop independent*.[2] On the other hand, consider a slightly different loop.

```
                    DO 100 I = 1, N
S₁                      A(I) = . . .
S₂                      . . . = A(I - 1)
        100     CONTINUE
```

In this case, we may *not* safely transform the loop to execute in parallel, because the dependence crosses loop iterations, that is, values created on one iteration are used on a later iteration. A dependence of this sort is said to be *loop carried*, because it exists only by virtue of the iteration of the loop — if the loop body is executed only once, the dependence ceases to exist.

It is useful to observe that a loop-carried dependence arises because of the iteration of a particular loop. For instance, in the following nest of loops

```
                    DO 200 I = 1, M
                      DO 100 J = 1, N
S₁                        A(I,J) = . . .
S₂                        . . . = A(I - 1,J)
        100         CONTINUE
        200     CONTINUE
```

iteration of the outer loop (on I) gives rise to the dependence. So long as the outer loop is iterated sequentially, the dependence will be satisfied. In light of this observation, we classify carried dependences by indicating the particular loop that creates the dependence.

Loop-carried and loop-independent dependence provide a precise characterization of the execution orders that are important within a program. So long as these dependences are preserved, the results of a computation will also be preserved. This fact is extremely important in vectorization as performed in PFC, an automatic vectorization system written at Rice.[13,16] PFC constructs a complete dependence graph for the program to which it is applied in order to distinguish the statements that can be executed as vector operations from those that cannot. This graph can also be used to determine parallel loops, as the next section shows.

From the parallel programming paradigm presented above, it should be clear that there are two tasks to be performed in converting sequential code: (1) identifying local and global variables, and (2) finding loops suitable for parallel execution. These two tasks are highly interdependent as the following loop illustrates:

```
                    DO 100 I = 1, N
                      T =
                      . . . = T
        100     CONTINUE
```

If T is a variable that can be kept in the local memory of individual processors, as above, then there is no dependence that prevents parallel execution of the loop. If, on the other hand, T must be globally available (as would happen if there were a use of the last value of T after the loop), then the loop would not execute correctly in parallel. Once T becomes a global variable, different processors may intermix intermediate computations involving T, with unpredictable results.

Identifying the variables that must be global to a parallel loop turns out to be a relatively straightforward process. Simply stated, a variable must be globally available if it holds a value that is used outside the loop, or, in the symmetric case, if a definition from outside the loop reaches into the loop. Given the dependence graph described above, recognition of variables that must be globally allocated becomes a fairly trivial task. The system need only analyze the true dependences coming into and going out of a loop; any variables that give rise to such a dependence must be made globally available. In actual practice, the problem is somewhat more complicated, because full dependence graphs are normally constructed only over loop bodies. However, definition-use chains[15] are quite commonly constructed over an entire program (as they are in PFC) and can be used in the absence of stronger dependence information.

The second aspect of parallel programming is the identification of loops suitable for parallel execution or, more correctly, determination of loops which cannot be executed in parallel. Recall that parallel execution, as used in this discussion, means execution of different iterations of a loop on different processors. We can achieve this without synchronization only so long as one processor does not compute a value needed by another processor (or one processor does not store on top of a value needed by another processor, etc.). Accordingly, the key restraints to parallel execution are dependences carried by the candidate loop that involve a global variable. For instance, in the following example

```
COMMON  A(100,80)
DO 100 I = 1, 80
        DO 100 J = 1, 100
             A(J,I) =
             = A(J - 1, I) +
100     CONTINUE
```

the I loop carries no dependences and may be safely executed in parallel. The J loop, however, is a different matter. If it is executed on multiple processors, each processor will create values needed by a different processor. Hence, a loop may be executed correctly in parallel if, and only if, that loop carries no dependences based on a global variable.[4]

These two principles are the fundamental considerations in converting a sequential code for parallel execution. Because both conditions are tedious to verify, a human converting a code for parallel execution may easily overlook an important problem. More specifically, a human programmer might not notice that a variable must be globally allocated or might miss a loop-carried dependence. Since the resulting faults may manifest themselves as errors only under specific schedules, they can be extremely difficult to locate.

PTOOL divides quite naturally into two subcomponents: one to construct the dependence graph, and the other to answer queries about potential parallelism in the program. The first of these components, called PSERVE, is a modified version of PFC. When PTOOL is invoked, the user submits the program source, including all the subroutines, to PSERVE for analysis. PSERVE constructs the dependence graph and saves it in a database of two files. This database is shipped back to the user for interactive analysis.

The user then conducts an interactive dialogue with the display facility, called PQUERY. PQUERY shows the actual program source on the screen and permits the user to select a loop for analysis. It uses the database constructed by PSERVE to identify the global variables in the proposed parallel region and it presents any dependences that might impede parallel execution.

While the division of PTOOL's functions into two relatively independent components prohibits incremental reanalysis of the program as the user eliminates problems, it does permit PQUERY to be run on a variety of machines. In fact, we have already implemented versions for an IBM PC and a SUN workstation.

Before computing a dependence graph, PFC employs a number of preliminary transformations to enhance the precision with which it can determine dependences. These transformations can radically change the structure of a program. Since the purpose of PTOOL was to display the dependences as they exist in the *user's* source program (and not in a transformed program), it was necessary to carefully alter the transformations so that dependences could be accurately calculated for the original program.

The two main preliminary transformations in PFC are "induction variable substitution"[3,24] and "IF conversion".[1] Induction variable substitution is the process of replacing auxiliary induction variables in a loop with a direct expression based on the loop induction variable. For instance, in the following example

```
            DO 100 I = 1, 100
                IX = IX - 1
                A(IX) = A(IX) - 1
    100     CONTINUE
```

IX is used as an auxiliary induction variable to run the loop "backwards".The difficulty with such variables is also illustrated above — an automatic system cannot easily determine whether the references to A are independent. This difficulty can be removed as follows:

```
            DO 100 I = 1, 100
                A(IX - 1) = A(IX - 1) - 1
    100     CONTINUE
            IX = IX - 100
```

This process, which is essentially the inverse of the classic optimization *strength reduction*,[15] is normally performed in PFC so that the array dependences of A may be accurately calculated. The problem with this transformation is that it eliminates some scalar dependences that inhibit parallel execution.

In order to correctly handle induction variables, PTOOL must use multiple passes to add dependences to the graph. Dependences for induction variables must be added before induction variable substitution is performed. After induction variable substitution, array dependences can be accurately calculated and added to the graph. As PTOOL adds dependences for the induction variables, it flags the edges so that it can inform the user that these dependences can be removed by substitution.

Control dependences are transformed into data dependences by a process known as "IF conversion".[1] The basic idea is to transform a statement whose execution is affected by a transfer of control to a conditional statement, controlled by a Boolean "guard" that exactly represents the conditions under which control flow would have reached the statement.

GOTO's can be quite naturally classified into two categories: forward GOTO's and backward GOTO's. Each of these categories may possess the further property of being an *exit* branch, meaning that the branch exits one or more loops. Although nonexit forward branches cannot directly cause any problems with parallelization, they can cause indirect problems by forcing some variables under their control to be global. Thus, while not strictly necessary, forward branches are converted because of this problem.

Backward branches are converted, via IF conversion into DO WHILE loops, which permit more effective analysis by PFC. Because of the conversion, the user can select a backward GOTO when analyzing the program in exactly the same manner that he selects a DO loop. IF conversion as implemented in PFC is powerful enough to convert any sequence of branches into an equivalent branchless program.

Transformations are only one of the ways in which PFC attempts to produce a precise dependence graph. Another technique that has proved extremely successful is interprocedural analysis.

Most automatic vectorizers make no attempt to trace dependence edges across procedure boundaries. As a result, they either ignore loops containing procedure calls (declaring them unvectorizable) or assume that all possible variables (i.e., parameters and COMMON) are modified. PFC employs a more sophisticated approach: it uses iterative data flow analysis on the program call graph to determine the side effects of procedure calls.[5] This additional information permits a substantially more precise dependence graph in the presence of subroutines.

The most significant advantage of this approach is the reduction in size of the dependence graph. On GAMTEB, a 504-line program written at Los Alamos, the dependence graph computed without interprocedural analysis contained 22,998 edges. With the application of interprocedural analysis, the graph was reduced to only 2605 edges. The smaller graph is beneficial to PTOOL's user both directly and indirectly. Directly, the user no longer sees spurious edges associated with variables in COMMON; hence, he is able to narrow his focus to real problems. Indirectly, the user experiences a performance improvement in PQUERY because the database is substantially smaller.

A second benefit of the interprocedural information has been the identification of COMMON variables actually used in calls. A fairly standard programming practice for large programs is to have the same COMMON blocks across all subprograms, thereby avoiding the problems of determining which variables have to be passed as parameters and of determining how actual parameters line up with formal parameters. Since all COMMON variables have to be in global memory, it is possible that a temporary value placed in such a variable at the beginning of a loop iteration will prevent parallel execution. Additionally, access to global memory is usually slower than access to local memory. Because PTOOL is able to pick out precisely the variables that are used and modified across procedures, it can aid a user in moving these variables to local memory.

In those situations where the source code of a procedure is not available, PFC assumes that all parameters and COMMON variables are both used and modified.

Recall that the original goal of PFC was to vectorize Fortran programs. It proceeds by generating a dependence graph and then producing a vector Fortran equivalent of the input Fortran program. For PSERVE, PFC has been modified to execute up to the point that the dependence graph and definition-use chains are assembled. Loop-independent edges are then filtered out of the dependence graph and each loop-carried edge is annotated with additional information (such as nesting level, whether the variable is in COMMON, etc.) before being added to PSERVE's dependence graph file. In addition, a second file, containing information about loop structure and definition use chains, is built. This information allows the display process to discern which lines make up a loop and identify the parameters of a given loop.

The primary novelty of PTOOL is the interactive display of a dependence graph in PQUERY. This section illustrates the power of PTOOL's display through the example in Figure 1, which captures a number of important characteristics from scientific applications. The code itself can be viewed as solving a wave equation, or computing an unknown function at a mesh of points given only its partial derivative in one direction. The first loop initializes boundary conditions at the border of the mesh. The next loop nest sweeps the calculation across the array. The inner loop moves up a column, calculating the value at any particular location j by using the value calculated at location $j - 1$. It also performs some auxiliary computations, based on the value at the location. The outer loop sweeps the computation across all the columns. The calculations moving up the columns are all dependent upon previous values and must proceed sequentially. However, nothing prevents parallel computation of different columns.

In addition to the code that is shown in Figure 1, the source for all of the function and subroutine calls was included when the code was run through

```
      PROGRAM MAIN
C
      COMMON M,H(10,10),P(10,10)
      DATA N/10/
C
      DO 10 K=1,N
         H(K,1)=FUNC(K,1)
         H(1,K)=FUNC(1,K)
   10 CONTINUE
C
      DO 30 K=2,N-1
         DO 20 J=2,N-1
            T=DERIV(H(J-1,K))
            H(J,K)=H(J-1,K)*T
            IF(H(J,K).EQ.0) GO TO 20
            E(J,K)=H(J,K)*PSI(J,K)
            P(J,K)=(E(J,K)**2)/(2*M)
   20    CONTINUE
   30 CONTINUE
C
      WRITE(6,*X'The resulting H array is',((H(J,K) J=1,N) K=1,N))
      WRITE(6,*X'The resulting E array is',((E(J,K) J=1,N) K=1,N))
      WRITE(6,*X'The resulting P array is',((P(J,K) J=1,N) K=1,N))
      WRITE(6,*X'The last DERIV is',T)
      STOP
      END
```

PSERVE. The only important property of these procedures was the fact that they were "pure"; that is, they did no READs or WRITEs and did not access global memory.

Once PSERVE has constructed the dependence database and downloaded it to the PQUERY machine, the system is ready for an interactive session. The user begins by browsing the complete Fortran source file. The first task is to identify loops that are likely candidates for parallelization. For example, in Figure 1, we can see that either of the two outer loops (DO 30 and DO 20) should execute correctly in parallel. A cursory examination of the code reveals no obvious problems with this conclusion.

The PQUERY browser provides a typical set of editor commands for moving about the file (e.g., search, go-to-line, page-forward). The user can thus proceed to the first loop to be checked. Once the DO statement (or backwards GOTO) defining this loop is visible on the screen, the loop can be selected by placing the cursor on the appropriate line (using either a mouse or cursor keys) and hitting a selection key.

When a loop is selected, PQUERY displays a list of variables not in COMMON that need to be global to the loop. As stated earlier, these variables must be parameters to the resulting parallel subroutine call. A user would see after selecting the DO 30 loop (from the code in Figure 1) for parallel execution that PTOOL has detected two variables that must be shared — T and N. T is somewhat surprising, since it appears to be a local temporary.

Since it is natural for the programmer to question PTOOL's decisions, we provided a simple explanation facility. For example, if the programmer questions the global allocation of T, he will be shown that T is used on line 32 after being defined in the selected loop (lines 12 to 20) — a use outside the loop. The variable, if present in the source (it is possible that COMMON variables will not appear at a call site), is highlighted and the reason that the variable must be global is given.

After reviewing the parameters, the user can add or delete any parameters. This permits PTOOL to solve a number of problems. For instance, on one pass over a loop, a user can allow all of the parameters to remain and discover the scope of the overall problem. Then he can delete all parameters to investigate dependences on variables in COMMON. If dependences are present, the user can attempt to eliminate them by transforming the program source. When no dependences arise from COMMON variables, the user can examine parameters singly, using transformations to eliminate each of the problems. PSERVE can then be invoked on the transformed source to verify that the process has been successful.

A number of studies have observed that there are two general classes of errors introduced in converting programs for parallel execution.[7,20]

1. Errors due to unintentional data sharing or access to shared variables in an improper order.
2. Errors due to incorrect synchronization code.

PTOOL is effective in helping the programmer identify the causes of errors in the first class. By displaying dependences in an understandable manner, PTOOL can immediately pinpoint problems that might require enormous amounts of unaided human time to find. In fact, in the first demonstration it found a problem that had consumed three man-months of effort at Los Alamos. Interprocedural analysis has been an essential element in the success of PTOOL, because it eliminates most of the spurious dependences caused by large COMMON blocks. Furthermore, it makes it possible to analyze complete Fortran programs of substantial size.

In this context, it is useful to contrast PTOOL with the DAPP (Data Flow Analysis for Parallel Programs) system of Appelbe and McDowell.[7] When completed, DAPP will accept a Fortran program in which synchronization code has already been inserted and produce a report of *parallel access anomalies* — pairs of statements that can access the same location simultaneously during a legal schedule. DAPP has the advantage over PTOOL of applicability to a variety of concurrent structures. On the other hand, it uses an analysis technique that is potentially exponential in the size of the program (the single-procedure analysis of PTOOL is proportional to the square of the procedure size, and the interprocedural analysis runs in time almost linear in the size of the call graph). Another drawback of DAPP is that it presents an exhaustive report of potential anomalies. Our experience with PTOOL and the experience of other researchers[9] is that exhaustive reporting produces an enormous amount of information in which the important facts can be lost. The selective display provided by PTOOL permits the programmer to focus on one problem at a time. For these reasons, we think that PTOOL will be more useful in analyzing large programs.

On the other hand, PTOOL is of no help whatsoever for problems of the second class — errors in synchronization code. It is our belief that writing synchronization code is inherently difficult and that programmers should be able to deal with parallelism at a higher level of abstraction. There are several systems of language extensions, implemented through preprocessing, that provide a higher level interface[8,11,14]

Another approach would be to extend PTOOL to support parallel programming in a more direct fashion. For example, once the programmer has dealt with all dependences that prohibit parallel execution of a given loop, he might request that the system generate the code required to initiate and synchronize the parallel execution. Such a facility would be the first step in evolving PTOOL from a debugging aid to a programming system. Dependence analysis can be a very powerful tool during the program development process. When displayed in an informative way, dependences can provide information about how effectively the program is making use of parallel hardware.

One major drawback to using PTOOL as a programming support system is that it is not sufficiently interactive, since it cannot redisplay dependences immediately after a change. Redisplay requires that PSERVE be invoked again on the whole program, an expensive process. If PTOOL is to be a truly effective programming support tool, it must be converted to an interactive system. However, an interactive programming system must also support other functions, such as editing, compilation, and execution. In short, the system must be a complete programming environment.

An interactive programming environment is appealing for other reasons. One of the main hindrances to accurate dependence analysis is variables whose values are known only at run-time. A compile-time analysis must make worst-case assumptions about the values of such variables, resulting in dependences that may not be present at run-time. An interactive system can query the programmer regarding values of such variables, and can embed the information provided into the dependence graph and into the resulting code (in case a programmer's assertion turns out to be false). Additionally, a programming environment makes reasonable an enhanced form of interprocedural analysis, in which the effects of procedures on portions of arrays and vectors can be summarized. Such analysis can be especially beneficial on a parallel multiprocessor, since the ability to run procedure invocations in parallel is one of its primary advantages.[23]

REFERENCES

1. **Allen, J. R., Kennedy, K., Porterfield, C., and Warren, J.,** Conversion of control dependence to data dependence, in Conf. Record of the Tenth Annu. POPL, Austin, TX, January 1983, 177.

2. **Allen, J. R.,** Dependence Analysis for Subscripted Variables and Its Application to Program Transformations, Department of Mathematical Sciences, Rice University, Houston, April 1983.

3. **Allen, J. R. and Kennedy, K.,** PFC: a program to convert Fortran to parallel form, in *Supercomputers: Design and Applications*, Hwang, K., Ed., IEEE Computer Society Press, Seattle, 1985, 186.

4. **Allen, J. R. and Kennedy, K.,** A parallel programming environment, *IEEE Software* 2:4 (July 1985), pp. 22-29.

5. **Allen, J. R., Callahan, D., and Kennedy, K.,** An Implementation of Interprocedural Analysis in a Vectorizing Fortran Compiler, Department of Computer Science, Rice University, Houston, December 1985.

6. **Allen, J. R., Callahan, D., and Kennedy, K.,** Program Transformations for Parallel Machines, Department of Computer Science, Rice University, Houston, February 1986.

7. **Appelbe, W. F. and McDowell, C.,** Anomaly detection in parallel Fortran programs, in Proc. Workshop on Parallel Processing Using the HEP, May 1985.

8. **Babb, R. B., Jr.,** Programming the HEP with large-grain data flow techniques, in *Parallel MIMD Computation: HEP Supercomputer and Its Applications,* Kowalik, J. W., Ed., MIT Press, Cambridge, MA, 1985.

9. **Conradi, R.,** Static Flow Analysis of Large Programs — Some Problems and Results, Technical report 22/83, Division of Computer Science, University of Trondheim, Norwegian Institute of Technology, N-7034, Trondheim, Norway, September 1983.

10. **Cytron, R.,** Compile-Time Scheduling and Optimization of Asynchronous Machines, University of Illinois, Urbana, August 1984.

11. **Darema-Rogers, F., George, D. A., Norton, V. A., and Pfister, G. F.,** VM/EPEX — A VM Environment for Parallel Execution, IBM Research Report RC11225, January 1985.

12. **Flatt, H. and Kennedy, K.,** The Performance of Parallel Processors, Department of Computer Science, Rice University, Houston, June 1985.

13. **Hood, R. and Kennedy, K.,** A Programming Environment for FORTRAN, Department of Computer Science, Rice University, Houston, June 1984.

14. **Jordan, H. F.,** Structuring parallel algorithms in an MIMD shared memory environment, in *Proc. 18th Hawaii Int. Conf. on System Sciences,* January 1985.

15. **Kennedy, K.,** A Survey of Data Flow Analysis Techniques, in *Program Flow Analysis: Theory and Applications,* Muchnick, S. S. and Jones, N. D., Eds., Prentice-Hall, Englewood Cliffs, NJ, 1981, 1.

16. **Kennedy, K.,** Automatic Translation of Fortran Programs to Vector Form, Rice Technical Report 476-029-4, Rice University, Houston, October 1980.

17. **Kuck, D. J.,** *The Structure of Computers and Computations* Vol. 1, John Wiley & Sons, New York, 1978.

18. **Lubeck, O. M., Fredrickson, P. O., Hiromoto, R. E., and Moore, J. W.,** Los Alamos experience with the HEP, in *Parallel MIMD Computation: HEP Supercomputer and Its Applications,* MIT Press, Cambridge, MA, 1985.

19. **Lubeck, O. M. and Simmons, M.,** An Approach to Partitioning Scientific Computations for Shared Memory Architectures, Computing and Communications Division, Los Alamos National Laboratory, Los Alamos, NM 87545.

20. **McGraw, M. R. and Axelrod, T. S.,** Exploring Multiprocessors: Issues and Options, UCRL-91734, preprint, Lawrence Livermore National Laboratory, Livermore, CA, October 1984.

21. **Ottenstein, K. J.,** A brief survey of implicit parallelism detection, in *Parallel MIMD Computation: HEP Supercomputer and Its Applications,* Kowalik, J. S., Ed., MIT Press, Cambridge, MA, 1985.

22. **Padua, D. A.,** Multiprocessors: Discussion of Some Theoretical and Practical Problems, University of Illinois, Urbana, 1980.

23. **Troilet, R. J.,** Contribution a la Parellisation Automatique de Programmes Fortran Comportant des Appels de Procedure, L'Universite Pierre et Marie Curie (Paris VI), December 1984.

24. **Wolfe, M. J.,** Techniques for Improving the Inherent Parallelism in Programs, Report 78-929, Department of Computer Science, University of Illinois, Urbana, July 1978.

Part III
MIMD Issues

Exploiting parallel processing is harder than sequential processing, and MIMD is the hardest form of parallel processing. Moreover, to move into the realm of parallel supercomputing, one must embrace massive parallelism, typically more than 1000 processors. Developing applications for a massive collection of processors, each with its own process, synchronized with all the others, and keeping processor utilization high, is a daunting challenge. However, the perceived payoff is so high that large numbers of researchers, engineers, and companies have been trying to overcome the obstacles for almost 20 years.

In this part of *Parallel Supercomputing in MIMD Architectures* we examine four of the major issues of the technology: scalability, i.e., what impediments are encountered as one attempts to enlarge an application; partitioning, i.e., how does one break an application efficiently into multiple processes that can be executed in parallel; utilization, i.e., keeping all of the hardware gainfully occupied; and heterogeneous networks, i.e., is the competing but related technology of local area nets of stand-alone uniprocessors ganged together to solve a single application more effective than a single MIMD parallel machine?

There are, of course, many issues untouched here. The literature is rich, but these are the most critical topics now.

20 A SCALABILITY ANALYSIS OF THE BUTTERFLY MULTIPROCESSOR

OVERVIEW

The availability of a variety of parallel computers has led to a serious concern about the scalability of their architectures. Scalability is an important consideration for the designers of multiprocessor systems.[5] Two aspects of scalability of an architecture are resource scalability and application scalability.[3] The resource scalability measures the growth rate of architectural properties and associated costs, whereas the application scalability measures the utilization of the resources and the efficiency of execution of an application. The purpose of this chapter is to present our experiences with an application scalability analysis on the BBN Butterfly multiprocessor.[1]

The test problem used for the performance analysis is the solution of a tridiagonal system (TDS) of linear equations. TDS of linear equations form a very important class of linear algebraic equations. Such systems arise frequently in many areas of large-scale scientific and engineering computing. For example, the core of finite-difference methods of solving partial differential equations characterizing continuous fields are the solutions of linear TDS. Consequently, fast and efficient solutions for solving such systems become crucial for many numerical methods. As a result, several TDS solvers suitable for vector or multiprocessor machines have been proposed.[7,8] For our purposes, we used a parallel version of the block partitioning method of Kowalik and Kumar.[4] The method of Reference 4, designed for general-purpose MIMD computers is very well suited to a shared memory multiprocessor, such as the Butterfly computer, and is specifically efficient in cases of $p < < n$ (p = number of processors, n is number of equations). This parallel algorithm provides opportunities for applying full force of parallelism, and it also includes data dependencies forcing sequential execution. The problem is decomposed in to p subproblems where p is supplied as an input parameter. Hence, the algorithm does not need a separate coding for p = 1 and caters to the availability of a limited number of processors. For experimental results, the method is implemented on the Butterfly computer using different problem sizes and different processor configurations ranging from 1 to 80. The criterion used for performance analysis is the amount of execution speed-up that results from applying successively more processors to the problem being solved.

Consider the tridiagonal system of linear equations:

$$b_j x_{j-1} + a_j x_j + c_j x_{j+1}, \ 1 \le j \le n, \ b_1 = c_n = 0$$

which in matrix-vector notation can be written as shown in Figure 1A.

It is assumed that the number of equations n is much larger than the given number of processors p and that pivoting for numerical stability is not required. The partition method of Kowalik and Kumar[4] is based on the scheme of "divide and conquer". The method solves Figure 1A by simultaneous elimination of blocks of b's and then blocks of c's and then using back substitution to derive the final solution. The computational process consists of the following phases.

Phase 1
Segmentation
The given system (Figure 1A) is divided into p, $2 \le p \le [n/2]$, subsystems so that each subsystem is of size k = n/p. Since we deal with a MIMD machine, it is not necessary that p divides n, but this choice simplifies our notation. Given p processors and n = k*p unknowns, reorder the original ordering 1, 2, 3, . . ., n of unknowns as the new ordering 1, 2, . . ., k – 1, k, k + 1. . . ., (p – 1) k + 1, (p – 1) k + 2, . . . pk. Now divide the whole system into p blocks each of size k. An example of such a partition for n = 9 and p = 3 is shown in Figure 1B.

Elimination
In this phase, coefficients in the lower diagonal (b's) and then the ones in the upper diagonal (c's) are eliminated by Gaussian elimination in the following order ("reduction to zero" in the following is by Standard Gaussian elimination):

1.   Perform k – 1 Gaussian eliminations, each of p parallel reductions, i.e., reduce $(b_2, b_{k+2}, b_{2k+2}, \ldots, b_{(p-1)k+2})$ to zero in parallel, then $b_3, b_{k+3}, b_{2k+3}, \ldots, b_{(p-1)k+3)}$ in parallel, . . . and at last, $(b_k, b_{2k}, b_{3k}, \ldots, b_{pk})$ to zero in parallel.
2.   First reduce $(c_{k-2}, c_{2k-2}, c_{pk-2})$ to zero in parallel, then $c_{(k-3)}, c_{2k-3}, \ldots c_{pk-3})$, . . . and at last, $(c_1, c_{k+1}, \ldots, c_{(p-1)k+1k})$ to zero in parallel.

We define the code segment for the elimination steps of the program BPTRD as the computational tasks T1$_i$, $1 \le i \le p$, for each of the p blocks where the task T1$_i$ for a given i is defined by:

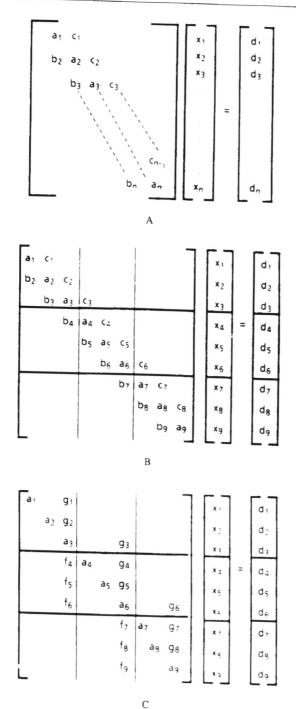

FIGURE 1. Scalability.

```
TASK T1_j
  If i ≠ 1 then f_(i - 1) k + 1 = b_(i - 1) k + 1
  For j = (i - 1) k + 2 to ik do
  m_j = b_j/a_j - 1
      a_j = a_j - m_j c_j - 1
      if i ≠ 1 f_j = -m_j f_j - 1
      d_j = d_j - m_j d_j - 1
  Endj
  g_ik - 1 = c_ik - 1
  For j = ik - 2 down to (i - 1)k + 1 do
      m_j = c_j/a_j + 1
      g_j = -m_j g_j + 1
      if i ≠ 1 then f_j = f_j - m_j f_j + 1
      d_j = d_j - m_j d_j + 1
  Endj
  If i ≠ 1 then do
      m_i = c_(i - 1)k/a_(i - 1)k + 1
      g_(i - 1)k = m_j g_(i - 1)k + 1
      a_(i - 1)k = a_(i - 1)k - m_i f_(i - 1 k + 1
      d_(i - 1)k = d_(i - 1)k - m_j d_(i - 1)k + 1
  End If
  End T1_j
```

The tasks $T1_1, T1_2, T1_3, \ldots T1_p$ are mutually noninterfering[5] and hence can be executed in parallel. For example, in the system shown in Figure 1B, first (b_2, b_5, b_8), then (b_3, b_6, b_9), then (c_1, c_4, c_7), and last (c_3, c_6) can be eliminated (in this order) in parallel. These eliminations, of course, are going to introduce some fill-ins. For Figure 1B, these fill-ins are shown in Figure 1C, as fs due to bs elimination and gs for those of cs. The reduced system after this step is shown in Figure 1C.

Phase 2
Solving the Reduced System of Size p
After the elimination, the resulting system Figure 1C contains p block-connecting equations (shown adjacent to the block boundaries) which, again, form a TDS of size p. Hence, this step solves the p equations:

$$f_{ki}x_{k(i-1)} + a_{ki}x_{ki} + g_{ki}x_{k(i+1)} = d_{ki}$$
$$i = 1, 2, \ldots, p \text{ and } f_{k1} = g_{kp} = 0$$

The solution is obtained by first eliminating the lower diagonal coefficients and then using back-substitution. The code segment for this computation of the program BPTRD is defined by the following task T2.

```
TASK T2
  For i = 2k to pk by k do
```

```
m  = f  / a
 i    i    i - 1
a  = a  - m  g
 i    i    i  i - 1
d  - d  - m  d
 i    i    i  i - 1
End i
For j = pk down to k by k do
  x  = (d  - g  x    )/a
   j     j    j  j + 1    j
End j
End T2
```

For small values of p it is more efficient to execute T2 sequentially by one processor (almost negligible computation time). If p is larger, then parallel recursion can be used on T2.

Phase 3
The Final Solution
The system (Figure 1C) is now decoupled into p independent subsystem which can be solved for the remaining $n - p$ x's by back-substitutions. This computation is defined as tasks $T3_i$, $i = 1, 2, \ldots, p$,

```
TASK T3
        i
For j = ik - 1 down to (i - 1)k + 1 do
  x  = (d  - f  x        - g  x   / a
   j     j    j  (i - 1)k    j  ik    j
End j
End T3
       i
```

The p tasks $T3_1$, $T3_2$, $T3_3$ . . ., $T3_p$ can be executed simultaneously by p processors.

The BPTRD program outlined above was coded in the Butterfly-C language and was run on the 32-node Butterfly computer of Boeing Computer Services and on the 80-node configuration system at the University of Maryland, College Park. The Boeing machine is equipped with 32 processor nodes, each with 4 Mbytes of memory and MC68020 microprocessor with MC68881 floating-point coprocessor. The program development was done on a Sun workstation as a host. The machine at the University of Maryland is equipped with MC68000 processors (no hardware floating point) and 1 Mbyte of memory with each processor.

The Uniform System approach of parallel programming was used to run the experiments. The Uniform System software creates an environment in which tasks may be distributed to processors without regard to the physical location of the data associated with the tasks.

For storage allocation purposes, the input and output vectors (total of t: a's, b's, c's, d's, and x's) were spread across the physical memory of the machine by storing the data as $n \times 5$ matrix and then scattering them across the memories of the machine in order to reduce memory contention.

The model of fork and join for generating and synchronizing the computational tasks of BPTRD described above is shown by the task graph[5] of Figure 2. The Uniform System matches available processors to the task generated. The

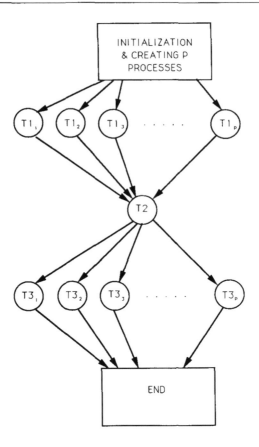

FIGURE 2. The task graph of BPTRD (K = N/P).

processor that calls a generator will always be available to work on the task it generates, and perhaps other processors will be as well. The tasks are distributed to processors on a first-come first-served basis. Two methods of task generations were tested:

• Create the number of parallel tasks equal to the number of processors available.
• Create the number of parallel tasks greater than the number of available processors (to avoid processor starvation).

The experiments were run for values of n and p, $96 \leq n \leq 16000$, and $1 \leq p \leq 80$. For simplicity reasons, n was taken to be divisible by p so that each block in the segmentation phase is the same size (not a restriction).

The coefficient matrix A (three diagonals) was randomly generated for a given value of n. The solution vector x was initialized to 1.0, and then the right-hand side vector d was obtained by matrix vector multiplication Ax. Thus, the solution was known in advance for each generated system.

Table 1
Performance of the BPTRD Program When the Number
of Generated Processes is More than the Available
Processors (N = 16000, P = Processors in Use)

P	T_p	S_p	E_p
1	133.88	1.0	1.0
2	77.25	1.7	0.8665
4	39.16	3.4	0.8456
8	20.60	6.4	0.8122
16	10.52	12.7	0.7951
32	5.80	23.0	0.7202
64	3.85	34.4	0.5433
80	3.39	39.4	0.5193

Table 2
Performance of the BPTRD Program When the Number of Generated
Processes is More Than the Available Processors

P	T_p	S_p	E_p
1	207.19	1.0	1.0
2	101.65	2.0	1.0
4	51.37	4.0	1.0
8	26.14	7.9	0.9904
16	14.47	14.3	0.8943
32	8.66	23.9	0.7475
64	4.78	43.2	0.6762
80	4.78	43.2	0.5408

The actual performance results of the program runs by BPTRD are given in Tables 1 to 5. The following definitions hold for different parameters used in these tables:

N	Problem size
P	Number of processors
T_p	Execution time of BPTRD using P processors
$S_p = T_1/T_p$	Speed-up of the parallel code using p processors
$E_p = S_p/p$	Efficiency of the parallel code using p processors
Overhead (p)	$= T_p - T_1$
VAR	$= N, P,$ or T_p
Ratio (VAR)	$= \dfrac{\text{Current Value (VAR)}}{\text{Previous Value (VAR)}}$

Table 1 presents the total speed-up and efficiency of execution for a fixed problem size and various numbers of processors. In this case the parallel tasks generated in phases 1 and 3 of BPTRD were exactly same as the number of

TABLE 3

The Scalability Performance of BPTRD with Fixed
N (= 16000) and Various Configuration of P (≥ 2)[a]

P	T_p	Ratio (P)	Ratio (T_p)
2	77.25	2.0	2.0
4	39.16	2.0	1.97
8	2.60	2.0	1.90
16	10.52	2.0	1.96
32	5.80	2.0	1.81
64	3.85	2.0	1.51

[a] N = 16000, NP = 100.

TABLE 4

Scalability Ratio Example of the BPTRD for
Varying the Problem Size N and Fixed P (8)

N	T_p	Ratio (P)	Ratio (T_p)
2,000	3.21	1	1
4,000	5.71	2	1.8
8,000	10.72	2	1.9
16,000	20.13	2	1.9

TABLE 5

The Scalability Analysis (Execution and Speed-Up Performance) for
BPTRD with Increase in the Problem Size N and the Corresponding
Change in the Number of Processors P

N	P	T_p	Overhead	Speed-up (%)
100	1	0.94	—	100
500	5	1.23	0.29	76
1000	10	1.27	0.37	74
2000	20	1.35	0.41	70
4000	40	1.51	0.57	62
8000	80	1.84	0.90	51

processors, i.e., p = Procs In Use. The speed-up S_p for $2 \leq P \leq 32$ is within 72 to 85% of the linear (best speed-up) and decreases as the number of processors increases thereafter, reaching up to 50% of the linear speed-up.

Table 2 provides the same parameter values as in Table 1, except for one. In this case the number of tasks generated was greater than the processors being used. This was mainly to avoid processor starvation. As it is clear from the numbers of Table 2, this improved the corresponding value of S_p and E_p, bringing them between 75 to 100% of linear speed-up and efficiency for $2 \leq p \leq 32$ and worked 20% more efficiently as compared to Table 1 results (from 50 to 70%) with $p = 64$. The results for 80 processors were also improved. This gain, of course, is achieved at the cost of process creation overhead, which decreases (see value of T_p in Table 2) as the number of processors used is increased. The number of tasks generated in phases 1 and 3 were $NP = 100$, and increasing this number more than 100 did not help.

Tables 3 and 4 show the changes in execution time as one of the parameters N and P is fixed and the other varies. The ratios of the two consecutive values of these variables are compared with the corresponding ratios of the execution time T_p. The results are encouraging and fully support scalability of the Butterfly on this application.

Table 5 presents the scalability results when the architecture and the application are equally increased in size, i.e., the values of N and P are increased by the same size. The results show that Butterfly is well scalable with respect to the presented application (it is the application characteristic that it is more efficient when $p < < n$). Of course, a drop in execution efficiency, in some cases, is the result of increased hardware delay.

It is noted that the difference between the expected linear speed-up and efficiency and the actual performance of this BPTRD could be attributed to the sequential work (task T2) and the overhead of parallelization.

The sequential part of the program BPTRD is by no means restricted to be executed on one processor. In fact, for a large value of p, the program can be parallelized recursively until T2 reduces to size 2 system. However, it was found that for values of p used in our experiments, T2 took so little time as compared to T1 and T3 (as shown in Table 4), that it was considered not worthwhile to parallelize this part.

As for the MIMD computation overhead part, for the presented implementation, task generation and task scheduling and data communication appear to be the main sources. Since the task granularity (block size) decreases as the number of tasks increases, the advantage of creating many more tasks may not be supported by the associated task creation overhead.

The application scalability characteristics of the Butterfly multiprocessor system are analyzed by mapping an application onto its architecture. A block partitioning method of solving a linear TDS equation is implemented and analyzed in terms of its performance on the 80-node Butterfly parallel computer. It is shown that when the architecture and the application are equally increased in size, the achieved execution efficiency is within 50 to 100% of the desired linear (best performance) results. From this point of view, the architecture is well scaled. The shown closeness between the linear and achieved performance is sufficiently good to give credibility to the presented work. Further, the achieved results reflect that the cost of the overhead of a MIMD computation becomes less significant as the ratio between the problem size n and the number of processors p increases.

REFERENCES

1. BBN Inc., Butterfly Parallel Processor, BBN Laboratory, Cambridge, MA, May 1986.
2. BBN Inc., Benchmark Results for a 256-Node Butterfly Parallel Processor, BBN Laboratory, Cambridge, MA, August, 1985.
3. **Deshpande, S. R. and Jenevein, R. M.,** Scalability of a binary tree on a hypercube, in Proc. of the Int. Conf. on Parallel Processing, 1986, 661.
4. **Kowalik, J. S. and Kumar, S. P.,** Parallel algorithms for recurrence and tridiagonal equations, in *Parallel MIMD Computations; HEP Supercomputer and Its Applications,* Kowalik, J. S., Ed., MIT Press, Cambridge, MA, 1985.
5. **Lipovski, G. J. et al.,** Inductive Computer Architectures: A Class of Supercomputer Architectures, MCE report, Parallel Processing Group, Microelectronics and Computer Technology, Austin, TX, 1983.
6. **Kumar, S. P.,** Parallel Algorithms for Solving Linear Systems of Equations on MIMD Computers, Ph.D. diss., Department of Computer Science, Washington State University, Pullman, 1982.
7. **Stone, H. S.,** Parallel tridiagonal equation solvers, *ACM Trans. Math Software*, 1, 289, 1975,.
8. **Wang, H. H.,** A parallel method for tridiagonal equations, *ACM Trans. Math. Software*, 7, 170, 1981.

21 MATHEMATICAL MODEL PARTITIONING AND PACKING FOR PARALLEL COMPUTER CALCULATION

Multiple processors, operating together to solve a single problem, can in many cases decrease the time of calculation. This is important in time-critical applications, such as real-time simulation, where this technique can provide computational rates unachievable on a single processor or allow the use of lower cost hardware to provide the necessary computational capabilities. For certain classes of problems it is possible to configure a network of microcomputers to achieve the same throughput rate as a large mainframe computer at a lower initial and ongoing maintenance cost.

The parallel processing concept has opened new areas of research and development in hardware, software, and theory. Some efforts sponsored by NASA Lewis are described in References 1 to 6. Techniques for developing mathematical models that can be solved efficiently on parallel processors is a key area of study. The first step in developing these multiprocessor models is to identify parallelism within the mathematical formulation of the problem. This requires a data flow analysis of the equations of the problem and will identify the "critical path" and the minimum achievable calculation time. The next step is to arrange, or "pack" the noncritical path computations on the minimum number of processors so as to make maximum use of the available computing resources.

This section presents a method for partitioning and packing equations for multiprocessor solution. Reference 6 gives a more detailed discussion of these techniques, including a comprehensive example of applying them to a turboshaft engine model.

A mathematical model of a physical system consists of a set of equations which describe, to some degree of accuracy, the response of that system to external influences (driving functions) over a limited range of operation. This range is defined in terms of the maximum and minimum values of the driving functions and, if time dependent, the maximum frequency or maximum rate of change of these functions. Generally, the object of this modeling effort is to provide a simulation of the physical system.

347

The prerequisite to developing parallel processor simulations is to be able to identify the parallel computational paths contained in the model. In general, a dynamic model can be programmed on a digital computer as a set of N computationally sequential equations of the form

$$X_3(ih) = f_K\left[X_m(ih), X_m((i-1)h),\dots,u(ih)\right]$$

where $X_K(ih)$ is the result of the Kth equation at time ih. Here, h denotes the simulation time step or update interval of the model calculations. The arguments $X_m(ih)$ are the current values of the results of preceding equations in the model (i.e., $m = 1$ to $K - 1$), and $X_m((i-1)h)$ are the past values of the results of all equations in the model. The argument u represents values obtained from sources external to the model which are always available at the start of the model computation sequence. The functional relationship between X_K and its arguments is represented by f_K. Assuming an equation is an indivisible computational unit, then the parallelism in the model is determined by the arguments of each equation. That is, two equations, or sets of equations, can be computed in parallel within an update interval if their arguments are independent of the results of the others computed in that interval.

For example, a model of the form

$$X_1(ih) = f_1\left[X_3((i-1)h)\right]$$
$$X_2(ih) = f_2\left[X_1(ih)\right]$$
$$X_3(ih) = f_3\left[X_2(ih), X_3((i-1)h)\right]$$

contains no parallelism since $X_3(ih)$ requires $X_2(ih)$ and $X_2(ih)$ requires $X_1(ih)$. These calculations must be done serially. However, the model

$$X_1(ih) = f_1\left[X_3((i-1)h)\right]$$
$$X_2(ih) = f_2\left[u(ih)\right]$$
$$X_3(ih) = f_3\left[X_1(ih), X_2(ih), X_3(i-1)h\right]$$

does contain parallelism since X_1 can be computed at the same time as X_2.

Parallelism due to decoupled or loosely coupled equation sets is easily identified from the physical nature of the model. A more difficult task is the identification of parallelism in a set of closely coupled equations, where the process dynamics dictate the use of current arguments in solving for equation results. For instance, suppose a model contains the following set of equations:

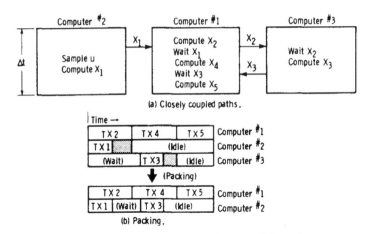

(a) Closely coupled paths.

(b) Packing.

FIGURE 1. Partitioning and packing closely coupled equations.

$$X_1(ih) = f_1\left[X_5\big((i-1)h\big), u(ih)\right]$$

$$X_2(ih) = f_2\left[X_5\big((i-1)h\big)\right]$$

$$X_3(ih) = f_3\left[X_2(ih)\right]$$

$$X_4(ih) = f_4\left[X_1(ih), X_2(ih)\right]$$

$$X_5(ih) = f_5\left[X_3(ih), X_4(ih), X_5\big((i-1)h\big)\right]$$

The variable X_1 can be computed at the start of the calculation interval, since it is a function of the past value of X_5 and the external variable u. X_2 may also be computed at the start of the interval. However, the calculation of X_3 must be delayed until X_2 has been determined, the calculation of X_4 must be delayed until both X_1 and X_2 are determined, and the calculation of X_5 must be delayed until both X_3 and X_4 have been determined. As shown in Figure 1a, three computational paths can be identified which can be assigned to three different computers in the simulation. Note that "wait states" have been inserted to ensure the currency of the arguments. That is, equation calculation is delayed until current argument values become available. The X_3 calculation is shown to be delayed slightly for the transfer of X_2. The shaded areas (time slots) in Figure 1b indicate the time available for result transfer to computer number 1. The calculation of X_1 and X_3 can take place anywhere in the time slot.

The detection of this type of computational parallelism can become burdensome when the equation set becomes large. The technique, however, can be automated. Related to this problem of partitioning is the problem of allocation (i.e., packing these paths into a minimum number of computers without extending the update time). Figure 1b demonstrates packing of the paths defined in

Figure 1a. Arbitrary calculation times of TX_1, TX_2, TX_3, TX_4, and TX_5 have been assigned to the equations producing results X_1 through X_5, respectively. The time TX_1 includes the time required to obtain the value of u. Note in Figure 1b, that because of the calculation times, the $X_2 - X_4 - X_5$ path is critical in that it contains no idle states. This path, therefore, dictates the minimum possible update time ($\Delta T = TX_2 + TX_4 + TX_5$). The paths X_1 and X_3 are assigned to separate computers. Packing in this example is a trivial task, since the X_3 calculation can be moved onto computer number 2 to be calculated during the idle period, as shown in the figure.

In many cases, efficient packing requires shifting equations in their time slots. This causes a ripple effect on the time slots of other equations which can complicate the packing problem. Because of the nature of the packing problem, a unique solution to the development of a packing algorithm does not exist. There are many ways to pack most parallel models.

In the following, partitioning and packing algorithms that have been developed at NASA Lewis are discussed. These algorithms were tested with a model of a turboshaft engine and detailed results are presented in Reference 6.

To begin the discussion of the partitioning algorithm, certain terms should be defined. A mathematical model is a set of equations, written to define the characteristics of a physical system to some desired degree of accuracy. A program is a sequential set of digital equations and supporting information (e.g., variable and constant definition) which define the mathematical model within the constructs of a programming language. A path is a subset of these equations which, because of interrelationships between arguments and results, contains no parallelism. Partitioning is the transformation of the program equations into a number of paths which may be calculated if parallel. Packing is the combination of paths into a minimum number of processors (computers) which provide computation of the model within a prescribed update interval. The critical path is the longest path, and the prescribed update interval must be greater than or equal to the calculation time of the critical path.

In this discussion of partitioning it is assumed that a program is given. That is, these equations, when executed serially, provide the required results. No assumptions are made concerning the parallelism of computational operations contained in the program equations.

$$x = a * y + b * z$$

contains parallelism (i.e., $a * y$ can be calculated in parallel with $b * z$) which will be ignored if we are concerned with partitioning at the earliest level and not parsing. For purposes of this discussion, the above equation will be considered as

$$X = f(a, y, b, z)$$

where f is some single operation. Therefore, equations will be assigned to paths in their entirety and not broken up into more primitive result-argument relationships.

As indicated in the last section, partitioning requires the establishment of result-argument relationships for the serial set of equations in order to develop computational paths. It is also necessary to know the calculation time of each equation. The program must be processed to provide this information. For this effort, the result-argument relationships and the calculation time information are outputs of the multiprocessor programming utility RTMPL.[2,3] The primary function of this utility is to translate a structured program of the mathematical model into assembly language for the simulation processor(s). As an option, the utility also provides information on the result and arguments of each equation, the processor operations necessary to obtain the result, and the processor calculation time for each operation. For the equation

$$X = y + z$$

The processor operations to compute the equation are load register R1 with z (requires 8 time units), add variable y to R1 (16 time units), and store R1 as the value of variable X (8 time units). This type of information is generated for each equation in the program.

Consider the close-coupled example in the previous section. The first step in the partitioning process is to convert utility-generated information into the form needed for partitioning. Dependent arguments are those which are the results of previous program equations calculations in the update interval (e.g., X_1 is a dependent argument of X_4). These are the drivers for partitioning since their current values are required before the computation sequence can continue. The independent arguments u and X_5 do not affect partitioning, since only past values are used. The calculation time for each equation is determined by adding the calculation times of the given processor operations.

The time at which an equation can start is determined by the arguments and calculation time of each equation. The first equation of a set only has independent arguments and, thus, can always start at time 0 (measured from the beginning of the calculation update interval). It can never require results from calculations in the current update interval since none are yet available. An equation can end at the time obtained by adding its calculation time to the time it can start. The general formula for obtaining this time is

```
CANSTART(RESULT) = MAX(CANEND(ARG 1),
CANEND(ARG 2), . . ., 0)
CANEND(RESULT) = CANSTART(RESULT) +
CALCTIME(RESULT)
```

where ARG1 is the first dependent argument, etc. This formula is applied sequentially to each equation in the program.

Once these attributes have been established for each program equation, the

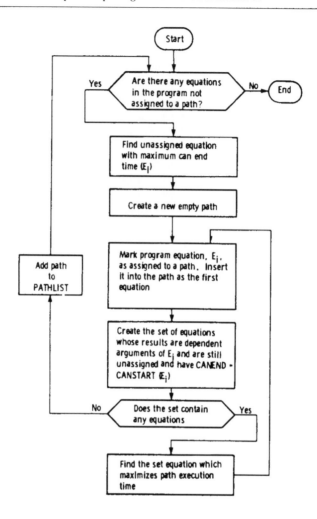

FIGURE 2. Path identification algorithm.

identification of computational paths contained in the program can begin. The
algorithm used for path identification is shown in Figure 2. Its purpose is to
identify all sequences of equations which contain no parallelism and which must
be computed serially. These paths are organized into a linked list called
PATHLIST. The paths in PATHLIST are ordered in terms of decreasing path
calculation time. Therefore, the first path in PATHLIST is the critical path. To
form a path, the algorithm selects the program equation, having the maximum
CANEND time, and which has not already been assigned to a path. This is the
result equation of the path. The next equation selected is the one which produces
a result used as a dependent argument of the result equation. If more than one
equation result is used as a dependent argument, then the one with maximum
CANEND time is selected. The selected equation then is inserted in front of the

result equation in the path. The path formation process continues until an equation is inserted that has no dependent argument equations which are not already assigned to a path. Paths are formed until all program equations have been assigned.

Partitioning has been discussed in terms of equations that produce values of variables. Often, mathematical models contain statements that do not produce values. Two common examples are conditional statements (e.g., IF . . . THEN . . . ELSE) and command statements (e.g., I/O operations). The calculation time of such a statement must be combined with a preceding or following equation. This could impose limitations on program structure and is a subject for future study.

The partitioning process produces a number of paths consisting of equations which must be computed serially and a table of information on each equation. The final task in the process of formulating a multiprocessor model is to pack the paths for assignment to a minimum number of processors. The first path in our list has the largest calculation time due to the partitioning algorithms. This is called the critical path, and its calculation time is the minimum time in which the model can be computed, no matter how many processors are used. The number of paths identified through partitioning is usually greater than the minimum number of processors necessary.

The minimum number of processors necessary to implement a multiprocessor simulation depends on how fast the simulation must be computed. This update time must be specified prior to packing. The simulation time step h is usually based on stability and dynamic accuracy requirements. For real-time applications, the update time ΔT must be equal to h. The update time also specifies when the computations must end. The first step in packing is to determine when each equation must end using the specified update time.

To determine when an equation must end, we begin with the state variables (defined here as those variables in which *current values* are not used as arguments in the model, but appear as results of model equations). The state variable computations will be the last computations performed and, thus, must end at the prescribed update intervals. The calculation of equations which are dependent arguments of these variables must end no later than the time at which the state variable calculations must start. The times when subsequent equations in the result/argument string must end are similarly determined. Since a variable can be used as a dependent argument in more than one equation, care must be taken that the earliest time, arrived at after all paths are analyzed, is used to specify when that equation must end.

We now have determined when an equation can start, can end, and must end. These are termed equations attributes. Since the paths are serial they can also be assigned these attributes: a path can start when its first equation can start, a path can end when its last equation can end, a path must end when its last equation must end, and additionally, the calculation time of a path is the summation of the calculation times of its equations. This is sufficient information to pack the paths.

The solution to the packing problem is not unique in that many arrangements of paths in processors can result in a satisfactory solution.

The packing algorithm, shown in Figure 3, was designed to achieve the minimum number of processors. Other requirements which may be imposed, such as memory size limitations, and interprocessor data transfer limitations were not imposed on the algorithm.

As input, the algorithm requires (1) that all paths be specified in a linked list called PATHLIST in order of decreasing calculation time; and (2) that the required update time of the simulation, ΔT, is specified and that the attributes of each equation and path (CANSTART, CANEND, MUSTEND, CALCTIME) have been determined as described above.

The packing algorithm creates processors as needed and inserts paths from PATHLIST according to a hierarchy of relationships between existing equations in a processor and the equations in the unpacked paths. When a processor is created, the path with the longest calculation time in PATHLIST is inserted. Next, the paths which are related to paths already in a processor are tested to see if they fit (see discussion of TESTFIT algorithm, below). If so, they are inserted if not, they are placed in a carry-over list.

Then, paths in PATHLIST which are unrelated to the equations in the processor are tested. If one of these is inserted, unrelational testing is ended and relational testing begins again. When no other paths can be inserted into a processor, another processor is created. This process continues until all paths in PATHLIST are inserted into a processor.

Relational testing is prioritized. All unpacked paths which provide critical arguments are tested first. (A path is considered to provide a critical argument if the result of the last equation in the path (EL) is an argument of a processor equation (EP) and

```
MUSTEND (EL) = MUSTEND (EP) - CALCTIME (EP)
```

Next, other related paths are tested. Then paths in the carry-over list (which was formed from paths which were related to equations packed into previously formed processors, but not yet packed) are tested.

Paths are tested for insertion on an equation-by-equation basis, using the test fit algorithm shown in Figure 4. First the attributes (CANSTART, CANEND, MUSTEND, and CALCTIME) of all program equations are saved. This is necessary because inserting an equation into a processor can cause a ripple effect on the attributes of other equations. If the whole path does not fit, any equation of the path inserted into the processor must be removed and the attributes of affected equations restored.

The ripple effect is illustrated in Figure 5. Assume a processor contains two equations (A and B) and that it has been determined that equation (C) can be inserted between them. The calculation time of each equation is shown as the shaded areas. For packing purposes, the calculation of each equation can take place anytime between its CANSTART time and its MUSTEND time.

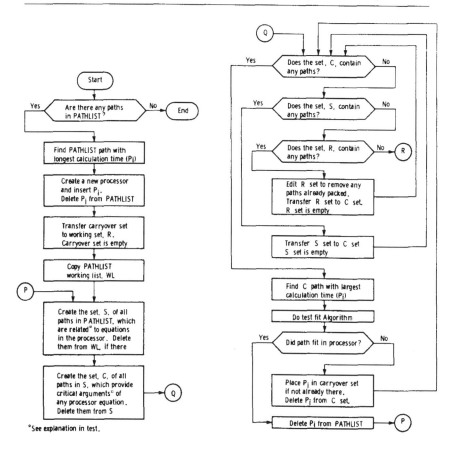

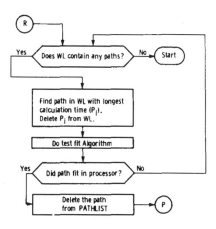

FIGURE 3. Packing algorithm.

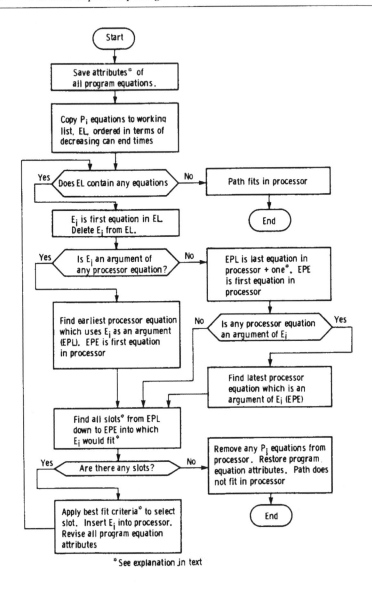

FIGURE 4.   Test fit algorithm.

The space available for C is the difference between the time at which B must end, and A can end minus the calculation time of B. Equation C will be inserted to start directly after A can end. The calculation of B will be delayed until C can end. Note that the difference between the MUSTEND and CANEND times of A and C have been reduced to zero by the positioning of C and that the time difference for B has been reduced. The primary impact of these changes is to reduce the space in the processor available for packing other paths. There is

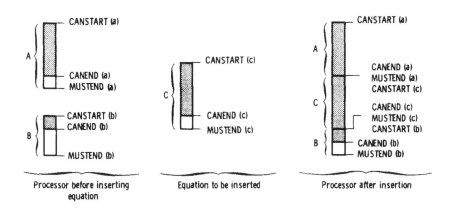

FIGURE 5. The effect of insertion on equation attributes.

also a secondary impact of equal importance. By increasing the times at which C and B can end, any unpacked equations which use these equations as arguments have their starting times delayed. Similarly, by reducing the MUSTEND times of A and C, the MUSTEND time of any unpacked arguments of these equations are moved up. These effects tend to reduce the slot sizes of unpacked equations, restricting the range of time into which they can be packed into a processor. Also, these ripple effects may introduce computational gaps within unpacked paths.

After the attributes are saved (Figure 4), the equations within a path are ordered in terms of decreasing CANEND times for insertion testing. That is, the latest equation will be tested first and the earliest last.

The processor equations are arranged in sequential order, where EP(1) is the earliest equation and EP(n) is the latest equation. Testing to determine if a path equation (E(i)) can be inserted into the processor involves the identification of all slots between any two processor equations (EP(j − 1), EP(j)) where the equation fits. The processor end points (i.e., EP(j) = EP(1) and EP(j − 1) = EP(n)) must also be considered. Because of the argument and result relationships between E(i) and the processor equations, it is required that the range of processor equations be limited for testing purposes. Let the end points of the range be designated by EPE and EPL (the earliest and latest processor equations, respectively, before which E(i) may be inserted). This range is established as follows: if E(i) is an argument of a processor equation, then EPL is the earliest processor equation of which E(i) is an argument (EPE = EP(1)); if any processor equation is an argument of E(i), then EPE is the one following the latest of these and EPL is the last processor equation plus one (end point); if E(i) is unrelated to any processor equation, then EPE = EP(1) and EPL is the last processor equation plus one.

Once the range of testing has been established, all slots within that range are tested to determine if E(i) fits. The fit criterion is as follows:

1. If EP(j) = EP(1) then CANEND* (E(i)) = CALCTIME (E(i)) else CANEND* (E(i)) = CANEND (EP(j – 1)) + CALCTIME (E(i)).
2. If CANEND* (E(i)) < CANEND (E(i)) then CANEND* (E(i)) = CANEND (E)i)).
3. If EP(j – 1) = EP(n) then MUSTEND* (E(i)) = ΔT else MUSTEND* (E(i)) = MUSTEND (EP(j)) – CALCTIME (EP(j)).
4. If MUSTEND* (E(i)) > MUSTEND (E(i)) then MUSTEND* (E(i)) = MUSTEND (E(i)).
5. If [[CANEND* (E(i)) < = MUSTEND (E(i))] and [CANEND* (E(i)) < = MUSTEND* (E(i))]], then E(i) fits.

The asterisk indicates the attributes of E(i) if it were inserted into the slot between EP(j – 1) and EP(j).

If it is established that E(i) can fit in more than one slot, the test fit algorithm proceeds to select the slot into which E(i) best fits. The best-fit criterion is as follows.

If a slot exists such that

```
CANEND*  (E(i))  -  CALCTIME  (E(i))  -
CANEND  (EP(j  -  1))  =  0
```

then this slot is selected. Otherwise, the latest slot which maximizes

```
[MUSTEND*  (E(i))  -  CANEND*  (E(i))]
```

is selected.

This criterion provides for efficient packing by eliminating processor idle time if possible, and if not, then the ripple effect from insertion is minimized.

Once a slot is selected, the equation is inserted and the attributes of all program equations are updated to reflect the insertion. If any path equation cannot be inserted into the processor, path equations which have already been inserted are removed from the processor, the original attributes are restored to the program equations, and the test fit algorithm ends.

The packing algorithm was programmed in Pascal, along with the partitioning algorithm. It was then tested on a turboshaft engine model. The results, in terms of percent processor utilization, are presented in Table 1. The first column is the update time ΔT specified prior to packing. It is given in terms of machine cycles. In this case, four processors were required. The second column gives the number of processors required for packing. The remaining columns show the percent utilization of the update time in calculating the assigned equations of each processor. The first specified update time (5666 cycles) was the minimum possible time and corresponds to the critical path calculation time. The second update time selection (10,000 cycles) required two processors. Note that the percent utilization of the last processor in each case exceeds the summation of time available on the other processor(s). The algorithm, therefore, functioned satisfactorily.

Table 1
Packing Algorithm Results for Turboshaft Engine Model

Update time	Processors required	Processor percent utilization			
		P number 1	P number 2	P number 3	P number 4
5,666	4	100	98	97	53
10,000	2	99	98	—	—
19,568	1	100	—	—	—

Since the packing algorithm does not account for data transfer times, it is possible that the available time between when a variable is computed on one processor and when its value is required for computation on another processor will be less than the time required to transfer the variable between the processors. The effect of this will be to increase the effective calculation time of the packed simulation and, therefore, to increase the minimum achievable update time. Data transfer effects may be significant for multiprocessor systems with inefficient data transfer mechanisms or for simulations that require large volumes of data transfer between processors. While these effects will increase the critical path time and, therefore, the minimum update time, proper consideration of data transfer will minimize this increase and provide far more efficient packing.

REFERENCES

1. **Blech, R. A. and Arpasi, D. J.,** Hardware for a Real-Time Multiprocessor Simulator, NASA TM-83805, 1984.
2. **Arpasi, D. J.,** RTMPL — A Structured Programming and Documentation Utility for Real-Time Multiprocessor Simulations, NASA TM-83606, 1985.
3. **Arpasi, D. J.,** Real-Time Multiprocessor Programming Language, (RTMPL) — Users Manual, NASA TP-2422, 1985.
4. **Cole, G. L.,** Operating System for a Real-Time Multiprocessor Propulsion System Simulator, NASA TM-83605, 1985.
5. **Milner, E. J. and Arpasi, D. J.,** Simulating a Small Turboshaft Engine in a Real-Time Multiprocessor Simulator (RTMPS) Environment, NASA TM-87216, 1986.
6. **Arpasi, D. J. and Milner, E. J.,** Partitioning and Packing Mathematical Simulation Models for Calculation on Parallel Computers, NASA TM-87170, 1986.

22 INCREASING PROCESSOR UTILIZATION DURING PARALLEL COMPUTATION RUNDOWN

INTRODUCTION

General purpose parallel computations are usually divided into phases that must execute sequentially in order to guarantee algorithmic integrity. For instance, the checkerboard approach to the successive overrelaxation solution of the potential field problem divides into two such phases: the "odd" locations phase and the "even" locations phase. On the parallel phase level, the iterated values of the previous phase must be complete before the new values of the next phase can be correctly computed.

In the checkerboard algorithm, the execution time of each location is definite (nominally, the time for four additions and a divide). Thus, the distribution of work among processors can be accurately planned. Under ideal conditions (involving the number of checkerboard locations in comparison to the number of processors), the distribution of work can be arranged so that each processor shares an exactly even portion of the work and, as a consequence, each processor completes its work at exactly the same time. Perfect computation resource utilization is realized (at least in a practical sense), since the next computational phase can begin immediately.

Unfortunately, ideal conditions are infrequently found in real applications. Continuing with the checkerboard algorithm, consider the situation when the potential grid is 1024 points on a side (2**20 grid points) and 1000 processors are available. Each computational phase will provide 524,288 individual computations, or 524 computations for each of the 1000 processors; however, 288 computations will be left over for distribution among the 1000 processors. This will leave 712 processors with nothing to do while the final 288 computations are carried out; this is an example of rundown.

The burden of experience gained by the author suggests that even this example is optimistic. Most computations carried out by the author's parallel

Navier-Stokes solver (the Combined Aerodynamic and Structural Dynamic Problem Emulating Routines, or CASPER,[1] which was controlled by the Parallel, Asynchronous Executive, or PAX[2] could not even be ascribed with definite execution times. In some instances, whether or not the computation was even to be carried out in a particular instance was a conditional part of the algorithm. No control over the computation count-to-processor ratio was attempted — processors were allocated as they became available on a the-more-the-merrier basis. Also, shared information access times were unpredictable and unrepeatable from instance to instance. As a result, there was no assurance that individual processors could be kept busy as a particular computational phase drew to a close.

The PAX/CASPER project provided the experience base cited later in this chapter. PAX/CASPER was focused on a parallel, general purpose, Navier-Stokes solver. Thus, this experience base is presented not as a grand generalization for all of parallel processing, but as a specific example in practical parallel processing.

Certain other situations that might seem of interest in the overlapping of computational phases (for instance, the possibilities for overlapping in a tight iterative loop) are not treated for the simple reason that they did not occur in the PAX/CASPER project. PAX/CASPER was not so much a research project in parallel processing as an exploratory development of a far-term aerodynamic tool. Thus, the motivation was to solve the problems that occurred rather than to solve the problems that one could imagine.

It has been suggested that scheduling and overhead problems will be a particular problem in PAX/CASPER. So far, this has not been the case. Operational experience shows that the ratio of computation to management has been running at something in the neighborhood of 200. There are additional strategies which have been identified for development, which include a middle management scheme to parallelize the serial management function, a direct worker-to-worker lateral communication scheme, and a data-proximity work assignment algorithm. These strategies combined with the overlapping of computational phases should enhance the management overhead situation.

Various solutions to the computational rundown problem may be acceptable. Some parallel processing schemes for general purpose computation may choose simply to accept the lower processor utilization as a minor design flaw. Another alternative is to create a multi-parallel-job-stream environment that allows computational work of one job stream to fill in when another job stream enters a computational rundown situation. This will bring processor utilization up; however, it should be recognized that the primary goal of parallel processing is to reduce elapsed wall-clock time for a given job. The introduction of such a "batch" environment will inevitably distribute processor resources among the several job streams and, thus, reduce the total processing power on any particular job and lengthen its elapsed wall-clock time.

The goal, then, is to find more ready-to-compute work from the parallel algorithm that is being computed. As mentioned previously, this is not possible at the parallel phase level: each phase must be completed before the next is begun in order to guarantee algorithmic integrity; however, if an examination is made at a deeper (subphase or, in the terminology of the author, task) level, it is frequently discovered that the completion of portions (tasks) of one phase will allow the correct computation of portions (tasks) of the succeeding phase.

Consider again the checkerboard algorithm. If all the "odd" locations adjacent to a particular "even" location have been updated with new values from the current computational phase, then the new value for that particular "even" location for the next computational phase can be correctly computed. Additionally, since all the computations requiring as an input the current value of that particular "even" location have been completed, the value for that "even" location can be updated without affecting the results of the current computational phase.

At this point, it is necessary to make certain assumptions (or, alternatively, set certain system design constraints) about the nature of computational-phase rundown. Two basic situations arise: one in which task assignments and releases are statically determined and one in which such matters are dynamically determined.

The static situation is much simpler from the standpoint of next-phase task release timing, since everything is determined ahead of time. In this case, it can be acceptable for computational rundown to begin almost immediately, since the scheduling of the next-phase task has already been statically determined. No completion processing of current-phase tasks is required to schedule the release of the next-phase task. (In fact, work in this area for the purposes of real-time simulation has been conducted for some years at NASA Lewis.[3,4])

The dynamic scheduling situation is substantially more interesting. Some time delay must be available between the completion of the first current-phase tasks and the onset of computational rundown. This delay is needed to provide time to process the completion of the early current-phase tasks and, in so doing, schedule the next-phase tasks that are thus enabled. During this delay, there must be enough current-phase tasks to keep the processing resources busy in order to avoid a computational load dip while the next-phase tasks are scheduled.

In the dynamic scheduling situation, enablement relationships between the current-phase tasks and the next-phase tasks (i.e., the relationship that enables a next-phase task based upon the completion of a current-phase task) may be either static or dynamic. That is, the completion of a particular current-phase task may always enable the same next-phase task (the static enablement case) or it may enable some next-phase task that can only be identified at the time of execution (the dynamic enablement case). The nature of the enablement relationship is important because it is involved in setting the time delay from the completion of the first current-phase tasks to the availability of the first enabled next-phase tasks.

Considering these characteristics of the dynamic scheduling situation (i.e., the time to process current-phase task completion, the time to recognize enablement relationships, and the time to schedule enabled next-phase tasks), it can be observed that the number of tasks should substantially outnumber the number of processors. Certainly, there should be at the outset of the current-phase work at least two tasks for each processor so that at least one task execution time will be available to process the completion of the first task assigned to the processor and to schedule the enabled next-phase task. This presumes that completion processing and task scheduling time is small with respect to task execution time. In particular, it assumes that one such completion, enablement, and scheduling cycle for each of the processors in the system can be completed in a single task execution time. (The author's experience with PAX suggests that this is reasonable even for dynamic, managerial-style parallel processing systems. Systems that use hardware-level synchronization primitives presumably would be at even greater advantage in this area.)

The conditions under which this overlapping of computational phases can correctly occur are the same as those that allow parallel computations within a particular phrase. Let the logical predicate PARALLEL(x,y) return the condition TRUE when x and y are such that parallel computations are allowed. Clearly, PARALLEL(n,m) must always be TRUE if n and m are distinct computational granules of the same parallel computational phase. Let q be an uncompleted granule of the current phase and r be a granule of the next phase that has been enabled by some completed granule, p, of the current phase. If PARALLEL(q,r) necessarily returns the value TRUE, then the current phase and next phase can be correctly overlapped.

The exact nature of the logical predicate PARALLEL(x,y) is, of course, of substantial practical interest; however, it has no direct impact upon the ability to overlap phases, as outlined above. Different parallel systems may identify different logical predicates.

The first challenge to be met is to find a way of identifying enabled next-phase granules for overlapping. It is easy to postulate that some mapping function exists either to map from the set of completed granules, p, to the set of enabled granules, r, or to map from the set of uncompleted granules, q, to the set of enabled granules, r. It is very difficult to establish what this mapping function might be in any general way. Fortunately, this mapping function is much more easily identified when each concrete situation is faced.

First, consider the simplest imaginable case as represented by the following Fortran code segment:

```
      ...
      DO 100 I=1,N   ;        First computational phase
      B(I)=A(I)      ;
100   CONTINUE       ;
      DO 200 I=1,N   ;        Second computational phase
      D(I)=C(I)      ;
200   CONTINUE       ;
      ...
```

Assuming that there are not shared output area constraints, it can be observed that these two parallel computational phases can be computed in parallel with each other. This represents what might be called a universal mapping function, wherein any granule of the second computational phase is enabled by any granule or set of granules (including the null set) of the first computational phase.

PAX/CASPER experience shows that 6 out of 22 (or 27%) of the parallel computational phases allow universal mapping enablement of the succeeding phases. This represents 266 out of 1188 lines (or 22%) of the code that is executed in parallel in PAX/CASPER.

This universal mapping usually occurs in PAX/CASPER when the nature of the larger computational process is changing. For instance, the change over from power of compression computations to interpolator matrix generation is one such character change. The two computations do not involve shared information of any kind; thus, they can be entirely overlapped. Of course, the two phases could be merged into one by a preprocessor of the parallel control stream; however, since the mechanisms necessary to handle this case would be a subset of those needed for the following case, it might well be simpler to support this enablement mapping.

For the next case, consider the following Fortran fragment that is to be computed in parallel as two succeeding computational phases:

```
      . . .
          DO 100 I=1,N   ;        First computational phase
          B(I)=A(I)      ;
  100     CONTINUE       ;
          DO 200 I=1,N   ;        Second computational phase
          C(I)=B(I)      ;
  200     CONTINUE       ;
      . . .
```

Again, assuming that there are not shared output area constraints, it can be observed by inspection that the identity mapping function $(I = I)$ maps from completed granules, p, to enabled granules, r. This is also a simple and easily identified mapping.

PAX/CASPER experience indicates that it applies in 9 out of 22 (or 41%) of the parallel computational phases (representing 551 of 1188 code lines, or about 46% of the parallel code in PAX/CASPER). Combining this direct mapping with the simpler universal mapping above indicates that (at least in PAX/CASPER experience) 68% of the parallel computational phases and 68% of the code executed in parallel can be easily overlapped to defeat computational rundown. These two enablement mapping possibilities are the most frequently occurring situations in PAX/CASPER experience.

The next most frequently occurring enablement mapping in PAX/CASPER experience is what could be called null mapping, that is, the situation in which no overlapping is possible. This occurs in 4 out of 22 (or 18%) of the computational phases and represents 262 out of 1188 (or 22%) of the lines of code executing in parallel. In all cases the cause was not that such an overlapping did not exist between the parallel computations, but was, in fact, that serial actions and decisions had to occur between the phases. This is important since it allows one to assess how often the extra effort of supporting overlapping features will be entirely defeated, regardless of the sophistication of the overlapping phase support features.

Another enablement mapping occurring in PAX/CASPER experience is a reverse indirect mapping. Consider the following Fortran fragment:

```
      . . .
      DO 10 I=1,N            ; Set up source mapping
      DO 10 J=1,10           ;
      IMAP(J,I)=IRAND()      ; IRAND produces an integer
10    CONTINUE               ;  in the range 1 to N
      DO 100 I=1,N           ; First computational
      A(I)=FUNC(I)           ;  phase generates some
                             ;  number in A(x)
100   CONTINUE               ;
      DO 200 I=1,N           ; Second computational
      DO 200 J=1,10          ;  phase sums subsets of
      B(I)=1(IMAP)(J,I))     ;  the results of the
                             ;  first computational
                             ;  phase
200   CONTINUE               ;
      . . .
```

Clearly, this computation can be overlapped; however, determining the enablement mapping is very difficult. This is because knowing that a particular first-phase granule is complete does not directly identify any distinct second-phase granule as computable; however, a reverse mapping from desired second-phase granule to required first-phase granules is possible.

In PAX/CASPER experience, this situation occurs in 2 of 22 (or 9%) of the computational phases, representing 78 out of 1188 (or 7%) of the lines of code executing in parallel. While this is not a frequently occurring situation in PAX/CASPER experience, it cannot be ignored out of hand. Some engineering judgment must be made to weigh the cost (in terms of management overhead, computational resource transferred from workers to management, etc.) of some reverse enablement mapping solution against the cost of computational rundown in 9% of the parallel computational phases.

Certainly, a solution exists for the reverse, indirect enablement mapping. Once the values of the information selection map (represented in the code fragment by the array IMAP) have been determined, it is a simple matter to produce a composite map of first-phase granules that must be completed in order to enable a particular second-phase granule. The executive can then use this map upon each first-phase granule completion to determine the computability of particular second-phase granules. This map could also be used to direct a preferred order of first-phase granule dispatching so as to enable a known second-phase granule as early as possible.

Two important facts about this reverse enablement mapping must be included. First, both occurrences of this situation involved a dynamically generated information selection map. Thus, the composite granule map would have to be generated by the executive at or after first-phase initiation, but before any second-phase enablements. Second, the impact of the executive computation must be considered. In the PAX/CASPER test bed, executive computation was done at the direct expense of worker computation. Thus, extensive composite granule map generation could be self defeating. Some parallel machines may provide separate executive computing resources, in which case the generation and use of composite granule maps would not be out of the question.

A final enablement form was observed in PAX/CASPER that could be characterized as a forward, indirect mapped situation. Consider the following Fortran fragment:

```
      DO 10 I=1,M            ;    Generate forward
      IMAP(I)=IRAND()        ;      map
10    CONTINUE               ;
      DO 100 I=1,M           ;    Use forward map
                             ;      to operate on a
      B(IMAP(I))=A(IMAP(I))  ;      subset of the
100   CONTINUE               ;      arrays
      DO 200 I=1,N           ;    Perform some further
      C(I)=B(I)              ;      operation on the
200   CONTINUE               ;      complete arrays
      ...
```

This situation is somewhat easier than the reverse, indirect mapping in that the identification of a particular granule in the first phase can be directly mapped to an enabled granule in the successor phase; however, much of the complication of a mapped enablement remains. This form was the least frequently occurring situation in PAX/CASPER, showing up only once (5% of the phases) and accounting for only 31 of 1188 lines of code executed in parallel.

No other forms of enablement mapping were observed in PAX/CASPER. Certainly, extensions of the forms already presented can be imagined. Additionally, a seam mapping problem (such as would be appropriate for the checkerboard approach to the successive overrelaxation problem) can be foreseen. These other forms are beyond the scope of the present discussion.

The developing PAX/CASPER language is simple and requires the user to make specific statements concerning choices for the management of each parallel computational phase. Statements involving the enablement of a succeeding phase could be made at two times: during the definition of a computational phase to the management system, and during the invocation of the phase for actual computations. The difficulty to be faced is that the statements no longer apply solely to the phase being referenced, but rely also on the characteristics of the succeeding phase.

The simplest approach is to require the user to specify the appropriate enablement mapping method when the phase is invoked. It might appear as in the following PAX parallel language fragment:

```
...
DISPATCH        phase-name
                ...
                ENABLE/MAPPING=option
                ...
```

This is simple and explicit; however, it leaves the door wide open to user mistakes. There is no interlock between this phase and the next that can be verified by the executive. A simple solution to this would be to identify the name of the enabled next phase so that the executive system (or language processor) can verify that, in fact, that phase is following. This might appear as follows:

```
...
DISPATCH        phase-name
                ...
                ENABLE [phase-name/MAPPING=option]
                ...
...
```

This allows the desired verification, but also brings up a new possibility. Occasionally, a conditional branch that is not dependent on the computational phase separates that phase from two or more succeeding phases, each of which may (or may not) be overlappable. If each of these phases were identified in the above construct, the executive could preprocess the branch and overlap the appropriate phase. This could look as follows:

```
...
DISPATCH        phase-name
                ...
                ENABLE/BRANCHINDEPENDENT
                        [phase-name-1/MAPPING=option
                         phase-name-2/MAPPING=option]
                ...
IF              (IMOD(LOOPCOUNTER,10).NE.0)
THEN            GO TO branch-target
DISPATCH        phase-name-1
                ...
    GO TO   rejoin
branch-target:
    DISPATCH phase-name-2
rejoin:
    ...
```

Finally, the matching of mapping selections and phases and the invocation of the appropriate overlapping services is something that could be done when the parallel phase is defined to the system; however, it would still be necessary to identify preprocessable branches at the computation invocation site. This could appear as follows:

```
DEFINE PHASE phase-name
               . . .
               ENABLE          [
                               phase-name-1/MAPPING=option
                               phase-name-2/MAPPING=option
                               phase-name-3/MAPPING=option
                               ]
   . . .
   DISPATCH        phase-name
                   . . .
                   ENABLE/BRANCHDEPENDENT
   . . .
```

The ENABLE/BRANCHINDEPENDENT would be deleted when branch preprocessing was either not appropriate or not needed. The executive system could perform the appropriate look-ahead to see whether any of the named succeeding phases was actually following and apply, as appropriate, the specified enablement mapping.

Control strategies for enabling and scheduling overlapped parallel computational phases are, of course, highly dependent upon the overall parallel processing strategies. As alluded to earlier, some approaches to parallel processing may do all of this before any computations are begun. Indeed, the entire process may be done manually by a human being when the pattern of parallel processing is fixed for the life of the system.

Within the PAX system, the opposite is true: the identification and scheduling of computable granules is entirely automatic. A scheduling mechanism for enabled computational granules already exists within the PAX system. It was developed to schedule dynamically created computations that conflicted (usually in terms of shared data access) with preexisting computational granules.

Within PAX, each internal description of one (or more) computational granule(s) included a queue head for a double circularly linked list of computable but conflicting computational granules. Upon completion of the described computation, all the queued conflicting computations became unconditionally computable and were placed in the waiting computation queue. The waiting computation queue was kept in a known order, and, for the purposes of the conflicting computation problem, it was determined that such conflicting computations would be placed ahead of the normal computations in the queue and, thus, be given higher priority.

The scheduling of universally mapped successor phases within this system is very easy indeed. At the time of phase initiation, the successor phase is also initiated and the resulting computation description placed in the waiting computation queue behind the current phase description.

The scheduling of directly enabled successor phases is similarly easy at first sight. At the time of phase initiation, the successor phase is also initiated and the resulting computation description is placed in the conflicted computation queue of the current-phase description. Thus, when the current-phase computation is completed, the now-enabled successor computation will be placed in the waiting computation queue to be considered for scheduling.

The above approach for directly enabled successor phases is fine if each indivisible granule of computation is described separately. Unfortunately, this is usually not economical (in terms of storage space and task search times, among other things) and was not the choice taken in PAX design. Computations were, instead, described as large, contiguous collections of granules. The descriptions were split apart as necessary to produce conveniently sized tasks for workers and then merged back into single descriptions when the work was completed. This splitting of descriptions requires that queued computation descriptions also be split so that each queued description will accurately reflect the enablement relationship between the computation and its queued successor computation.

While this is certainly possible, it forces a further design decision for the executive software. PAX computation splitting was demand driven by the presence of an idle worker. It was felt that the delay while splitting a task description was acceptable; however, the additional delays of splitting queued successor computation descriptions may represent an unacceptable situation. Two possible solutions exist. One possibility is to presplit the tasks before idle workers present themselves to the executive. This would allow the executive to work ahead in otherwise idle time. Alternatively, the splitting of a computation could generate a successor-splitting task that could be quickly queued for later attention when the executive would again be idle.

The successor computation description could be removed from the current computation description and included in the successor-splitting task information. When the successor-splitting task is executed, the successor computation could be split and requeued to the appropriate current computation descriptions.

Management of indirectly (both forward and reverse) mapped successor computations is a good deal more interesting. The description of the successor computation cannot simply be queued to the description of the current computation, since there is no guarantee of the enablement relationship. Additionally, it would seem wise to get the current phase into execution without the delay of constructing the necessary information for enabling successor computations. Both forward and reverse indirection would seem well handled by much the same mechanisms since the only significant difference is the direction of the indirection. Each leads naturally to a list of current-phase granules that must be completed to enable a particular successor-phase granule.

It would seem appropriate to identify a subset group of successor-phase granules that are to be the subject of the enablement operation, so as to avoid solving an unnecessarily large enablement problem. Once this subset has been identified, the current-phase granules that enable the successor subset can be identified. Since these are not necessarily the current-phase granules that would be naturally selected by the scheduling mechanism, they should be split into individual descriptions and placed in the waiting computation queue in such a manner as to elevate their computational priority.

It is important to note that the description of the successor subset cannot simply be queued to any one of the identified current-phase granules, since it is enabled not by the completion of any one such granule, but by the completion of all the identified granules. This enablement on completion of all identified current-phase granules can be handled by any number of simple mechanisms. For instance, during completion processing, a status bit (set when the current-phase granules were identified and split into individual descriptions) can be checked and, if it is set, an enablement counter decremented. When the enablement counter reaches zero, it can be taken as a signal that the successor-phase granules are computable.

This discussion has explored the possibilities for overlapping parallel computations in a general-purpose, parallel-computation environment so as to minimize loss of computational resources. Practical experience with PAX/CASPER, a parallel Navier-Stokes solver, suggests that simple and plausible steps could provide such overlapping in 68% of the computational phases and that, with extended effort, more than 90% of the computational phases are amenable to some form of phase overlapping.

REFERENCES

1. **Jones, W. H.,** Combined Aerodynamic and Structural Dynamic Problem Emulating Routines (CASPER): Theory and Implementation, NASA TP-2418, 1985.
2. **Jones, W. J.,** Parallel, Asynchronous Executive (PAX): System Concepts, Facilities, and Architecture, NASA TP-2179, 1983.
3. **Arpasi, D. J.,** Real-Time Multiprocessor Programming Language, (RTMPL) User's Manual, NASA TP-2422, 1985.
4. **Arpasi, D. J. and Milner, E. J.,** Partitioning and Packing Mathematical Simulation Models for Calculation on Parallel Computers, NASA TM-87170, 1986.

23 SOLVING COMPUTATIONAL GRAND CHALLENGES USING A NETWORK OF HETEROGENEOUS SUPERCOMPUTERS*

Wide-area computer networks have become a basic part of the computing infrastructure of today. These networks connect a variety of machines, representing an enormous computational resource. We** describe two software packages we are developing to facilitate the use of a network of heterogeneous computers. The first, Parallel Virtual Machine (PVM), allows utilization of the network of machines as a single computational resource. The second, Heterogeneous Network Computing Environment (HeNCE), is a graphical tool to assist the user in writing and analyzing parallel programs. We describe the parallelization of a large materials science application code including modifications required to run on heterogeneous supercomputers. Finally, we present results from our initial tests of this software.

PVM is a software package that enables concurrent computing on loosely coupled networks of processing elements. PVM may be implemented on a hardware base consisting of different machine architectures, including single CPU systems, vector machines, and multiprocessors. These computing elements may be interconnected by one or more networks, which may themselves be different (e.g., Ethernet, the Internet, and fiber optic networks). These computing elements are accessed by applications via a library of standard interface routines. These routines allow the initiation and termination of processes across the network as well as communication and synchronization between processes.

*Adapted with permission from the *Proceedings of the Fifth SIAM Conference on Parallel Processing for Scientific Computing*, edited by Danny Sorenson. Copyright 1992 by the Society for Industrial and Applied Mathematics, Philadelphia. All rights reserved.

**Adam Beguelin, Jack Dongarra, Oak Ridge National Laboratory and University of Tennessee; Al Geist, Oak Ridge National Laboratory; Robert Manchek, University of Tennessee; and Vaidy Sunderam, Emory University.

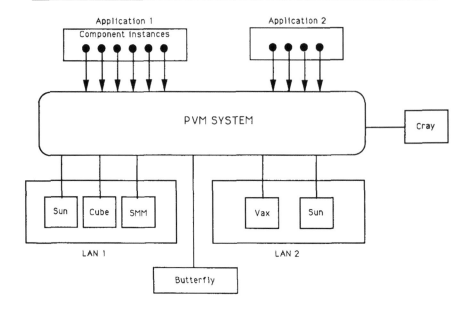

FIGURE 1. PVM architecture model.

Application programs are composed of *components* that are subtasks at a moderately large level of granularity. During execution, multiple *instances* of each component may be initiated. Figure 1 depicts a simplified architectural overview of the PVM system.

Application programs view PVM as a general and flexible parallel computing resource that supports an MP model of computation. This resource may be accessed at three different levels: the *transparent* mode, in which component instances are automatically located at the most appropriate sites, the *architecture-dependent* mode, in which the user may indicate specific architectures on which particular components are to execute, and the *low-level* mode, in which a particular machine may be specified. Such layering permits flexibility while retaining the ability to exploit particular strengths of individual machines on the network. The PVM user interface is strongly typed; support for operating in a heterogeneous environment is provided in the form of special constructs that selectively perform machine-dependent data conversions. All communications done between PVM processes use the external data representation standard (XDR).[15] Interinstance communication constructs include those for the exchange of data structures as well as high-level primitives such as broadcast, barrier synchronization, and rendezvous.

Application programs under PVM may possess arbitrary control and dependency structures. In other words, at any point in the execution of a concurrent application, the processes in existence may have arbitrary relationships between each other, and, further, any process may communicate and/or synchronize with any other. This is the most unstructured form of crowd computation, but in practice, a significant number of concurrent applications are more structured.

Two typical structures are the trees and the "regular crowd" structure. We use the latter term to denote crowd computations in which each process is identical; frequently such applications also exhibit regular communication and synchronization patterns. Any specific control and dependency structure may be implemented under the PVM system by appropriate use of PVM constructs and host language control flow statements.

PVM is available through *netlib*. To obtain a copy, send mail to *netliboornl.gov* with the message: *send index from pvm.*

While PVM provides low-level tools for implementing parallel programs, HeNCE provides the programmer with a higher level abstraction for specifying parallelism. The HeNCE philosophy of parallel programming is to have the programmer explicitly specify the parallelism of a computation and to aid the programmer as much as possible in the tasks of designing, compiling, executing, debugging, and analyzing the parallel computation. Central to HeNCE is a graphical interface which the programmer uses to perform these tasks. This tool supports visual representations of many of its functions.

In HeNCE, the programmer is responsible for explicitly specifying parallelism by drawing graphs that express the parallelism. The user directly inputs the graph using tools which are part of the HeNCE environment. Each node in a HeNCE graph represents a procedure written in either Fortran or C. There are three types of arcs in the HeNCE graph. One represents dependency; a second represents looping; and the third represents pipelined sections. An arc from one node to another represents the fact that the tail node of the arc must run before the node at the head of the arc.

Once the graph is complete, HeNCE will automatically write the parallel program, including all the communication and synchronization routines using PVM calls. HeNCE tools assist the user in compiling this program for a heterogeneous environment.

Graphical tools in HeNCE allow the user to dynamically configure a parallel collection of machines into a parallel virtual computer. During execution, HeNCE dynamically maps procedures to the machines in the heterogeneous network based on a user-defined cost matrix. HeNCE also collects trace and scheduling information, which can be displayed in real time or saved to be replayed later.

During the last few years, ORNL material scientists and their collaborators at the University of Cincinnati, SERC at Daresbury, and the University of Bristol have been developing an algorithm for studying the physical properties of complex substitutionally disordered materials. A few important examples of physical systems and situations in which substitutional disorder plays a critical role in determining material properties include metallic alloys, high-temperature superconductors, magnetic phase transitions, and metal/insulator transitions. The algorithm being developed is an implementation of the Korringa, Kohn and Rostoker coherent potential approximation (KKR-CPA) method for calculating the electronic properties, energetics, and other ground-state properties of substitutionally disordered alloys.[3] The implementation allows the treatment of materials having complex atomic lattices and having any number of disordered sublattices. The algorithm also solves the underlying quantum mechanical equations within a fully relativistic framework.[1] This is necessary for materials containing heavy elements such as barium in which even the outer, bonding electrons are traveling at a significant fraction of the velocity of light.

The KKR-CPA method extends the usual implementation of local density approximation to density functional theory (LDA-DFT)[4] to substitutionally disordered materials.[2] In this sense it is a completely first-principles theory of the properties of substitutionally disordered materials, requiring as input only the atomic numbers of the species making up the solid.

The KKR-CPA algorithm contains several locations where parallelism can be exploited. These locations correspond to integrations in the KKR-CPA algorithm. Evaluating integrals typically involves the independent evaluation of a function at different locations and the merging of these data into a final value. The two most obvious locations for parallelization in the KKR-CPA algorithm are in the integration over the Brillouin zone and the integration over energy. Each location was evaluated in terms of the available parallelism and the required communication overhead incurred by splitting the algorithm at that point. The Brillouin zone integration is the main step in each CPA iteration. The disadvantage of parallelizing the Brillouin zone integration is the large amount of communication volume that would be required. For this reason the integration over energy was parallelized. Typically, this integration involves the evaluation of the single-site Green's function for between 200 and 1000 energies in order to determine the charge density for the next self-consistency iteration. Each of these tasks involves the iterative solution of the CPA equations for the given energy and requires significantly more computation than communication.

The parallel implementation is based on a master/slave paradigm to reduce memory requirements and synchronization overhead. In our implementation, one processor is responsible for reading the main input file, which contains the number of processors to be used, the problem description, and the location of relevant data files. This master processor also manages the LDA-DFT charge self-consistency iteration. The slave processors require only enough memory to solve the CPA equations for a single energy, which presently requires 7 Mbytes. Memory reduction was a crucial consideration for porting the KKR-CPA code to the Intel iPSC/860, because the individual nodes only contain 8 Mbytes of RAM. If only one processor is requested, then a subroutine is called that calculates the tasks serially one after another. When more than one processor is requested, a pool-of-tasks scheme is used to accomplish dynamic load balancing. In this scheme, the tasks are arranged in a queue in approximate order of decreasing difficulty and assigned to idle slave processors as they become available. Thus, all processors are busy as long as there are tasks in the queue.

Additional modifications were required to allow this KKR-CPA code to execute on multiple heterogeneous supercomputers. On a Cray YMP, two options are now available in the code. The user can specify a number of tasks to initiate simultaneously on the Cray multiprocessor (ideally one per processor).

A second option is to set a Cray multitasking switch in the input file and then initiate just one multitasking task. This second option is now under investigation. To use an iPSC/860 we had to write an intermediate routine that runs on the Intel host and performs three functions. This routine allocates the number of node processors specified by the master process. It receives messages from other PVM processes, converts these messages to Intel node messages, then passes the messages to the appropriate nodes (and vice versa). Finally, it releases the nodes when the problem is finished. Because of the size of the KKR-CPA code (16,000 lines of Fortran 66), we have not ported the code to run on Thinking Machines CM2, which requires applications to be written in Fortran 90.

So far, the results on a network of heterogeneous supercomputers show the viability of using such a system. Test programs of simple tasks like solving the Mandelbrot problem have been run on a DECStation, Cray XMP, iPSC/860, and CM2 configuration.

The KKR-CPA code was initially run on a network of IBM RS/6000s. Using six model 530s and four model 320s, an execution rate of 207 MFLOPS was measured. This performance is comparable to running it on a single processor of a Cray YMP.

The KKR-CPA code was also run on a virtual machine consisting of an RS/6000, a Cray XMP, and 16 nodes of the iPSC/860. Performance was not measured during these initial runs, which were designed to test the viability of the system. Performance tests will be done when a dedicated Cray YMP/8 and 128-node iPSC/860 are simultaneously available. Experiments are now underway that involve running the KKR-CPA code on a virtual machine composed of three dedicated Cray YMPs.

These experiments are the initial steps in showing how several Cray and Intel supercomputers can be configured into a virtual super-parallel computer in order to achieve the computational power necessary to solve Computational Grand Challenge Problems.

REFERENCES

1. **Ginatempo, B. and Staunton, J. B.,** The electronic structure of disordered alloys containing heavy elements — an improved calculational method illustrated by a study of CuAg alloy, *J. Phys. F*, 18, 1827, 1988.

2. **Johnson, D. D., Nicholson, D. M., Pinski, F. J., Györffy, B. L., and Stocks, G. M.,** Total energy and pressure calculations for random substitutional alloys, *Phys. Rev. B*, 41, 9701, 1990.

3. **Stocks, G. M., Temmerman, W. M., and Györffy, B. L.,** Complete solution of the Korringa-Kohn-Rostoker coherent potential approximation: Cu-Ni alloys, *Phys. Rev. Lett.*, 41, 339, 1978.

4. **von Barth, U.,** Density functional theory for solids, in *Electronic Structure of Complex Systems*, Phariseau and Temmerman, Eds., NATO ASI Series, Plenum Press, New York, 1984.

5. SUN Network Programming Manual, Part Two: Protocol Specification, 1988.

LIST OF SOURCES

Background

Introduction to Parallel Programming, Steven Brawer (Encore Computer Corporation, Marlborough, MA), Academic Press Inc., 1250 Sixth Ave., San Diego, CA 92101, 1989.

PART 1: MIMD Computers

COMMERCIAL MACHINES

Thinking Machines CM-5

The Connection Machine CM-5 Technical Summary, Thinking Machines Corp., 245 First Street, Cambridge, MA 02142-1264, Oct. 1991.

NCUBE

Development of Parallel Methods for a 1024-Processor Hypercube, John L. Gustafson, Gary R. Montry, and Robert E. Brenner, Sandia National Labs, Albuquerque, NM 87185, SIAM Journal of Scientific and Statistical Computing, Vol. 4, No. 9, July 1988, P609; 3600 University City Science Center, Philadelphia, PA 19104-2688.

iWarp

iWarp: An Integrated Solution to High-Speed Parallel Computing, Shekhar Borker, Robert Cohen, George Cox, Sha Gleason, Thomas Gross, H. T. Kung, Monica Lam, Brian Moore, Craig Peterson, John Pieper, Linda Rankin, P. S. Tseng, Jim Sutten, John Urbanski, Jon Webb, Dept. of Computer Science, Carnegie-Mellon University, Pittsburgh, PA 15213 and Intel Corp., 5200 N. E. Elam Young Parkway, Hillsboro, OR 97124, Proceedings of Supercomputing '88, Orlando, FL, Nov. 14–18, 1988, Computer Society of the IEEE, 1730 Massachusetts Ave., Washington, DC 20036. ©1988 IEEE.

iPSC and iPSC/2

The Scalability of Intel Concurrent Super Computers, David S. Scott and Justin Rattner, Proceedings of Supercomputing '88, Computer Society of the IEEE.

Using Givens Rotations to Solve Dense Linear Systems on the Hypercube, TIBA Porta, Dept. of Computer Science, Yale University, P. O. Box 2158 Yale Station, New Haven, CT 06520, Proceedings of the Third International Conference on Supercomputing, International Supercomputing Institute Inc., 1988, P443; 127 9th Street East, Tierra Verde, FL 33715-2204.

Paragon XP/S Product Overview, Intel Corp., Supercomputing Systems Systems Division, 15201 N.W. Greenbrier Parkway, Beaverton, OR 97006, 1991.

Encore Multimax

Parallel Block SOR Methods for Solving Poisson Equation on Shared and Local Memory Multiprocessors, Xiaodong Zhang, Computer Science Dept., Univ. of Colorado at Boulder, Boulder, CO 80309, Third International Conference on Supercomputing, International Supercomputing Institute Inc., 1988, P473.

AT&T DSP-3

High Performance Reconfigurable Parallel Processing Architecture, R. R. Shively, E. B. Morgan, T. W. Copley, and A. L. Gorin, AT&T Bell Laboratories, Whippany, NJ.

Digital Signal Processor Evaluation, K. Barnes and W. Schaming, General Electric Advanced Technology Laboratories, Moorestown, NJ 08057, CMAT-90-TR-005, 6 April 1990.

Meiko Computing Surface

A Parallel Architecture for Flexible Real-Time SAR Processing, C. J. Oliver, Royal Signals and Radar Establishment, Malvern, Worcs., WR14 3PS, United Kingdom, and S. P. Turner, Meiko Scientific, Bristol, BS12 4SD, UK.

Inside a Heterogeneous Parallel Computer, Dick Pountain, Byte Magazine, February 1991, One Phoenix Mill Lane, Peterborough, NH 03458.

The Edinburgh Concurrent Supercomputer: Project and Applications, Ken Bowler, Richard Kenway and David Wallace, Physics Department, University of Edinburgh, The King's Buildings, Mayfield Road, Edinburgh EH9 3JZ, Scotland, Proceedings of the Third International Conference on Supercomputing, International Supercomputing Institute Inc., 1988.

BB&N Butterfly

Butterfly Parallel Processor Overview, BBN Report No. 6148, 6 March 1986, BBN Advanced Computers Inc., 10 Fawcett St., Cambridge, MA 02138.

Parallel Vision with the Butterfly Computer, Christopher M. Brown, Computer Science Dept., Univ. of Rochester, Rochester, NY 14627, Third International Conference on Supercomputing, International Supercomputing Institute Inc., 1988, P54.

Sequent

The Measured Performance of Parallel Dynamic Programming Implementations, Kenneth Almquist, Richard J. Anderson, and Edward D. Lazowska, Dept. of Computer Science, Univ. of Washington, Seattle, Washington 98195, 1989 International Conference on Parallel Processing, P III-76; IEEE Computer Society, 1730 Massachusetts Ave., N.W., Washington, DC 20036. ©1989 IEEE.

Teradata

DBC/1012 Data Base Computer, Concepts and Facilities, C02-0001-05, Release 3.1, 1988, Teradata Corporation, 12945 Jefferson Blvd., Los Angeles, CA 90066.

RESEARCH MACHINES

J-Machine

The J-Machine: A Fine Grain Concurrent Computer, William J. Dally, Andrew Chien, Stuart Fiske, Waldemar Horwat, John Keen, Michael Larivee, Rich Lethin, Peter Nuth, and Scott Wills, Artificial Intelligence Laboratory for Computer Science, Massachusetts Institute of Technology, Cambridge, MA 02139, Paul Carrick and Greg Fyler, Intel Corporation, Santa Clara, CA 97006, Information Processing 89, G. X. Ritter (Ed.), Elsevier Science Publishers B.V. (North Holland), IFIP 1989; Elsevier Science Publishers, P. O. Box 882, Madison Square Station, New York, NY 10159.

PAX

The PAX Computer and QCD PAX: History, Status and Evaluation, David Kahaner, Office of Naval Research-Far East.

Concert

The Concert 1.0 Execution Environment, MIT Laboratory of Computer Science, 545 Technology Square, Cambridge, MA 02139 and Advanced Technology

Department, MS 3A/1912, Harris Government Systems Sector, P. O. Box 37, Melbourne, FL 32902, August 6, 1988.

Associative String Processor

Computer Vision Applications with the Associative String Processor, Anargyros Krikelis, Aspex Microsystems LTD., Brunel University, Uxbridge, UB8 3PH, U.K., Journal of Parallel and Distributed Computing, 13, 170-184 (1991), Academic Press Inc., 1 East First St., Duluth, MN 55802.

PART 2: MIMD Software

Operating Systems: Trollius

All about Trollius, Greg Burns, Vibha Radiya, Raja Daoud & Raghu Machiraju, The Ohio State University, Trollius Project, Research Computing, 1224 Kinnear Rd., Columbus, OH 43212-1154, OCCAM User's Group Newsletter, 1990.

The Performance/Functionality Dilema of Multicomputer Message Passing, Gregory D. Burns & Raja B. Daoud, Research Computing, The Ohio State University, Proceedings of the Fifth Distributed Memory Computing Conference, ©1990 IEEE.

Parallel Programming Languages: Apply

Apply, A Programming Language for Low-Level Vision on Diverse Parallel Architectures, Leonard G. C. Hamey, Jon A. Webb, and I-Chen Wu, Dept. of Computer Science, Carnegie-Mellon University, Pittsburgh, PA 15213.

Translating Sequential Programs to Parallel: Linda

Translating Sequential Programs to Parallel, Dan Weston, Cogent Research Inc., 1100 Compton Dr., Beaverton, OR 97006, Electronic Design, March 8, 1990, P81, 611 Route #46, West Hasbrouck Heights, NJ 07604.

Semiautomatic Parallelizing: PTOOL

PTOOL: A Semiautomatic Parallel Programming Tool, Randy Allen, Donn Baumgartner, Ken Kennedy & Allan Porterfield, Dept.of Computer Science, Rice University, Houston, TX 77251, Proceedings of the 1986 International Conference on Parallel Processing, IEEE Cat. No. 86CH2355-6, P164.

PART 3: MIMD Issues

Scalability

A Scalability Analysis of the Butterfly Multiprocessor, Swarn P. Kumar and Stuart M. Stern, Boeing Computer Proceedings of the Third International

Conference on Supercomputing, International Supercomputing Institute Inc., 1988, P101, Vol. 1.

Partitioning

Mathematical Model Partitioning and Packing for Parallel Computer Calculation, Dale J. Arpasi and Edward J. Milner, NASA/Lewis Research Center, Cleveland, OH 44135, Proceedings of the 1986 International Conference on Parallel Processing, IEEE Cat. No. 86CH2355-6, P67.

Utilization

Increasing Processor Utilization During Computation Rundown, William H. Jones, NASA/Lewis Research Center, Cleveland, OH 44135, Proceedings of the 1986 International Conference on Parallel Processing, IEEE Cat. No. 86CH2355-65, P139.

Heterogeneous Networks

Solving Computational Grand Challenges Using a Network of Heterogeneous Supercomputers, Adam Beguelin & Jack Dongarra, Mathematic Sciences Section, P. O. Box 2008-6012, Oak Ridge National Laboratory, Oak Ridge, TN 37831-6367 and University of Tennessee; Al Geist, Oak Ridge National Laboratory; Robert Manchek, University of Tennessee, Computer Science Dept., Knoxville, TN and Vaidy Sunderam, Math and Computer Science Dept., Emory University, Atlanta, GA.

INDEX

Printed and bound by CPI Group (UK) Ltd, Croydon, CR0 4YY

22/10/2024

01777622-0015